Hillslope materials and processes

Hillslope materials and processes

M. J. SELBY

OXFORD UNIVERSITY PRESS · OXFORD · 1982

Oxford University Press, Walton Street, Oxford OX2 6DP

London Glasgow New York Toronto
Delhi Bombay Calcutta Madras Karachi
Kuala Lumpur Singapore Hong Kong Tokyo
Nairobi Dar es Salaam Cape Town
Melbourne Auckland

and associates in
Beirut Berlin Ibadan Mexico City Nicosia

Published in the United States
by Oxford University Press, New York

British Library Cataloguing in Publication Data

Selby, M. J.
Hillslope materials and processes.
1. Slopes (Physical geography)
I. Title
551.4'36 GB448
ISBN 0-19-874126-X
ISBN 0-19-874127-8 Pbk

Library of Congress Cataloging in Publication Data

Selby, Michael John.
Hillslope materials and processes.
Bibliography: p.
Includes index.
1. Slopes (Physical geography) 2. Weathering.
I. Title.
GB448.S44 551.4'36 81-22458
ISBN 0-19-874127-8 (pbk.) AACR2
ISBN 0-19-874126-X

Typeset by Hope Services, Abingdon
Printed in Great Britain
at the University Press, Oxford
by Eric Buckley,
Printer to the University

Preface

The study of hillslopes is of concern to many scientists — geologists, pedologists, hydrologists, engineers, and geomorphologists. As a result the literature on the subject is dispersed through many books and journals and the approaches to the subject are varied. In writing this text I have attempted to draw on all the relevant disciplines and particularly to use the contributions of the exponents of rock and soil mechanics. Work by geomorphologists during the last twenty years has concentrated on the study of processes: I hope to modify this concern by emphasising the importance of rock and soil resistance to erosion and in so doing to re-assert the importance of geological influences on hillslope forms and development.

Considerable attention has been paid to the study of landslides because they are important as processes modifying the landscape; they are of economic significance — direct and indirect costs of landslides in the United States alone have been estimated to exceed $1 000 000 000 per year at 1976 prices; and landslides also provide an opportunity for quantitative study of the forces acting on hillslopes.

I am deeply indebted to a number of people for their help and tolerance during the preparation of this book: my wife drew some of the diagrams and took more than a fair share of our domestic responsibilities allowing me to concentrate on writing; of my colleagues Dr A. W. P. Hodder spent many hours critically reading and commenting on the whole text and I also received valued advice and comments from Professor H. S. Gibbs on Chapters 2 and 3; Drs R. J. Eyles and M. J. Crozier of Victoria University commented on Chapters 4, 5, 6, and 7; Dr P. J. Hosking of Auckland University also read much of the text and Dr R. J. Wasson of the Australian National University commented on the section on talus deposits and alluvial fans. Any errors or misrepresentations which remain are entirely my responsibility. I also wish to thank Mrs Margaret McLean for typing and retyping the text, Frank Bailey and Ken Stewart for draughting most of the diagrams, and Rex Julian for preparing the photographic prints. All photographs not otherwise acknowledged are my own.

The fieldwork on which many of my observations are based was carried out during leave from the University of Waikato and on expeditions of its Antarctic Research Unit. I am most grateful to the Council of the University for making this possible.

June 1980   M. J. Selby
Hamilton, New Zealand

Contents

Acknowledgements

Thanks are due to the following publishers and learned bodies for permission to reproduce copyright material:

American Geological Institute, Figs. 7.24, 7.25

American Scientist, journal of Sigma Xi, The Scientific Research Society, Fig. 2.7

Cambridge University Press, Figs. 1.3, 5.34

Clarendon Press, Oxford, Figs. 3.1, 3.5, 3.6

Department of Geography, University of Guelph, Fig. 4.30

Freeman Cooper and Co., Figs. 7.1, 8.12

Gebrüder Borntraeger, Figs. 8.2, 8.14, 9.9, 9.10

Geological Society of America and the authors, Figs. 2.10, 2.11, 5.31, 8.11

Geological Society of Australia, Inc. Fig. 3.4

International Association of Engineering Geology and the authors, Figs. 4.26, 4.28, 4.29

John Wiley and Sons Inc., Fig. 4.12

Macmillan Journals Ltd, Fig. 2.9

McGraw-Hill Book Co., Fig. 5.17

New Zealand Department of Scientific and Industrial Research, Figs. 2.15, 5.4, 6.22

New Zealand Hydrological Society, Figs. 5.6, 5.7

Oliver and Boyd, Figs. 2.1, 11.3, 11.4

Pergamon Press, Fig. 4.4

Soil and Water and the author, Fig. 5.32

Soil Conservation Society of America, Fig. 5.25

South African Institution of Civil Engineers, Fig. 4.31

The Institution of Civil Engineers, London, Fig. 4.20

The Institution of Engineers, Australia, Fig. 6.19a

The Institution of Mining and Metallurgy, London, and the authors, Figs. 4.2, 4.32, Table 4.2

Twidale, C. R., Figs. 8.9, 8.10

Symbols used

A	area
A_j	total surface area within a rock joint
$*c$	cohesion
c'	cohesion with respect to effective stresses
c_r	residual cohesion
e	void ratio, kinetic energy
E	potential energy
F	factor of safety of a slope against landsliding
g	the acceleration due to gravity
G_s	specific gravity
H, h	height of slope, depth
h	piezometric height
H_c	critical height for slope stability
h_c	depth of crack
i	roughness angle of asperities in a rock joint
k	hydraulic conductivity
LL	liquid limit
l	length
m	mass, decimal part of depth (as in mz)
n	porosity
PL	plastic limit
PI	plasticity index
Q	total discharge per unit of time
R	hydraulic radius
r	the radius of the arc of a plane of sliding
S	slope
SL	shrinkage limit
s	shear strength at failure
u	pore-water pressure
V	water pressure in a tension crack, velocity
v	voids, velocity

W	weight
z	depth, thickness of overburden
z_w	depth of water
α (alpha)	angle of inclination of a failure plane
β (beta)	slope angle
γ (gamma)	unit weight of soil or rock at natural moisture content (the symbols γ_d and γ_{sat} may be used for dry and saturated soil respectively.)
γ_w	unit weight of water
ϵ (epsilon)	linear strain
ρ (rho)	bulk density
σ (sigma)	total normal stress
$\sigma_1, \sigma_2, \sigma_3$	major, intermediate and minor principal stresses respectively
σ'	effective normal stress
τ (tau)	shear stress
$*\phi$ (phi)	angle of internal friction
ϕ_r	residual angle of internal friction
ϕ'	angle of internal friction with respect to effective stresses
ϕ_j	joint friction angle
ϕ_{jr}	residual angle of friction along a rock joint
ϕ_p	maximum or peak value of internal friction

* In order to distinguish the parameters c and ϕ for the various shear strength test conditions the subscripts u, cu, and d are used to denote undrained, consolidated-undrained, and drained respectively, e.g. c_u.

1

Introduction

Hillslopes occupy most of the land surface with the exception of terraces and plains formed by river deposits. Even extensive surfaces, like the Great Plains of North America and the plateaux of Africa, are largely formed of hillslopes of low angle between crests of interfluves and valley floors. The most spectacular hillslopes are the great rock cliffs of the high mountain chains and the extensive valley slopes of deep gorges in uplifted plateaux.

Because of their extent the study of slopes has always been close to the heart of geomorphology although most geomorphological work on slopes, carried out before the middle of the twentieth century, was related to the classical models of slope evolution proposed by W. M. Davis, Walther Penck, and L. C. King. Only since the 1950s has much attention been paid to processes of hillslope denudation, and the study of slope material strength and resistance has been brought into the subject only in the 1970s.

Slope systems

Weathering and removal of rock and soil on hillslopes is not a uniform process in either time or space: it is episodic and depends upon the availability of energy and a transporting medium. As a result hillslopes can be regarded as a system of stores (Figs. 1.1, 1.2) which are periodically unlocked by processes. Very resistant stores, such as are provided by massive hard rock outcrops, may only yield material at very infrequent intervals. Soil slopes in a humid tropical climate may yield solutes almost continuously, but solids by landslide processes much less frequently. Each process, therefore, has its own magnitude and frequency of operation which is controlled by the resistance of the hillslope rock and soil, and by the intensity of the denudational processes.

By tracing and measuring the movement of material from hillslopes by different processes, and by measuring the modification of hillslope form produced by them, it is possible to evaluate short-term changes of slope profiles. In theory this should eventually provide an understanding of how hillslopes evolve. In most environments hillslopes change too slowly for the progression from long steep slopes to slopes of lesser angle to be observed. That such changes occur is indicated from geological sections, in which erosional surfaces have been cut across complex structures to produce unconformities which may be preserved in the depositional record. Attempts to compensate for the lack of observations of slope change usually involve one of two possible procedures. In a few rather rare situations space may be substituted for time, as where Savigear (1952) was able to measure slope profiles along a cliff which had been protected from wave attack at its base for varying periods, so that a sequential development of hillslope forms was assumed to have been produced. Less secure methods involve the measurement of different slope profiles in one area and the assembly of these profiles into a sequence. Such methods are extremely uncertain because the underlying assumption of sequential change cannot be verified.

The second method – that of the development of process–response models – forms the basis of most modern hillslope geomorphology (Fig. 1.3). Analysis involves the measurement of the resistance of rock and soil to change, the force and mode of action of a process causing change, and variation in the rate of change through time and in space. A model seeks to link variations in the rate of change over the hillslope to development of slope profiles (frequently by using differential equations). Field or laboratory methods often involve the establishment of a statistical relationship between a slope change and measurements of soil or rock properties and a process

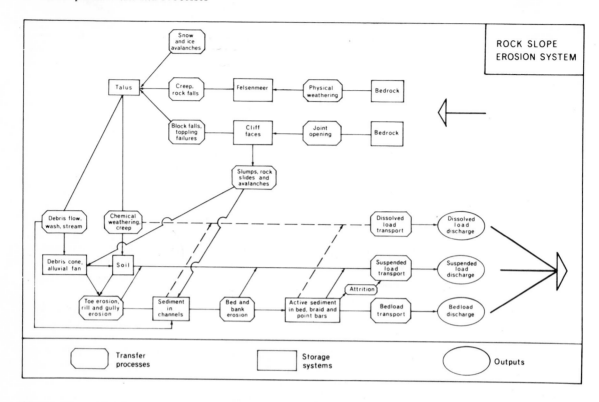

Fig. 1.1 The system of stores and transfer processes on a rock slope.

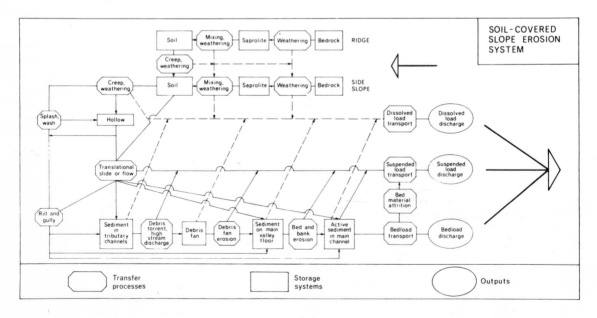

Fig. 1.2 The system of stores and transfer processes on a soil-covered slope.

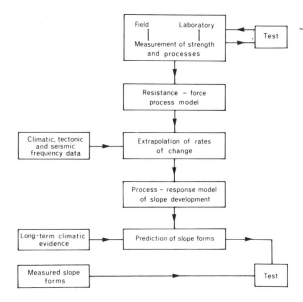

Fig. 1.3 The pattern of hillslope studies (modified from Carson and Kirkby, 1972).

such as rainfall. The primary difficulty in such methods is that of adequately sampling slopes, rock and soil types, and process types and rates. A second difficulty arises from the uncertain degree of inherited influences upon current hillslope forms, so that when a certain suite of slope forms is predicted from a sequence of processes the prediction may not be tested against natural slopes with a known history. There is at present little hope of developing predictive slope-form models which incorporate the multitudinous climatic, and hence process intensity, changes of the Quaternary. The emphasis of this book is therefore upon those components of hillslope studies which can be studied effectively — the strength of slope materials, the effect and intensity of processes acting on those materials and the resulting short-term changes on slopes.

Energy available for slope processes

The energy available for slope processes is derived from three sources: solar radiation, gravity, and endogenetic forces. Solar radiation directly promotes weathering processes but much of it is effective by driving the circulation of water in a hydrological cycle between the atmosphere, pedosphere, lithosphere, and ocean (see Fig. 5.1). Raindrops striking the ground, water flowing, and boulders falling or rolling downslope, have energy provided by the gravitational force that attracts them towards the centre of the earth. Endogenetic forces are generated

by radioactive decay of natural isotopes producing heat. Geothermal heat drives volcanic activity and creates the stresses which are released in earthquakes. Over most of the land surface the energy available from solar and gravitational sources for geomorphic work is several thousand times greater than that available endogenetically. Only where volcanic activity, or sudden release of seismic energy, concentrates power can internal energy produce distinctive landforms or landforming processes.

Plan of the book

The book begins with a discussion of weathering because this process is responsible for the creation of soil and broken rock on which most hillslopes are formed, and because it is primarily responsible for the selective formation of many minor landforms on bare rock slopes, even though these landforms cannot develop until a transporting process removes the weathering products. Weathering also has the effect of producing residual crusts within the regolith, which can control the form of hillslopes under certain climatic regimes.

The resistance of rock masses and soils to imposed stresses is a neglected topic in geomorphology and has consequently been accorded more emphasis than is usual. In Chapter 4 the nature of strength, its measurement and assessment in the field and laboratory, is described so that the selective activities of erosional and transporting processes, discussed in the next three chapters, can be more readily understood. In Chapters 5 to 7 mass wasting is emphasised, partly because it is the dominant process of cliff retreat and partly because it is possible to show the relationship between resistance of materials and the forces acting to promote landsliding. The components of Chapters 2 to 7 and the linkages among them are, perhaps, most readily appreciated from a study of Figs. 1.1 and 1.2.

Tors and bornhardts are treated as a separate topic, rather than with other rock slopes, because much may be learnt from a study of them. Tors show us that structurally controlled weathering may pre-condition the shape of the land surface long before the surface of the soil is lowered by erosion to expose the results of differential weathering. Bornhardts provide examples of the control on landforms exerted by the spacing of joints and, more particularly, by the effect of sheeting on the form of the rock surface. The study of both tors and bornhardts provides a salutary warning that it is impossible to deduce the nature of the processes responsible for a landform from a study of only the shape of that landform.

The long-term evolution of hillslopes has always been a major topic in geomorphology and has traditionally been based upon changing slope morphology. The evidence for past processes is usually fragmentary and only decipherable from a close study of the deposits on, or at the base of, hillslopes. While stratigraphy may eventually assist us in deciphering the record of the past it is premature to attempt a detailed account of the methods and conclusions which may be reached. The nature of fluctuating climatic regimes throughout late Cenozoic time is only just being appreciated from studies of deep-sea cores and loess deposits. It is already evident that terrestrial deposits contain only a partial record of the past and that the relationship between slope forms and climatically controlled processes is far from adequately understood. The discussion of slope profiles, consequently, is confined to the types of models which may be used to study hillslope forms. The rates at which slopes change is discussed in Chapters 10 and 11.

2

Weathering processes

Weathering is the process of alteration and breakdown of rock and soil materials at and near the earth's surface by physical, chemical, and biotic processes. Igneous and metamorphic rocks, as well as deeply buried and lithified sedimentary rocks, are formed under a regime of high temperature and/or pressure. At the ground surface the environment is dominated by temperatures, pressures, and moisture availability more characteristic of the atmosphere and hydrosphere: thus rocks are altered by weathering to new materials which are in equilibrium with surface conditions. Weathering has three very important results: it is the process which renders resistant rock and partly weathered rock into a state of lower strength and greater permeability in which the processes of erosion can be effective; it is the first step in the process of soil formation; and during weathering the release or accumulation of iron oxides, lime, alumina, and silica takes place — where concentrated after initial solution these form indurated shells on rocks, or layers in the soil which may become hard and resistant to erosion.

The alteration of rock by weathering occurs in place, that is *in situ*, and it does not directly involve removal processes. It may be characterised by the physical breakdown of rock material into progressively smaller fragments without marked changes in the nature of the mineral constituents. This disintegration process leads to the formation of a residual material comprising mineral and rock fragments virtually unchanged from the original rock (Plate 2.1). By contrast chemical alteration may induce thorough decomposition of most or all of the original minerals in a rock, resulting in the formation of material composed entirely of new mineral species, particularly of clay minerals (Plate 2.2). Biological weathering induced by biophysical and biochemical agencies is largely confined to the upper few metres of the earth's crust in which plant roots are active.

It must be appreciated that physical, chemical, and biological processes usually operate together. Also erosion takes place from the surface of the ground and within the soil by solution almost continuously so that, although we speak of weathering as a process of decomposition *in situ*, transport of the residuum of weathering may be simultaneous and assists in the continuation of weathering.

Soil

The end-product of weathering is said to be soil, but this statement can be confusing for the word 'soil' is used in at least two senses. For the soil scientist (pedologist) soil is a material which results from both weathering and soil-forming processes and it is the material in which plants grow: weathering alone will not produce the horizons (layers) which are the fundamental features of soils in this plant-related sense.

To the engineer, by contrast, soil is a broken-down rock material of relatively low strength and he commonly uses the term 'soil' to include low strength materials, such as clay-rich sedimentary rocks and unlithified sands, which have not been altered by soil-forming processes and which may not have been weathered since exposure at the ground surface. The engineer usually wishes to dispose of that organic soil which is the focus of the soil scientist's concern and he usually removes it from any excavation or work site before placing a structure on the ground. Geologists usually use the term 'regolith' to refer to the whole profile of weathered rock and unconsolidated rock material of whatever origin: it thus corresponds to all that material comprising the weathering zone together with unconsolidated superficial deposits. This definition is close to the usage of 'soil' in engineering and soil mechanics.

2.1 A residual soil, derived from glacial moraine, in which there has been little or no chemical weathering and hence no clay mineral formation, Antarctica.

Factors affecting soil weathering

Few generalisations can be made about the rate of weathering of minerals because of the numerous factors which can influence the process. However, climate and the physical and chemical composition of the parent rock are of outstanding significance.

Climatic influences

Climatic conditions determine the temperature and moisture regime in which weathering takes place. Under conditions of low rainfall, mechanical weathering is dominant and, therefore, comminution of particles occurs, with little alteration of their composition. With an increase in precipitation, more minerals are dissolved and chemical reactions increase so that chemical decomposition of minerals and synthesis of clays becomes more important. In humid temperate climates silicate clays are formed and altered. Speed of chemical reactions is greatly increased by a rise of temperature: a rise of 10 °C

usually doubles or trebles the reaction rate. Increases of temperature also alter the relative mobility of minerals. Quartz, for example, is highly resistant to weathering in temperate climates, but fine-grained quartz particles are more easily weathered in tropical conditions, and in such climates iron and aluminium hydrous oxides are more resistant. The iron and aluminium oxides, therefore, tend to accumulate in tropical soils which get their red colour from the iron.

Another effect of climate is to control the vegetation and its production of litter. In humid tropical climates the production of organic matter is high – 3 300 to 13 500 kg/ha per year from tropical forests – compared with temperate forests that produce 900 to 3 100 kg/ha per year. This means that the supply of organic compounds to take part in chemical weathering is high in the tropical forests and low in the temperate ones. The appearance of the soils of these forests suggests the reverse because dark humus can accumulate in the

2.2 A soil with high kaolinite content formed on a sandstone. The presence of clay minerals is indicated by the shrinkage from drying of the upper part of the profile. The lower part of the profile is still moist.

cool forests but in the tropical ones organic matter is broken down very rapidly and much of the humus has a pale colour which makes it difficult to see. The turnover of tropical forest humus is about 1 per cent per day compared with 0.1 to 0.3 per cent in temperate forests.

The significance of climate has prompted the idea of weathering regions which approximately correspond to the distribution of major zonal soil groups (Figure. 2.1). This type of generalization is a useful model but it has to be qualified. Variations in soil type and weathering rates and depths depend not only upon differences in the kind of processes prevailing in climatic zones, but also upon the intensity of those processes. Broad schemes also have to be modified because they apply only to tectonically stable areas with adequate drainage. Uplands and depressions give rise to distinctive erosional and depositional processes which may mask zonal weathering processes. A further qualification is that zonal processes have been modified by climatic changes in large areas of the earth so that, although modern processes are occurring under modern climates, there may be relict weathering products in many areas derived from earlier climatic regimes.

In spite of these reservations we can detect areas of very thin weathering profiles in polar and desert zones where the absence of water and plants produces low weathering rates; intermediate rates occur in the temperate latitudes, and high rates in the humid tropics where weathering profiles are commonly

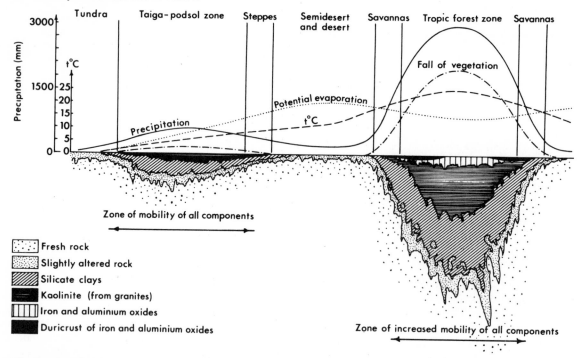

Fig. 2.1 The formation of weathering mantles in areas of tectonic stability and low relief. This scheme demonstrates a relationship between climatic factors, vegetation cover, depth of weathering and dominant profile horizons. It does not consider relict effects (after Strakhov, 1967).

deep and the formation of insoluble secondary clay minerals is at a maximum.

The physical characteristics of rocks

Particle size, hardness, permeability, and degree of cementation, as well as mineralogy all influence weathering processes.

Particle size is important because chemical weathering is mainly the result of surface reactions between solutions and mineral grains. The rate of weathering is therefore largely dependent upon the surface area of the grains (Fig. 2.2). Thus it is relatively slow in sands but fast in silts.

Hardness, mineralogy, and cementation affect the rate at which weathering can reduce the rock to smaller particles. A siliceous sandstone, because of its quartz constituents, hardness, and cement, will be more resistant than a calcareous sandstone. Permeability, however, is probably a more important characteristic for it will control the rate at which water can seep into the rock and the area of the surface on which it can act. Rocks display a very large range of permeabilities, as is shown by the approximate values in Table 2.1.

The chemical properties of rocks and soils are of fundamental importance in determining the rate

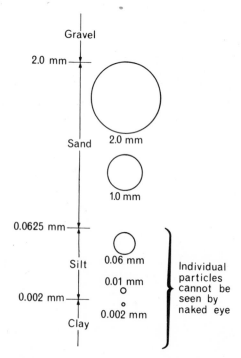

Fig. 2.2 The relative diameters of soil particles from which can be imagined the relative surface areas of the particles.

Table 2.1 The Porosity and Permeability of Selected Rocks

	Rock	Porosity (%)	Permeability m/day
Unconsolidated	Clay	45–60	10^{-6}–10^{-4}
	Silt	20–50	10^{-3}–10
	Sand	30–40	10–10^4
	Gravel	25–40	10^2–10^6
Indurated	Shale	5–15	10^{-7}–10
	Sandstone	5–20	10^{-2}–10^2
	Limestone	1–10	10^{-2}–10
	Conglomerate	5–25	10^{-4}–1
	Granite	10^{-5}–10	10^{-7}–10^{-3}
	Basalt	10^{-4}–50	10^{-5}–10^{-2}
	Slate	10^{-4}–1	10^{-9}–10^{-6}
	Schist	10^{-4}–1	10^{-9}–10^{-5}
	Gneiss	10^{-5}–1	10^{-9}–10^{-6}

Source: Gregory and Walling, 1973.

of processes and the products of weathering – these influences are discussed below in the section on chemical weathering. Other major factors which influence weathering are plants which produce CO_2 from their roots, and organic compounds from decay of plant tissue. Site factors – especially the rate of soil drainage – also have an influence on weathering.

The weathering profile

The depth of the regolith is extremely varied. In cold climates, and especially in areas which have been scoured by glacial erosion, bare rock may be at the ground surface or concealed by only a few millimetres of weathering products. Similarly, hot deserts may have a neglible soil cover; by contrast in some places the regolith is over 100 m thick and in parts of the Snowy Mountains of Australia completely weathered rock has been found at depths in excess of 300 m. Such great depths of weathering are, however, rare and 30–50 m is probably a common maximum depth of weathering on undulating surfaces in the humid tropics, where maximum depths might be expected. The rate of production of completely weathered material, often called saprolite, from rock is extremely variable and depends upon rock type, climate, and vegetation. Estimated rates vary from close to zero to 20 mm of rock weathered per 1000 years (e.g. Patterson, 1971; Owens, 1976).

A number of attempts have been made to characterise weathering profiles. Many of these attempts relate to regoliths on granitic rocks and these commonly display a series of zones. Ruxton and

Berry (1957) found that in the pedological soil at the surface in Hong Kong there was no trace of parent rock materials. In weathering zone V (Fig. 2.3) a residual debris of structureless sand and clayey sand may attain a thickness of 1-25 m. In this zone the residual material is largely of quartz and kaolinite. The zone IV residual material is a pallid silty sand with quartz particles in a matrix of chemically altered feldspars and kaolinite. The upper part of zone IV may be of entirely altered material but in the lower part small rounded stones and boulders may be embedded in a quartz-rich sand and angular gravel, known as grus. Zone IV may have a range of depth from 1 to 60 m (Plate 2.3).

In zone III the rounded boulders, known as core stones, increase in size and abundance and are progressively less rounded with depth. The degree of chemical alteration of the matrix around the boulders may also decrease with depth and some

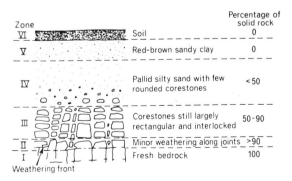

Fig. 2.3 Features of a full weathering profile developed on granitic rocks (modified from Ruxton and Berry, 1957).

2.3 Zone III of a profile formed on granite showing the large rounded corestones, Cape Coast, South Africa.

feldspars may be unaltered. The thickness of this zone is again variable but may be up to 10 m. Zone II is one of limited weathering along the joints. As much as 90 per cent of the rock is unaltered and decomposition is largely from the oxidation of the iron in biotites so that red staining is characteristic of the rims of joint blocks. The depth at which weathering is no longer detectable is commonly known as the weathering front. This is often a highly irregular feature and may itself be a zone rather than a surface.

Zones of weathering can be observed in many rock types as well as those of granite but the features are not always as distinctive as those described above. In some areas the regolith is composed almost entirely of transported materials and the break between the regolith and bedrock may be very sharp. Elsewhere changes may be very gradual and zones very difficult to distinguish (Plates 2.4, 2.5).

Core stones

Unweathered core stones form *in situ* within a regolith. They are common features of deep weathering profiles formed in granite but they are also found in basalts, andesites, and some sandstones. The feature of the parent rock which controls the production of core stones is the variation in permeability produced by jointing and the spacing of joints. Where joints are very close together or where the passage of solutions is through pores in the rock there are few zones of preferential water movement; where joints are widely spaced and the massive rock has low porosity and water movement

is along joints, weathering becomes concentrated there and works progressively into joint blocks. The unaltered core of the joint block thus becomes the core stone. In Fig. 2.4 the progressive breakdown of the joint block from its exterior towards the interior is illustrated. The formation of clay minerals through the breakdown of feldspars occurs close to the joint so that no trace of rock structure survives, then successively into the block are zones of decreasing alteration.

If during a phase of erosion the deeply weathered grus and clays are removed then core stones may be left at the ground surface as blockfields, boulder fields, or piles of core stones, known as tors. Such boulders may form distinctive features on some land surfaces, especially in the humid tropics (Plate 2.6). Other types of tor and residual features are discussed in Chapter 8.

Spheroidal weathering

One of the features of many core stones is that they are surrounded by concentric shells of rock and weathering products (Plate 2.7). There is still much controversy regarding the origin of such weathering forms. One group of hypotheses (e.g. Jocelyn, 1972) suggests that the concentric shells may be a result of residual internal stresses contained in the rock during contraction on cooling and that they were thus initiated at the same time as the major joints. A second group of hypotheses suggests that the shells are formed by chemical processes (e.g. Augusthitis and Otteman, 1966) which may be aided by the formation of microcracks along which chemical solutions can migrate (Bisdom, 1967).

2.4 Profile through the toe of a slope showing a sharp boundary between the calcareous sandstone bedrock and the soil above. Thickening of the soil profile downslope indicates the effect of erosion on upper slopes and accumulation near the slope base, Hawke's Bay, New Zealand.

The rounding off of corners of an original angular joint block (Plate 2.8) permits the gradual formation of spherical shells around a core stone. When this process is operating within a deep regolith it implies that the changes taking place must involve either contraction of the joint block or change of composition without increase in volume (i.e. isovolumetric weathering), for expansion of interlocking joint blocks is clearly impossible. A feature of many spheroidally weathered joint blocks is that the rock shells vary in colour sometimes with whitish and reddish-brown shells alternating. The reddish-brown colours are often found to be the result of deposition of limonitic iron between rock granules, and the whitish rings are depleted of iron but relatively enriched in silica, potassium, and aluminium with the formation of clay minerals. It seems probable that these migrations and sites of deposition are influenced by changes in pH of the migrating solutions so that once separation has started the precipitates create the pH which is suitable for further deposition of ions of their own species. Solution must be a very important part of the total process of change for, according to Blackwelder (1925), the chemical weathering of a granite to new mineral species would be accompanied by a

2.5 Indeterminate boundaries between horizons of a soil developed upon indurated sandstones. Slightly weathered rock extends in places nearly to the surface, but alongside this weathering has extended to a depth of more than 2 m, South Auckland, New Zealand.

GRANITE

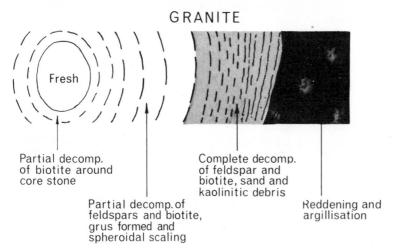

Partial decomp. of biotite around core stone

Partial decomp. of feldspars and biotite, grus formed and spheroidal scaling

Complete decomp. of feldspar and biotite, sand and kaolinitic debris

Reddening and argillisation

Fig. 2.4 The progressive development of weathering zones around a granitic corestone.

2.6 Granite boulders and a tor exposed at the ground surface in the savanna zone of Cameroon, West Africa.

50 per cent increase in volume if soluble products were not removed in the drainage water.

Spheroidal weathering is most common in igneous and metamorphic rocks containing a variety of minerals with variable resistances to weathering.

It is particularly common in granitic rocks, basalts, andesites, and dolerites. It is rare in sedimentary rocks which tend to have uniform compositions of relatively resistant minerals. Spheroidal weathering occurs within the regolith and has to be distinguished

2.7 Spheroidal weathering of basalt showing the successive rock layers.

2.8 The progressive rounding of a joint block by thickening of weathering skins at corners and by weathering along fractures at corners. The rock is a basaltic tuff.

from exfoliation of rock spalls from exposed surfaces.

Boulder fields may be the result of the stripping away of deeply weathered regolith materials to leave exposed core stones or they may result from exfoliation of boulders at the ground surface. It is only when a regolith containing core stones and spheroidal weathering forms is found in association with the boulder fields that their origin can be determined with confidence.

Processes of weathering

The formation of joints in rock is perhaps the most important single weathering process, even though it is seldom classed as such. It is by the formation of joints that stresses, produced by cooling and by the pressure of tectonism and an overburden, are released. The presence of stress in rocks is illustrated dramatically, and sometimes disastrously, in deep mines when shells of rock burst from the walls, floors, and roofs of galleries. Rock bursts are a response to the opening of the gallery which permits the rock to expand and so release its internal stress.

Once joints have developed in a rock physical

and chemical processes operate together in nearly all environments and at nearly all stages, although physical weathering may be dominant in early stages and chemical weathering dominant once a regolith has formed. The product of weathering has a composition which is determined by the dominant process. Physical disintegration produces a residuum of fractured and comminuted particles of the original rock minerals. Solution processes remove the soluble minerals and leave a residue of primary minerals which may be further altered by the chemical processes (Fig. 2.5). Chemical decomposition ultimately results in the production of clay minerals and complete alteration of original rock minerals. Stages in the weathering sequence are illustrated in Fig. 2.6.

Physical weathering processes

Physical weathering processes are most evident where rock is exposed at the ground surface, and are thus particularly obvious in hot or cold deserts and on cliff faces. The result of physical weathering is a chemically unaltered fragment of rock which may range in size from a fractured single rock

STAGES OF WEATHERING OF ROCK MATERIAL

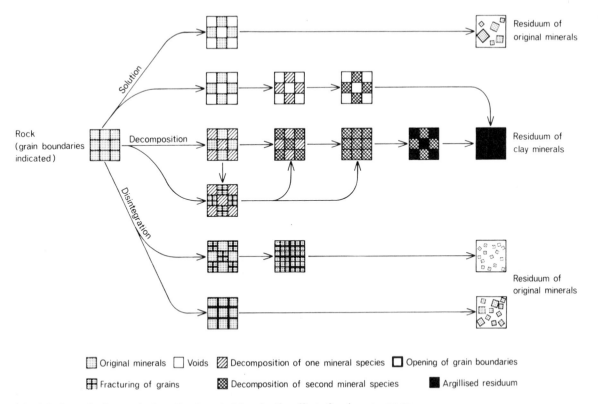

Fig. 2.5 Stages in the weathering of rock material under the effect of various processes.

crystal to a massive joint block. Stress release is almost certainly the only weathering process which can produce massive joint blocks, but small rock spalls, fragments, and separated crystals may be produced by a variety of processes, and because they leave the rock chemically unaltered it is often very difficult to determine which processes are responsible for a particular product. The most commonly recognised physical processes are those resulting from internal rock stresses, frost action, hydrofracturing, salt crystal growth, insolation, and wetting and drying.

Internal stresses exist in all rocks which have been subjected to high temperatures and pressures during their formation. In igneous rocks, such as granite, stresses are set up during cooling. Feldspar crystals form first, from a melt, and then quartz crystallises in a form known as high quartz, and with further cooling this converts to low quartz. The difference between these two types of quartz is dependent upon the arrangement of the atoms in the crystal. Solid silica of most forms is built from a tetrahedral unit consisting of one silicon atom surrounded by four oxygen atoms. This unit is connected to other similar units through the oxygen atoms, which allows for some variety in the possible structures. Low quartz differs from high quartz in one major respect: the bond angle at the oxygen is 180° in high quartz, but at about 150° in low quartz (Fig. 2.7). The transition from high to low quartz in a cooling melt occurs at about 573 °C and involves a change in crystal shape. As the quartz crystals are constrained by the surrounding crystals stresses develop in the rock and fractures form between crystals (Krinsley and Smalley, 1972).

Stress release in rocks which have been subjected to high tectonic or overburden pressures may produce small rock spalls or large sheet-like joint blocks (Gerber and Scheidegger, 1969) (see Chapter 8 for discussion of sheet jointing).

The microfissures, incipient joints, and other planes of weakness in rocks resulting from crystal-

STAGES OF WEATHERING OF A ROCK MASS

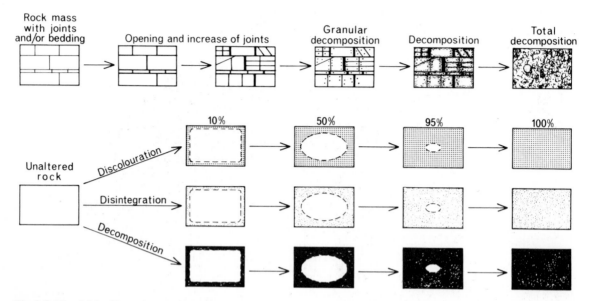

Fig. 2.6 The visible effects of weathering at successive stages.

lisation and stress release all provide lines of weakness which can be exploited by chemical and physical processes. The spalling of platey rock fragments — generally known as exfoliation — and the release of individual crystals or crystal aggregates — known as granular disintegration — may be entirely the results of release of internal stresses or they may result from the action of other physical

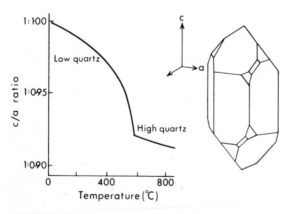

Fig. 2.7 During crystallisation from a melt high quartz changes to low quartz. These two forms of quartz have different ratios of the length of their *c* to *a* axes, different bond angles, and hence different volumes (after Krinsley and Smalley, 1972).

processes (Plates 2.9 and 2.10). It is seldom possible to determine the exact cause of a particular feature unless it has been created under controlled laboratory conditions.

Hydrofracturing and frost action are two of the most important and widely recognised processes of physical weathering. They operate by overcoming the tensile strength of rocks which may be as high as 30 MPa (see Chapter 4 for a discussion of rock strength). When water freezes at 0 °C it increases in specific volume by 9 per cent, as its molecules organise into a rigid hexagonal crystalline network. It was once thought that simple ice formation was itself capable of forcing open rock fissures, but it has been demonstrated that crystallisation pressures do not exceed 0.6 MPa and are hence far weaker than is required to disrupt rocks (Connell and Tombs, 1971). The formation of ice crystals, or frost action as it is commonly called, is thus limited in its direct weathering effects to the heaving of pebbles, soil aggregates, and clods. The freezing of water within rock pores and fissures produces fracturing by indirect effects.

If water is confined it does not begin to freeze until the temperature is below 0 °C (Fig. 2.8). In rock pores and fissures water at the entrance to a pore freezes first and confines water within the pore so that it will only freeze when temperature is

2.9 Exfoliation of thick plates from a granodiorite, Antarctica.

below 0 °C. Ice forming at normal freezing rates is not strong enough to completely seal a pore, as it becomes plastic at stresses in excess of 100 kPa and is extruded. The maximum theoretical confining pressure of 216 MPa at −22 °C (which would be capable of breaking any rock by tensile failure) is, therefore, not reached in natural conditions. Below −22 °C the pressure remains nearly constant and a denser kind of ice (Ice III) begins to crystallise. Furthermore, water in very thin films of the kind found in small pores and very narrow fissures will not freeze, even at very low temperatures, because of the strong capillary adhesion of water molecules to rock which prevents their reorganisation into ice crystals.

In very narrow fissures shallow freezing may aid in forcing capillary films of water towards the tips of fractures, so opening and extending the fracture. Such hydrofracturing may extend to depths of 12-15 m, even though the ground may freeze to a depth of only 1 m (Philbrick reported in Bloom, 1978), and can occur within the crystal lattices of

clays and other minerals. It is thus a potent weathering process at depths into rocks far in excess of the penetration depth of ice formation, and capable of dividing large rock masses and the smallest mineral grains (Dunn and Hudec, 1966; Hradek, 1977).

In spite of the extrusion of ice from pores, as temperature falls, there is some evidence that rapid freezing of pure water at temperatures a few degrees below zero may produce a shock wave of such energy that it will fracture rocks: calculated shock pressures range from 100 MPa to as high as 10 GPa. It has been suggested by Hodder (1976) that nucleation of ice and very rapid freezing may cause collapse of air cavities within water filling rock pores. Such collapse, known as cavitation, is implosive and the source of shock waves, which are heard as sharp cracks by travellers. Rapid freezing of rock containing supercooled water (at −5 °C or so), which is free from impurities or ice crystals, has been reported by Malaurie (1968) to produce high pressures as sudden conversion to ice takes place. Experimental work by Battle (1960) led him to

2.10 Granular disintegration of a marble boulder, Antarctica.

believe that the minimum rate of temperature change for the production of high pressures by ice growth is 0.1 °C/minute.

Most rocks have a thermal conductivity which will permit frost penetration to a depth of up to 20 m over a period of about 14 weeks, but such slow temperature changes are not likely to produce disruptive pressures from ice growth. Diurnal

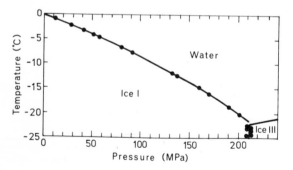

Fig. 2.8 The increase of pressure in confined pure water at lowered freezing temperatures (data from Bridgman, 1911).

changes across the freezing point are confined to depths of about 0.2 m, consequently shattering is most effective in producing relatively small rock fragments through which a temperature change can be transmitted rapidly.

Experimental work (Tricart, 1956; Potts, 1970) has shown that the number of oscillations across the freezing point is of greater importance to rock shattering than the intensity of freezing; thus an 'Icelandic' temperature regime with a range of +8 °C to −8 °C in 24 hours is far more effective than a 'Siberian' regime of +15 °C to −30 °C in a period of weeks; areas with low diurnal but large seasonal temperature changes, or even continuous low temperatures, are not beneficial to frost-induced shattering.

Frost-induced shattering is also dependent upon the presence of water. Total saturation of a rock may inhibit shattering, but partial wetting is most beneficial. Other factors which influence shattering are rock characteristics: planes of weakness, natural fissures, high porosity, and high permeability all

2.11 Frost-shattered dolerite. The joint blocks in this out-crop had a blade-like shape and fractured readily, Antarctica.

favour physical weathering. The alignment of rock fissures also influences the shape of the debris with rocks like schists producing elongated rod or plate-like fragments (Plate 2.11). In general then permeable fissured rocks in areas with frequent crossing of the freezing point and abundant moisture are likely to be severely frost weathered or hydrofractured. It must be recognised, however, that there is still much to be learnt about the processes of freeze–thaw. It has even been suggested that the widespread phenomenon of rock shattering is not caused by freezing and thawing but by hydration (White, 1976).

Shattering is largely confined to rock outcrops and a regolith greatly reduces or eliminates its effectiveness. In cold deserts or alpine areas with no regoliths the entire ground surface may shatter to produce a surface of broken rock fragments called a felsenmeer (a German word meaning 'rock sea') (Plate 2.12).

Salt weathering results from the crystallisation of salts in rock pores. Growing crystals can exert high pressures on the confining rock and cause exfoliation and rock fracturing. The salts involved in this type of weathering may be derived from the sea and blown inland in spray or carried by snow and rain, or they may be derived from chemical weathering of rock. In desert areas particularly, drainage waters may evaporate to leave a salt-rich sediment which will be deflated by wind and the salt redeposited on rock surfaces.

Salts may be carried in solution into pores and fissures in rocks by percolating water. Once in the rock the salts may contribute to rock failure by any of three processes which can create stresses resulting from: (1) the growth of crystals from solution; (2) thermal expansion; or (3) hydration.

(1) The growth of salt crystals from saturated solutions has now been studied in a number of experiments which have been reviewed by Evans (1970) and by Selby (1971a). It has been shown that, in a fissure large salt crystals will always grow at the expense of smaller ones and that they will continue to grow until they completely fill the pores in a rock. A crystal will grow in a large pore until the pressure builds up to such an extent that mechanical fracture occurs, or the confining pressure is sufficient to make the crystal grow down a capillary pore. Whether or not fracture occurs will depend upon the tensile strength of the rock. For rocks of equal inter-pore strength those with large pores separated from each other by microporous regions will be most liable to destruction by crystal growth.

(2) Thermal expansion of crystals which have already formed in rock fissures is a result of a rise in temperature causing the crystal volume to increase. Cooke and Smalley (1968) measured the thermal expansion of five salts ($NaNO_3$, $NaCl$, KCl, $BaSO_4$, $CaCO_3$) and found that in all cases except one ($CaCO_3$) the expansion was considerably greater than that of granite (Fig. 2.9). For a common salt such as sodium chloride a rise of 54 °C would give a volumetric change of 1 per cent, which is considerably greater than that in the surrounding rock. The stresses caused in the rock are concentrated at the inner extremity of the crack in which the salts

2.12 Felsenmeer of quartzite on a high plateau (2000 m), Antarctica.

occur and fissures may be progressively opened by this process even when there is a considerable confining pressure.

The temperature range of 54 °C may seem extreme but rock surfaces experience considerably higher and lower temperatures than the surrounding air and may well be subjected to such a range during a 24-hour period in extreme environments. The low thermal conductivity of rock probably limits this process to the outer few centimetres of any outcrop.

(3) Hydration of crystals occurs when some salts take up moisture within their lattices. The resulting expansion may be considerable and produce very large pressures. The hydration pressures of a number of salts have been calculated by Winkler and Wilhelm (1970). The hydration of thenardite (Na_2SO_4) to mirabilite ($Na_2SO_4 \cdot 1OH_2O$) following a change of relative humidity from 70–100 per cent at a temperature of 20 °C can exert a pressure of nearly 50 MPa, and hydration of bassanite ($CaSO_4 \cdot \frac{1}{2}H_2O$) to gypsum ($CaSO_4 \cdot 2H_2O$) with a change of relative humidity from 30 to 100 per cent at a temperature of 0 °C can exert a pressure of 200 MPa (Figs. 2.10 and 2.11). Such temperature and humidity ranges can occur in many arid regions and at a local scale on coastal cliffs and rock platforms which are frequently wetted by spray and then rapidly dried in the sun. Salt weathering is not likely to occur in humid environments where salt solutions are not saturated.

Salt weathering is most effective in porous and fissured rocks and least effective in fine-grained ones (Plate 2.13). The most obvious result is the exfoliation of individual crystals or groups of crystals from igneous rocks, or of sand and silt grains. In areas of considerable abundance of salts large hollows and even shallow caves may be formed (See chapter 3 for discussion of weathering forms).

Insolation weathering is the result of alternate warming and cooling of rock surfaces under the direct influence of solar heating. It has been calculated (see Garner, 1976) that a gneiss expands

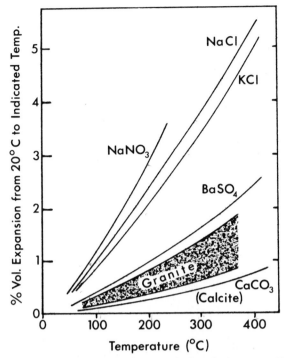

Fig. 2.9 The expansion in volume of selected salts compared with granite for given rises in temperature (after Cooke and Smalley, 1968).

approximately one part in 100 000 per 1 °C increase in temperature. Rock surfaces in direct sunlight in tropical regions may experience a range of temperatures of more than 80 °C and such a temperature range acting on gneiss could cause a 100 m long slab to expand by 4-5 cm. It has been suggested that such expansions and contractions could cause microfracturing, spalling and splitting of exposed rock.

Experimental work by Blackwelder (1925, 1933) and Griggs (1936) on small rock samples failed to

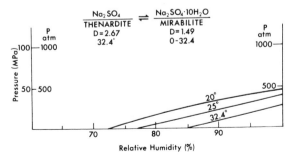

Fig. 2.10 Pressures exerted in the hydration of thenardite to mirabilite (after Winkler and Wilhelm, 1970).

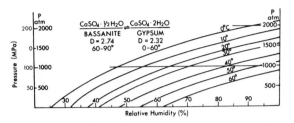

Fig. 2.11 Pressures exerted in the hydration of bassanite to gypsum (after Winkler and Wilhelm, 1970).

produce rock fracturing as a result of temperature changes, and it was suggested that small rocks are sufficiently elastic to absorb the change of volume caused by heating and cooling by solar radiation, although the greater heating in fires is well known to cause rock splitting. It may be that there is a critical size below which insolation weathering is ineffective for a particular type of rock (Rice, 1976).

In spite of the early experimental work the existence of clearly split boulders in desert areas (Plate 2.14) has promoted a continuing belief in the effectiveness of thermal expansion and contraction. Aires-Barros *et al.* (1975) repeated the early experiments and noted some microfracturing of rocks under dry heating, but substantially more fracturing and spalling when the temperature changes occurred in the presence of water. It seems possible that the volumetric expansion of water trapped in rock capillaries may disrupt rocks.

In desert areas rocks experience not only large temperature changes but may also suffer very large temperature gradients when, for example, internal temperatures are still dropping as a cold wave is conducted through the rock but the surface is being rapidly heated by the sun. Rapid quenching of the surface in a rainshower may also produce a shock effect which could cause fracturing. The production of sand-sized and other small grains from granular disintegration of rocks has been attributed to differential expansion of various minerals of different colours but there is little evidence in favour of this hypothesis (see Schattner, 1961).

The field evidence suggests that insolation weathering can occur on bare surfaces, but the experimental evidence is equivocal and the role of water is still uncertain.

Alternate wetting and drying has been suggested as a process responsible for the disintegration, or slaking, of fine-grained rocks (Ollier, 1969). The nature of the processes involved is not fully understood but it may involve the effects of both water

2.13 Weathering of dolerite, Antarctica. The coarse-grained variety is disintegrating but the fine-grained rock is nearly immune to the prevailing weathering processes.

and air (Taylor and Spears, 1970; Franklin and Chandra, 1972). Water molecules which are adsorbed on to the negatively charged surface of a crystal through their positively charged hydrogen ions may eventually force rock particles apart and then permit a collapse to a lesser volume in a subsequent drying of the rock; repeated adsorption and loss of moisture then finally disintegrates the rock (Plate 2.15).

Air breakage may occur when, during dry periods, evaporation from the surfaces of rock fragments promotes high suctions within the rock pores. At extreme desiccation the bulk of the voids will be filled with air which, on rapid immersion in water, becomes compressed by capillary pressures developed as the rock becomes saturated. High disruptive internal air pressures may cause cracks to be opened and extended from their tips.

The expansion and contraction of clays, with wetting and drying, is not normally included in a list of physical weathering processes, but, as clay minerals of the montmorillonite group can experience an increase in volume of up to 15 times on wetting, their expansion and contraction in clay-rich rocks such as mudstone can be a major weathering process, and may promote rapid losses of strength and landsliding (see Chapter 6).

Comparative studies of physical weathering processes indicate that, where conditions for the existence of saturated solutions are suitable, salt weathering is a most effective cause of rock breakdown and more effective than frost-induced shattering. Frost action and salt weathering in conjunction produce a higher rate of breakdown than salt action alone (Goudie, 1974). These two processes are far more effective than insolation and wetting and drying. Rock characteristics — especially water absorption capacity and bulk specific gravity — appear also to be highly significant controls on the rate of disintegration.

2.14 Possible insolation shattering of a quartzite boulder, central Sahara. The scale is 30 cm long.

Chemical weathering processes

The disintegration of rocks, such as granite, in the early stages of weathering is partly a physical and partly a chemical process. Studies using a scanning electron microscope (Baynes and Dearman, 1978) reveal that weathering is usually initiated along primary cracks, pores, and open cleavages. Such microfissures permit solutions to penetrate into the rock and commence chemical weathering. Quartz grains suffer microfracturing and pit etching. Microfracturing may occur when solution of neighbouring feldspar crystals causes an increase in porosity and permits expansion of quartz with the release of locked-in residual stresses produced during the early cooling history of the rock from a magma. Feldspars also suffer pit etching (Plate 2.16) and solution along structurally determined planes, while biotite undergoes decomposition to clay minerals and expansion of the crystal lattice. Clay mineral formation results from the removal of ions in solution; if only small amounts of cations are removed cation-rich clay mineral species, such as montmorillonite and illite, can form but intense

flushing and continued weathering leaves only kaolinite or gibbsite.

Chemical weathering processes require heat and water, consequently they tend to be at a maximum in the humid tropics and at a minimum in deserts — especially in polar deserts where the water is in the ice form and not available to take part in chemical reactions.

Solution is usually the first stage of chemical weathering. The amount of solution depends upon the volume of water passing the surface of a particle and the solubility of the solid being dissolved. Because nearly all water that comes into contact with rock is slightly acid, the solubility of the rock minerals has to be expressed in relation to the pH of the water. Minerals vary in their response to the attack of acid water: some are insoluble, some become gels, and some partly dissolve leaving the siliceous framework for the formation of clay crystals. During prolonged attack the siliceous frameworks consisting of silica tetrahedra and alumina octahedra are disintegrated. Silica is slightly soluble at all pH values, whereas alumina (Al_2O_3) is only

2.15 Weathering of a mudstone showing the effect of swelling and shrinking with wetting and drying.

2.16 A scanning electron micrograph of a feldspar crystal weathered in a soil profile. Removal of ions has left the crystal surface deeply pitted. (Micrography by Dr M. J. Wilson, Macaulay Institute for Soil Research, Aberdeen.)

readily soluble below pH4 and above pH9 (Fig. 2.12). Alumina therefore tends to accumulate in the clayey residuum during soil weathering and silica is slowly leached. Amorphous silica is nearly 20 times as soluble as quartz; this difference is important because amorphous silica is frequently a cement in sandstones and the resulting solution of the cement produces separate grains of quartz sand. At pH values above 9 the solubility of silica increases, but such pH values are rare in soils and confined to very alkaline desert conditions.

The solution processes which result in the formation of an alumina-rich residuum with varying amounts of silica are complex because they involve the initial release of ions into solution and then the reactions of these ions with other ions or minerals to form new mineral combinations. The addition of organic acids further complicates the processes so that even a simple order of solubility of ions is not applicable in all situations. The most commonly quoted order of solubility (Polynov, 1937) is:

$$Ca > Na > Mg > K > Si > Al > Fe,$$

but the only invariable rule is that Ca, Mg, Na, and K are all more soluble than Al, Si, and Fe.

Similarly, and because rocks have variable compositions, there can be no fixed order of rock

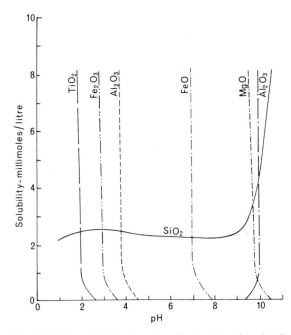

Fig. 2.12 The solubility in water of selected rock and soil minerals as a function of pH (after Siever, 1959).

resistance to chemical weathering although rocks rich in the ferromagnesian minerals and calcic plagioclase are less resistant than those rich in orthoclase, quartz, and muscovite (Table 2.2).

Table 2.2 Stability Order of Common Minerals under Acid Weathering

Most stable	Fe-oxides
	Al-oxides
	Quartz
	Silicate clay minerals
	Muscovite
	K-feldspar (orthoclase)
	Biotite
	Na-feldspar (albite)
	Amphibole
	Pyroxene
	Ca-feldspar (anorthite)
Least stable	Olivine

Hydration is the addition of water to a mineral and its adsorption into the crystal lattice. The adsorption may make the mineral lattice more porous and subject to further weathering. Iron oxides for example may adsorb water and turn into hydrated iron oxides or iron hydroxides:

$$2Fe_2O_3 + 3H_2O \rightleftharpoons 2Fe_2O_3 \cdot 3H_2O.$$
Hematite   Limonite

The hydration reaction is frequently reversible, and because it involves a considerable volume change it is important in physical weathering.

Hydrolysis is a chemical reaction between a mineral and water, that is between the H^+ or OH^- ions of water and the ions of the mineral. In hydrolysis, therefore, water is a reactant and not merely a solvent.

The concentration of hydrogen ions (measured as pH) is of fundamental importance in all weathering processes because: (a) it determines the solubility of silica, and metal oxides; (b) the H^+ ions combine with OH^- ions thus removing them from crystal surfaces and permitting further hydrolysis; and (c) H^+ ions replace other cations in the mineral crystals. The major sources of H^+ ions in the soils are acid clays (a clay with a high proportion of H^+ in its cation-exchange sites) and living plants. The plant roots exchange H^+ ions for nutrient ions (Fig. 2.13).

It is often said that hydrolysis is a major process in the production of clay minerals but it seems more probable, from the rarity of pure water in soil solutions and the abundance of bicarbonate solutions, that the allied process of carbonation is the cause of

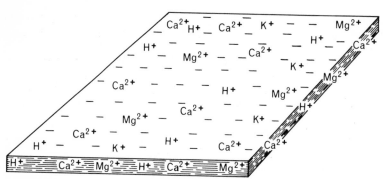

Fig. 2.13 Diagrammatic representation of a clay crystal, having sheet structure with a dominantly negative charge, and its adsorbed cations. No attempt has been made to show cations adsorbed between the plate-like crystal units nor the numerous adsorbed water molecules.

the decomposition of feldspars to clay minerals (Curtis, 1976).

Carbonation can occur readily because bicarbonate is nearly always present in weathering solutions and is the major solution component of drainage waters. The bicarbonate ion is derived from the photosynthetic fixation of carbon dioxide and its subsequent respiration from plant roots and the bacterial degradation of plant debris. The carbon dioxide in the soil atmosphere dissolves and dissociates to produce bicarbonate:

$$H_2O + CO_2 \rightleftharpoons H^+ + HCO_3^- \text{ (bicarbonate)}.$$

The supply of acid soil waters will thus be largely controlled by plant activity. The weathering of feldspars may thus be represented as:

$CaAl_2Si_2O_8 + 3H_2O + CO_2 \rightarrow$
(anorthite) $Al_2Si_2O_5(OH)_4 + Ca^{2+} + 2HCO_3^-$
                          (kaolinite)   (solution)

$2NaAlSi_3O_8 + 3H_2O + CO_2 \rightarrow$
(albite) $4SiO_2 + Al_2Si_2O_5(OH)_4 + 2Na^+ + 2HCO_3^-$
                          (kaolinite)   (solution)

$6KAlSi_3O_8 + 4H_2O + 4CO_2 \rightarrow$
(orthoclase)

$K_2Al_4(Si_6Al_2O_{20})(OH)_4 + 12SiO_2 + 4K^+ + 4HCO_3^-.$
(illite)   (solution)

In the humid tropics silica is rapidly leached away and the above equations should probably be rewritten to show not silica but soluble silicic acid in the soil solution:

$$SiO_2 + H_2O \rightarrow Si(OH)_4.$$
(solution)

The dissolution of limestone and dolomite follows the same pattern with half the bicarbonate being derived from the rock minerals:

$$CaCO_3 + H_2O + CO_2 \rightarrow Ca^{2+} + 2HCO_3^-.$$
(calcite)   (solution)

Other important reactions are those including sulphates which may be derived from the dissolution of gypsum and anhydrite, but more commonly they enter weathering profiles as a result of weathering of metal sulphides, especially of pyrite:

$4FeS_2 + 15O_2 + 8H_2O \rightarrow 2Fe_2O_3 + 16H^+ + 8SO_4^{2-}.$
(pyrite)   (sulphuric acid)

Shales containing pyrite are thus weathered rapidly and may provide drainage waters which will attack limestones if these are near by.

The terms *oxidation* and *reduction* are used in two different, but related, senses. Chemists use them to mean the removal of electrons from and their addition to the atoms of some elements involved in the chemical change, and oxygen need not be involved. For example, iron may be combined with sulphur to form FeS, in which case the iron is said to be oxidised.

The more common use of the term oxidation in earth sciences is to imply a simple reaction of a substance with oxygen to form oxides. Oxidation in this sense nearly always acts through the intermediary of water in which the oxygen is dissolved, although the presence of air is also required. One of the most readily oxidisable elements is iron, and the most easily recognised first alteration in weathering is the oxidation from the ferrous to ferric state:

$$4FeO + O_2 \rightarrow 2Fe_2O_3.$$
(ferrous oxide,   (ferric oxide,
reduced state)   oxidised state)

In thin sections of rocks this can be recognised by the presence of iron staining along cracks and between mineral grains. Oxidation is measured potentiometrically, on the Eh scale, where a positive

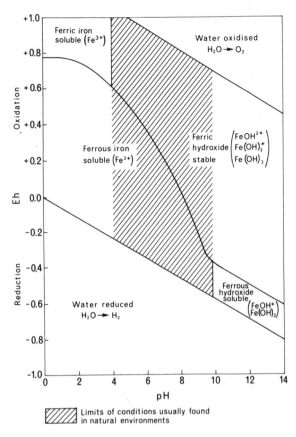

Fig. 2.14 The solubility of iron as a function of Eh and pH (after Hem and Cropper, 1959).

reading indicates a degree of oxidation and a negative reading indicates a degree of reduction. Water acting as the intermediary agent by which oxygen operates as an oxidant has an oxidising potential (Eh) of 810 mV at pH7. Where ferrous iron links silica tetrahedra in a silicate structure it may react with oxygen to form ferric oxide, the silicate structure then falls apart and alteration begins:

$$2FeSiO_3 + \tfrac{1}{2}O_2 + 5H_2O \rightarrow$$
$$Fe_2O_3 + 2Si(OH)_4 + 2H^+ + 2e^-.$$
$$\text{(solution)}$$

Thus the iron accumulates in oxide form within the soil or on an exposed rock outcrop. Eh and pH are important factors in biological and chemical processes and they interact to control the form in which iron exists in soils (Fig. 2.14).

Reduction is the reverse of oxidation and involves the release of oxygen from compounds. The reduction of iron to the ferrous form renders it more soluble and mobile. Much of the work of reduction

in waterlogged sites is carried out by anaerobic bacteria. In such acid conditions oxygen is rapidly used up and the resulting iron compounds are grey and green in colour — contrasting markedly with the red and brown colours of iron compounds in oxidising aerobic conditions. On well drained sites organic compounds derived from the leaves of plants are able to combine with many soil ions to form complex compounds which may be classed as *chelates*. Reduction may be aided, or may even require chelating agents, such as gallic acid (a tannin derivative), which act as electron donors, ferric iron acting as the electron receptor:

$$Fe(OH)_3 + e^- \rightarrow Fe^{2+} + 3OH^-.$$
$$\text{(ferric hydroxide)} \quad \text{(ferrous iron)}$$

Conditions in soils are not constant and, if there are seasonal fluctuations of the water table, alternate oxidation and reduction of iron are induced and the Eh varies: the soil becomes mottled where reduced iron is oxidised to red oxides. During the decomposition of organic matter the Eh remains above +400 mV, but in waterlogged conditions it falls below +200 mV.

Secondary mineral formation is the normal consequence of mineral weathering. It involves both the modification of crystal structures by cation-exchange and cation substitution to produce new minerals, and also the precipitation of new minerals from solution. The precipitation of calcium carbonate in soils of dry regions is an example of the latter process.

An example of the change of mineral species by ionic substitution is the alteration of mica or illite by the replacement of interlayer K^+ with other ions from the weathering solution. Such a reaction is the principal means of vermiculite formation in soils. Whereas the original mica structure is retained by vermiculite, it is different in that it has the capacity to expand when wet.

The stability of newly formed secondary minerals is a function of the environment. It depends to a large extent upon the continual release of soluble products from other weathering minerals, and if they cease to be supplied the secondary minerals will become less stable and may be replaced by other secondary minerals. Although the weathering sequence is from complex to simple clay minerals, different parts of the soil complex may be characterised by different clay minerals, and the minerals will be of mixed species. In general there is a progression of alteration:

feldspar → montmorillonite → kaolinite → gibbsite.

The clay composition of a soil therefore indicates the stage of weathering and this can be expressed generally by a diagram (Fig. 2.15). Knowledge of the weathering sequence also permits, to some degree, inferences to be made as to how soils will change their properties with time.

The formation of clay minerals from feldspars is probably the most important single weathering process of humid environments. The three basic reactions — hydration, carbonation, and oxidation-reduction — are the fundamental processes and the importance of carbonation, rather than simple solution, needs to be stressed.

Biotic weathering

Biotic weathering is a combination of chemical and physical weathering effects of which the following are the most important: (a) breakdown of particles by the action of roots and burrowing animals (Plate 2.17); (b) transfer of material by animals; (c) simple chemical effects as when solution of rock is increased by the CO_2 released into the soil during respiration; and (d) complex chemical effects such as chelation,

and the formation of complexes of organic-mineral substances.

Soil bacteria play a very important part in the weathering of minerals in many soils. Those chemotrophic bacteria which oxidise mineral substances such as sulphur, iron, and manganese for their metabolism are extremely active in reducing conditions and play a major part in the weathering of waterlogged soils.

Chelates are soluble and retain ions from iron and aluminium compounds and where the movement of the soil solution is downwards they thus permit the leaching of these ions. This washing down of chelates (i.e. *cheluviation*) is a process of major importance in podzolisation.

Organic acids that form chelates attack silicate minerals and may be very important contributors to soil weathering under certain conditions of climate and vegetation. Many coniferous trees and podocarps produce such organic compounds (polyphenols) and contribute to podzolisation. The decomposition products of grass leaves appear to have little or no chelating effect.

The role of lichens and mosses, which can grow

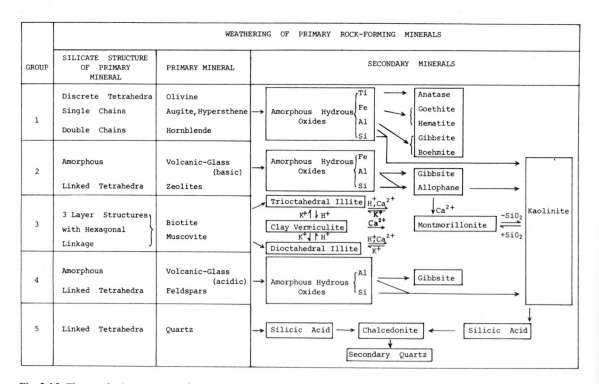

Fig. 2.15 The weathering sequence of primary rock-forming minerals (after Fieldes and Swindale, 1954).

2.17 Tree roots have grown in an open joint and prised loose the plate-like joint block so that it fell.

directly on rock surfaces, in weathering is not well understood. Some species release organic acids, such as citric acid, and can extract Al, Mg, and SiO_2 from rock, and so promote weathering and some produce organs capable of penetrating cracks (Winkler, 1966). It is not clear whether or not such plants produce a mechanical force which prises grains from rock surfaces.

Human interference with the environment, and especially by releasing industrial pollutants into the atmosphere, has had a marked effect upon rates of weathering. The release of sulphur compounds, which become sulphuric acid, and CO_2 into the atmosphere has greatly increased the acidity of rainfall close to industrial areas of Europe and North America. As a result soils have become more acid and rocks, especially limestones, are undergoing more rapid weathering. In cities, also, building stones are being attacked by acids, and by salt weathering resulting from the reaction of calcite in the rock with sulphuric acid to produce gypsum ($CaSO_4 \cdot 2H_2O$) (Winkler, 1970).

Sequence of weathering processes

Under a humid temperate climate the weathering of an igneous rock might be summarised as:

(1) mechanical fracturing of rock by stress release and opening of fissures between grains by physical processes;

(2) at the same time feldspars, micas, and ferromagnesian minerals will suffer carbonation to produce clay minerals, while releasing insoluble quartz to form an inert mineral skeleton of the soil; iron is oxidised and hydrated;

(3) porosity is increased and calcium, magnesium, sodium, and potassium compounds are removed in solution;

(4) the secondary hydrated silicate clay minerals will be further weathered and iron and aluminium oxides will increase in proportion in the residual soil mass.

Sedimentary rocks break down initially into particles which have a size determined by the original particle size and by the nature of the cement which binds the grains. Sandstones always weather to produce sand, but shales and mudstones may initially produce particles larger than clay-size if cement binds the clay. Soluble rocks like limestones lose their soluble components to leave an insoluble residue. Weathering always tends towards stability and insolubility.

3

Landforms resulting from weathering

The development of most landforms entails the reduction of rock by weathering to a more easily eroded material, and then the removal of weathering products by erosive agents such as running water, ice, wind, or waves. These agents create further rock destruction by abrading other rock surfaces with the debris they carry. The direct production of a distinctive landform by weathering is most readily seen in desert areas or on steep cliffs where little regolith can accumulate. Because continued weathering depends upon the removal of debris most weathering forms are relatively small. If attention is confined to these small forms, however, a false impression will be obtained of weathering significance and the variable resistance to it of different rocks. Much of the relief of the land which is not directly attributable to tectonic forces is due to differential weathering and erosion acting most effectively in weak rocks and along lines of structural weakness. Many examples of these processes are offered elsewhere in this book.

Pits, pans, caverns, and rills

Many outcrops have a microrelief attributable to a great variety of miniature depressions which have formed either along lines of noticeable weakness, such as joints, or on sites which appear to be very uniform. Many of the minor features occur on a variety of rock types although the readily soluble rocks, such as limestones, are particularly prone to various forms of etching. The silicate-rich rocks such as granite, granodiorite, schist, quartzite, sandstone, and gneiss also bear many minor weathering forms, as does basalt.

Although commonly reported from semi-arid and arid environments all forms are found in nearly all climates (Dahl, 1966). Most forms have a great variety of names in many languages so only the most frequently used terms are mentioned here.

Weathering pits of flat surfaces (German, opferkessel; Portuguese, oriçanga; Polish, kokiolki; Australian, gnamma) range in size from a few centimetres in diameter to several metres (up to 7 m have been reported) and from a few centimetres in depth to a metre or more. Pits may be hemispherical or flask-shaped hollows in which the walls are overhanging, or they may be flat pan-like forms; they may be circular or irregular in plan (Hedges, 1969) (Plate 3.1).

The walls of deep pits are often smooth and appear to retreat by spalling rock flakes; their smoothness is not like that produced by the swirling of debris-laden water, and the rock debris on floors of pits is not rounded, so these features are evidently not produced by the pot-holing action of streams or air. The walls of some shallow pans are irregular and, in granitic rocks especially, the rock is laminated.

Spillways lead away from some pans and pits and indicate that water is involved in the formation of many of these features. In a few pans on very soluble rocks small features that look like splash marks surround the pits and suggest that solution of the rock by water is the main process but, as will be seen below, this is only one of several possible processes contributing to the formation of these forms.

As the slope of the rock surfaces increases the shape and variety of weathering pits changes.

Honeycomb or alveolar weathering is characterised by numerous pits a few millimetres or centimetres in diameter and in depth. They may enlarge to produce a fretwork or honeycomb structure. Alveolar weathering occurs most frequently on granular rocks such as tuffs and sandstones and presumably involves selective granular disintegration of rock faces (Plates 3.2 and 3.3).

Tafoni (sing., tafone) are hollows which may be

3.1 Pan formed in quartzite, central Sahara. Note the pitting of the rim. The scale is 30 cm long.

3.2 Small tafoni in granite, Antarctica. Scale is given by the gloves and ice axe.

several metres in depth, width and height, although they are generally smaller. They are cut into steeply sloping rock faces and may have overhanging entrances, called visors. The floors of tafoni are often flat and may be littered with a debris of rock fragments. The shapes of caverns are sometimes influenced by structural features of the rock, but this is not always the case.

The origins of the various kinds of cavernous weathering are varied – they occur on a great variety of rocks; in all climates; on outcrops with a great variety of forms, and with both large and small exposed surfaces around them. Attempts to classify them according to shape, rock type, or climatic environment (e.g. Wilhelmy, 1964) have, therefore, not been successful: exceptions seem to be as common as the cases which fit the classification schemes.

Most pits are assumed to be initiated at some depression or weakness in a rock surface and to enlarge gradually, perhaps by consuming other pits. The rates at which they change are varied and range from deepening at a rate of 1 cm/year on some coastal cliffs in weak rock to an imperceptibly slow rate. In Antartica progressive development of tafoni, with increasing size of features and abundance on boulder surfaces, has been discerned in rocks on the surface of glacial moraines which can be ranked in order of age (Calkin and Cailleux, 1962). The processes of tafoni enlargement include granular disintegration and exfoliation of rock flakes. The removal of this debris, which is necessary for the continuation of weathering, may be by wind or

3.3 Sculpturing below a granite overhang showing exfoliation, central Namib desert. The cave is 2 m high.

under gravity, and for pans it may be aided by the overflow of water.

A variety of processes have been identified as causing cavernous forms. In some tafoni in the central Sahara it is obvious that granitic rocks are decaying by the decomposition of feldspars and mica to clay minerals, to leave a grus which is composed largely of quartz sand and fine gravel. Elsewhere, for example in Antarctica and the Namib desert, there is no trace of chemical decomposition in the rock spalls and the debris on the floor of each tafone is composed of unaltered quartz and feldspar crystal fragments (Selby, 1977b). In some cold climates frost action may contribute to rock weathering, for snow and ice will melt in contact with a rock surface warmed by insolation (perhaps to +20 °C) when the air temperature is at zero or below. Freezing will occur as soon as the rock is again in shadow.

Case hardening or the production of a weathering rind is a common feature of many rock outcrops. The minerals which produce the hardening are commonly iron and manganese hydroxides which may be associated with clay minerals such as goethite, although sometimes amorphous silica is the hardener. These minerals are dissolved from the rock interior, so weakening it, and drawn by capillary attraction towards the rock surface where they are precipitated and dehydrated, sometimes to a less soluble mineral form. The precipitates block the pores and fissures in the rock surface and so render it more resistant to weathering. Case hardening may be involved in the development of the visors above the openings of tafoni, although not all visors are so hardened, and it can also occur along joint planes at depth.

Organic action by algae, lichens, and fungi appears to be particularly important in the development of some cavernous forms. By releasing CO_2 these plants may encourage carbonation and particularly the decay of feldspars. Some lichens may also be involved in the extraction of iron, and perhaps silica, from rock minerals (Franzle, 1971) and are responsible for superficial rock staining to black and brown colours. This process weakens rock surfaces and does not cause case hardening. As the lichens shrink and swell with alternate dehydration and hydration they can also tear away rock particles and so enlarge hollows.

The high pH of water in many weathering pits

3.4 Rills on limestone, northern New Zealand.

suggests that alkaline waters, in which the solubility of silica is greatly increased, are responsible for much of the enlargement, and soluble salts may contribute to weathering both by salt weathering and by raising the pH of water. It has been pointed out by Twidale and Bourne (1975b) that some forms of cavernous weathering are initiated and develop within the regolith in the zone of ground water where chemical processes are almost certainly responsible for their production. Partly undercut or flared slopes may also originate in this way. Many forms of cavernous weathering, however, develop in areas where there is no trace of a deep regolith and no evidence that one ever existed on the present land surface — as in Antarctica and the Namib.

It is evident that many processes may contribute to cavernous weathering, although some are not active in certain environments, and even in one area the significance of such processes as hydration, alkaline water, and organisms will vary with slope, shade, and other factors. About the only valid

generalisation on the causes of tafoni is that water is nearly always involved in the weathering process (Dragovich, 1969).

In many environments cavernous weathering forms are present but no longer developing. This may be because they were formed under climatic conditions which no longer exist, as in parts of the northern Sahara where tafoni formed at the foot of some cliffs when water tables were higher than they are now (Smith, 1978), or tafoni may become inactive when they are so deepened that the diurnal fluctuations of moisture or surface heating, which were effective in shallow recesses, cannot reach the shaded and protected surfaces at the back of deepened hollows.

Rills, rillen, grooves, or gutters, as they are variously known (Plate 3.4), can also form in many different types of rocks and in many environments. Rills may form a regularly spaced pattern with individual rills being 5–30 cm deep and 20–100 cm wide, where they form the drainage network for evenly sloping rock surfaces, or they may be single features. Some are connected with lines of pits. Few rills have a large enough catchment with enough rock debris in it for them to be regarded as miniature stream beds. It seems much more probable that solution by alkaline waters, especially on silicate rocks, is an important process. The activity, in ways not well understood, of algal slimes from pits, and of lichens may also have some significance in the enlargement of rills.

Convergence of forms

It has been stressed in this section, and will be stressed again with reference to other landforms such as tors and bornhardts (Chapter 8), that similar landforms can be produced by a variety of processes in a variety of environments. This convergence of landforms of various origins towards a similar geometry is sometimes known as the principle of equifinality (Bertallanffy, 1950). It demonstrates that it is seldom possible to study the shape alone of the landform and so deduce its origin. It also implies, where detailed evidence is lacking, that it may be impossible to determine how a particular feature has evolved.

Duricrusts

An important group of landforms owes its origin to the presence and dissection of indurated crusts, of various mineral compositions, which have formed as a result of weathering processes or by the redistribution of weathering products. Such crusts frequently form resistant caps to hills and underlie extensive plains and plateaux, especially of the more stable continental platforms with tropical, subtropical, or temperate climates. The crusts, classed together as duricrusts, can also occur in other environments but seldom with as great a thickness or forming major features of the landscape.

The terminology of duricrusts is very confused and is still undergoing revision. The confusion results from disagreement over the origin of duricrusts and also because many original definitions have been changed as knowledge has extended. Many genetic definitions are now known to be misleading or wrong.

A duricrust is a product of processes acting within the zone of weathering to cause the accumulation of iron and aluminium sesquioxides, silica, calcium carbonate or, less commonly, gypsum and halite. The accumulation of these compounds may be either a result of removal of other materials to leave an enrichment of the crust-forming minerals, or it may be an enrichment caused by deposition from water or of windborne minerals which can then accumulate and harden to form a duricrust. This secondary or depositional origin can be of materials derived locally by material moving down a hillslope from a higher crust, or it may be a much more widespread redistribution by streams or wind.

In the twentieth century there has been a growing tendency to use terms which indicate the cementing agent or composition of duricrusts, and to add to the indicator of the cement the ending '-crete' or the word 'crust' to the composition term: thus calcrete or calcitic crust is used for a duricrust composed largely of calcium carbonate (see Table 3.1). This terminology avoids the very large number of local names which are in common use and also, for crusts rich in iron and aluminium oxides, the term 'laterite' which has caused much confusion. The application of such classifications is, however, not always simple because the chemical content of duricrusts can vary over short distances in response to change in rock mineralogy and site factors.

The word 'laterite' was originally used by Buchanan, in southern India in 1807, to refer to an indurated red clay containing much iron. The clay had many cavities or pores and was soft while excluded from the air and could be cut with a spade, but it hardened on exposure and was commonly used to make bricks for building. Hence Buchanan coined the word 'laterite' from the Latin *later*, a brick. Confusion in the use of the term developed because it was later applied to ironstone cappings found on the plateaux of southern India, to the zonal soils of the humid tropics, to the whole weathered profile beneath a laterite of Buchanan's meaning, and to iron-rich breccias and slope-wash accumulation. Because such a variety of materials with many types of composition and various origins have been called laterites it is now necessary to define the word whenever it is used. Some workers now prefer to use the terms 'ferricrete' for iron-cemented crusts and 'alcrete' for aluminium-cemented crusts, and distinctive soil names for soil types.

Among the local names for iron and aluminium-rich crusts are: canga (Brazil), murrum (Uganda), ferricrete (South Africa), cuirasse (France), plinthite (USA), and pisolite (Australia). Aluminium-rich crusts are widely called bauxite crusts. Silica-rich crusts have been called: surface quartzite (South Africa), porcellanite – especially of a very fine-

Table 3.1 Classification of Duricrusts

By cementing agent	By dominant content	Essential chemistry	Typical crystalline minerals
silcrete	silitic crust	SiO_2	quartz (90–95%)
	siallitic crust	SiO_2, Al_2O_3	quartz (aluminous compounds often amorphous to crypto-crystalline)
	fersilitic crust	Fe_2O_3, SiO_2	hematite, quartz
ferricrete	fersiallitic crust	Fe_2O_3, $FeO(OH)$, SiO_2, $Al_2O_3 \cdot nH_2O$, + $AlO(OH)$	hematite, geothite, quartz, gibbsite, +boehmite
	ferrallitic crust	Fe_2O_3, $FeO(OH)$, $Al_2O_3 \cdot nH_2O$, $AlO(OH)$	hematite, geothite, gibbsite, boehmite
	ferritic crust	Fe_2O_3 (up to 80%), $FeO(OH)$,	hematite, geothite
	fermangitic crust	Fe_2O_3, MnO_2	hematite, pyrolusite/psilomelane
alcrete	tiallitic crust	TiO_2, $Al_2O_3 \cdot nH_2O$	rutile/anatase, gibbsite
	allitic crust	$Al_2O_3 \cdot nH_2O$ (up to 60%), $AlO(OH)$	gibbsite, boehmite
calcrete	calcitic crust	$CaCO_3$	calcite (60–97%)
	calcsilitic crust	$CaCO_3$, SiO_2	calcite (silica often chalcedonic, etc.)
gypcrete	gypsitic crust	$CaSO_4 \cdot 2H_2O$	gypsum
salcrete	halitic crust	NaCl (usually impure)	rock salt

Note: gibbsite is $Al_2O_3 \cdot 3H_2O$; boehmite, $AlO(OH)$; geothite, $FeO(OH)$; hematite is $\alpha\text{-}Fe_2O_3$.
Source: based on Dury, 1969.

grained variety — and grey billy (Australia). Calcium carbonate rich crusts have been called: caliche (USA), croûte calcaire and carapace calcaire (France), travertine (Australia — where the $CaCO_3$ is deposited by a spring or from a stream this term is widely used outside Australia), tosca (Argentina), sheet limestone (South Africa), kankar and kunkur (India), tufa (South Africa — this term is also widely used for spring deposits).

Profiles

Duricrust profiles are generally of three types — those in which the crust is derived from the (a) overlying soil or (b) from underlying weathered rock, and those in which (c) the crust-forming material is detrital, that is, it has been transported and deposited or precipitated.

Ferricrete and alcrete profiles which are developed *in situ* usually have a number of horizons. The horizons vary in thickness, hardness and colour depending upon the degree of development and preservation of the profile, but the following features are frequently recognisable (Plate 3.5 and 3.6):

0-2 m thick — soil zone, often sandy and sometimes containing nodules or concretions; this may be eroded away;

1-10 m thick — crust of reddish or brown hardened or slightly hardened material, with vermiform (or vermicular) structures (i.e. having tube-like cavities 20-30 mm in diameter) which may be filled with kaolin; less cemented horizons may be pisolithic (i.e. formed of pea-sized grains of red-brown oxides);

1-10 m thick — mottled zone of white clayey 'kaolinitic' material with patches of yellowish iron and aluminium sesquioxides;

up to 60 m (but generally less than 25 m) thick — pallid zone of bleached kaolinitic material; the distinction between the mottled and pallid zone is not always apparent and they can be reversed; silicified zone which may be hardened;

3.5 An exposure of ferricrete crust above a pallid zone, northern Nigeria.

Up to 60 m thick	weathered margin of deeply weathered rock showing original rock structures.

(*Note*: mottled and pallid zones are not always present.)

In detrital ferricretes and alcretes, weathered zones beneath the crust (which are collectively sometimes called the lithomarge) are not present and the crust may have a variety of structures with inclusions of redeposited duricrust material and of 'foreign' rock debris which may become ferruginised. It is also apparent that many *in situ* ferricretes and alcretes do not have underlying deeply weathered profiles.

Ferricretes are widespread in the humid tropics and some are developing at the present. The formation of a crust may occur upon exposure of the iron-rich soil horizon and this has been aided by deforestation and soil erosion following excessive burning and agriculture (Plate 3.7). Laterisation (as it has been generally called) has thus been aided by human activity although most ferricretes are natural and often ancient. Alcretes occur in the rainforest zones of the humid tropics and are economically very important as a major source of bauxite.

Silcrete profiles consist of an indurated silicified layer which may be up to 3 m thick, with as much as 95 per cent SiO_2, with quartz grains set in a matrix of fine quartz or opaline silica. The silica-rich horizon commonly has columnar structure and is grey, yellow, or brown colour. The rock has a vitreous appearance and non-vesicular varieties have

3.6 A soil profile in the rainforest of central Zaïre. The pisoliths are pea-sized concretions of iron oxides.

3.7 An exposed ferricrete beneath a thin soil and sparse savanna woodland, northern Nigeria.

a conchoidal fracture. When struck with a hammer silcrete may emit a pungent smell. The silica crust usually overlies a kaolinised zone which merges downwards with the parent rock. Silica-rock and silica-cemented gravels are also referred to as silcrete. These are apparently derived from precipitates and are not part of the local weathering profile.

Silcrete is widespread in western and central Australia and in southeastern Africa. Isolated silicified boulders from parts of France, Central Otago (New Zealand) and southern England — where they are called sarsen stones — may be remnants of ancient silcretes. There is no conclusive evidence that silcretes are forming anywhere today. The extensive silcretes of central Australia developed during the Tertiary (Exon *et al.*, 1970), and thin silcrete horizons formed round the bases of many hills in the Adelaide to Lake Eyre region during the Pleistocene (Wopfner and Twidale, 1967).

Calcrete profiles are widely distributed in arid and semi-arid lands. On a world scale calcretes have about 80 per cent calcium carbonate (Goudie, 1973), and range in thickness from nodules in a soil

3.8 A desert soil with thin lenses of calcium carbonate shown by their pale colour, northern Sahara, Morocco.

profile to massive sheets over 10 m thick. Netterberg (1967) has suggested a sequence of calcrete types which is essentially an evolutionary sequence (Plates 3.8 to 3.10):

3.9 Nodular calcrete, near Kalkrand, Namibia.

3.10 Massive hardpan calcrete, near Kalkrand, Namibia.

calcified soil
↓
nodular calcrete
↓
honeycomb calcrete
↓
hardpan calcrete
↓
REWORKING
↓
boulder calcrete

Lime accumulation occurs within soils at the present time in semi-arid zones where evapotranspiration is in excess of precipitation for most of the year. Leaching through the soil profile is negligible and the carbonate is deposited at a level in the profile controlled by water removal by plants, by evaporation, or by the CO_2 content of the soil atmosphere (if CO_2 is released from solution $CaCO_3$ is precipitated). The removal of water by plant roots sometimes causes the carbonate to be precipitated around the roots where it forms casts called rhizomorphs (Plate 3.11).

All calcretes require a source for their calcium carbonate and this is usually locally derived from limestone, marble, or calcareous sandstones. Where the local rocks are deficient in lime, calcretes can form from deposition from flood waters, and such a fluvial origin probably accounts for the thick valley calcretes of the Kalahari and some of the calcrete cemented gravels of the inland Namib desert.

Both calcrete and silcretes may be formed, or at least added to, by volcanic ash or windblown dust which can accumulate in soils and also pass through the organic cycle before being precipitated in the soil (Fig. 3.1).

Gypcrete and salcrete profiles are exceedingly rare because of the high solubility of gypsum and halite. They are preserved only under arid climates and, because they are usually precipitates from the evaporation of sea water or saline desert streams, their profiles usually consist of a white or buff-coloured crust overlying lake or marine muds.

Origins of ferricretes, alcretes, and silcretes
The basic problem of understanding the genesis of ferricretes and alcretes is to determine how the chemical separation of iron, aluminium, and silicon

3.11 Calcium carbonate accumulations around fossil grass roots in ancient dune sands, central Namib desert. Such root casts are called rhizomorphs.

can occur. These three elements are normally relatively insoluble yet they are removed from their parent silicate and ferromagnesian rocks and then selectively concentrated. The commonly offered explanations for selective removal of Si and weakly soluble metals include considerations of: (1) rates of reactions; (2) preferential solubility of one element over a given pH range; (3) preferential

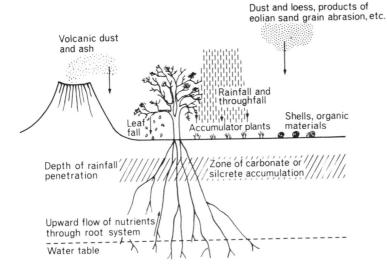

Fig. 3.1 Possible sources for accumulations of soil carbonate and silica. These sources do not include deposition from flood waters (after Goudie, 1973).

solubility of one element due to organic complex-ing; and (4) the availability of oxygen in ground water for oxidising processes – a condition usually equated with Eh. While any of these considerations may be relevant to particular situations it is most probable that (2) and (4) provide the most generally valid explanations of solubility.

Selective removal from the soil of alkalis, alkali earths, and silicon occurs at a certain combination of pH and Eh where the solubility of aluminium and metal oxides and hydroxides is less than that of quartz. Selective removal of iron from the soil with retention of aluminium requires unique Eh and pH conditions (Norton, 1973). At low pH aluminium is mobilised, but iron requires both low pH and low Eh before it is mobilised. At high Eh iron is im-mobile and becomes enriched, and at high pH and Eh both iron and aluminium are immobile and will be retained in soils. Given the availability of Fe and Al the formation of an aluminium-rich soil (bauxite) or an iron-rich soil will depend upon a critical combination of Eh and pH. A secondary enrichment of Al may be brought about by an increase in pH of groundwater, whereas secondary enrichment of Fe may be caused by an increase in pH or an indepen-dent Eh increase. The effect of chelating organic materials is to change Eh and pH conditions towards those favouring increased solubility of iron and aluminium.

Silica as quartz crystals has solubilities ranging from 6 to 14 ppm, but quartz dust and amorphous silica has solubilities of up to 140 ppm. This suggests that the form of silica is very important in influenc-ing its solubility, although the kinetics of pure silica solutions seldom apply in complex environ-ments within the soil.

Knowledge of the reactions which produce mobility of various elements in soils having organic materials within them is still limited and the dis-cussion above is, no doubt, greatly oversimplified. Our ignorance is increased by the certainty that many iron- and aluminium-rich soils and crusts are of considerable geological age – some in Australia are certainly as old as the Cretaceous and ferricrete formation there seems to have continued through Tertiary times (see Van de Graaff *et al.*, 1977 for a review). These crusts may have developed under vegetation covers unlike any existing today and in climates that were probably warmer, and perhaps with different rainfall regimes from those of the present humid tropics.

The greatest degree of iron and aluminium con-centration in modern weathering profiles appears to be occurring under tropical monsoon seasonal climates which encourage a fluctuating water table. This prompted the hypothesis that a seasonally fluctuating water table is essential for ferricrete formation. While there is no doubt that water table fluctuations promote considerable variations in soil Eh and pH there is much doubt that the fluctuations have a sufficiently large vertical range to cause precipitation of oxides over the great depth of many ferricrete profiles (up to 60 m). It is probable also that a rising water table would have an upper layer of relatively 'young' water overlying 'older' water with higher iron concentrations. It is unclear how young water with low iron concentrations could greatly increase ferricrete thickness from below. Furthermore, for a forest vegetation cover to survive, it is improbable that soil moisture levels ever fall far below wilting point, so drying of the soil is presumably limited. Water table fluctuations and capillary rise of ground water are now usually considered to be of limited significance in the formation of thick ferricretes and alcretes. The hypothesis of Lelong (1966) that migration of iron from depth in the profile may occur by ionic diffusion, during the early part of a dry season, has yet to be evaluated thoroughly. That of Trendall (1962) suggests that the most active weathering and mobilisation of iron occurs in the vadose zone, during the wet season, with concentration of iron at its upper surface to form an embryonic ferricrete horizon of minimal thickness. During the ensuing dry season salts in solution would move downwards with the falling water table, and the upper part of the profile would then settle. In this way, says Trendall, thick ferricretes would be formed as continued settling lowers the ground surface while the ferricrete layer thickens. This hypothesis requires prolonged periods of slow ground downwearing without marked changes in the form of the land surface.

Most early hypotheses stressed the relationship between the ferricrete and the underlying pallid zone of *in situ* crusts. It was generally believed that the iron of the crust must be derived from the pallid zone. Few quantitative assessments of the amount of iron in pallid zones have been carried out, but the available data suggest that pallid zones are not noticeably depleted in iron. Secondly not all ferri-cretes are underlain by pallid zones. These facts have led McFarlane (1976) to suggest that pallid zones are not the source of iron and that they develop separately from and often after a ferricrete.

Her hypothesis is that: (a) original iron and aluminium oxide precipitates form as pisoliths within the soil in the relatively narrow range of fluctuation of the ground water table; (b) as the land surface is lowered, by surface erosion and stream incision, the water table sinks and further pisoliths form below the existing precipitates; (c) when stream downcutting ceases the water table stabilises and a massive variety of vesicular or vermiform ferricrete develops; and (d) deepening of a ferricrete profile can continue because iron oxides are mobilised from the surface by the action of vegetation and are reincorporated at the base of the ferricrete horizon, and because additional oxides are available from newly weathered rock below the ferricrete. In this hypothesis pallid zones are neither necessary nor related to ferricrete formation, but they develop beneath ferricretes by leaching, especially when the land surface is incised and groundwaters can move downwards through permeable duricrusts, and then laterally through pallid zone materials towards the streams (Fig. 3.2).

The extensive silcrete deposits of central Australia were formed in mid-Tertiary to early Pleistocene times. It has been suggested by Stephens (1971) that the silica of which they are composed was derived from the selective leaching occurring beneath the ferricretes and alcretes of northern and eastern Queensland. He suggested that the drainage waters carrying the silica in solution, or as a gel, seeped and flowed towards central Australia where the silica was precipitated under a drier climate in areas of low relief, and restricted surface drainage with high pH. The silcretes were thus discontinuous, but because of the low relief, none the less extensive. This hypothesis certainly accounts for many silcretes which are clearly detrital in origin and explains the thick lenses of gypsum found in many silcrete profiles, for gypsum is commonly formed in desert salt lakes. The flood-deposit hypothesis is less satisfactory for silcretes above a related kaolinised profile unless it can be established that the silica-rich flood waters invaded old and deep weathering profiles over large areas. If this did occur then it is possible that some silcretes result from silicification

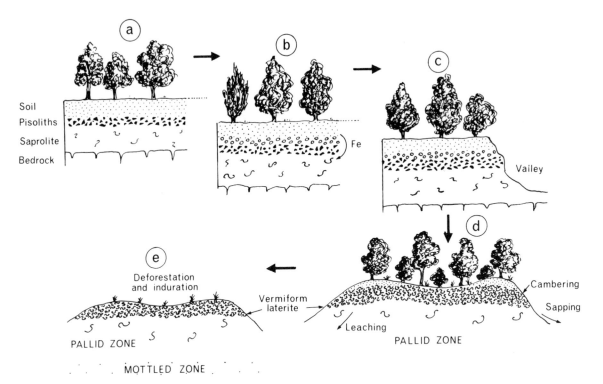

Fig. 3.2 McFarlane's (1976) model of laterite (ferricrete) formation below a downwearing surface a–c, followed by incision and the formation of a pallid zone, and then by deforestation, soil erosion, and induration of exposed ferricrete.

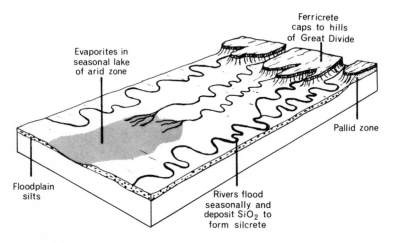

Fig. 3.3 A schematic representation of Stephen's (1971) hypothesis of the formation of silcrete in central Australia.

of ferricrete and others may be the hardened remnants of truncated ferricrete profiles. Gravels derived from silcrete now form the surface of extensive desert gibber plains, such as those of the Sturt Desert. The formation of opal in cavities in silcrete by rhythmic precipitation of silica has permitted the development of extensive mining in areas around Coober Pedy from where 90 per cent of the world's opals are derived.

The concentration of ferricretes and alcretes in the humid tropics, of silcretes in the subtropics, and calcretes in areas with a precipitation of less than

500 mm per year and temperate or tropical temperature regimes, has produced ideas that duricrusts may be classified on a zonal basis. Watkins (1967) suggested a scheme for part of Australia (Fig. 3.4). This scheme does not take into account the great age of many duricrusts nor the complexities of climatic changes they have suffered during their formation, although it does indicate an approximate degree of zonation which is commonly recognisable in tropical and subtropical zones. The dry margin of ferricrete formation in West Africa, for example, appears to be about coincident with the wet margin

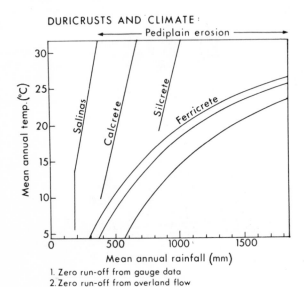

1. Zero run-off from gauge data
2. Zero run-off from overland flow
3. 50mm mean annual run-off
4. 250mm mean annual run-off

Fig. 3.4 (a) Watkin's (1967) model of the relationship between climate and various kinds of duricrust;

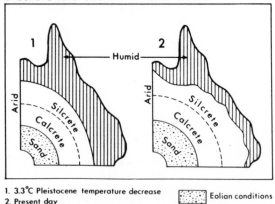

1. 3.3°C Pleistocene temperature decrease
2. Present day

▒ Eolian conditions

(b) Approximate zonation of duricrusts in Australia and the influence of climatic change. This concept assumes that duricrusts formed under climates like those of today. This concept may be false and does not consider the relict nature of many duricrusts.

of calcrete, at 500 mm of rain per year. The absence of obviously developing modern silcrete makes reliable comment upon the climate in which it evolved virtually impossible.

Rates of formation and hardening

Goudie (1973) suggests that rates of crust enrichment may reach about 1 m in 0.3 to 2.3 million years for ferricretes, and 1 m in 0.3 million years for calcrete. The figure for ferricrete is largely based upon the work of Trendall (1962) who calculated how much iron could be derived from the bedrocks in Uganda and concluded that 10 m of ferricrete could be derived from 200 m of bedrock over a period of about 35 million years.

It has been suggested by Persons (1970) that, in the Cameroons (West Africa) under a seasonal rainfall averaging 3250 mm per year, a 2 m of complete induration of ferricrete, after deforestation and exposure, would take about 100 years. Superficial hardening can occur in a few months or years. The process of hardening is, at least partly, due to dehydration of the iron and aluminium oxides and to the development of crystallinity in goethite and hematite (Maignien, 1966). Harder laterites are usually found to have higher iron contents and to have been exposed for greater lengths of time.

The process of reversal of hardening is very slow: it can usually occur only where there is an accumulation of superficial deposits and the establishment of vegetation on them.

Calcrete hardening is a result of cementation as precipitates form in soil voids, and by recrystallisation of the carbonates.

3.12 Calcrete cap to a small mesa, northern Sahara.

Landforms and duricrusts

Duricrusts form most readily on surfaces with gentle slopes. As a result they provide resistant caps to plateaux and hills (Plate 3.12) and act like cap rocks to buttes, mesas and, if warped by earth movements, can form the dip slopes of hills. Ferricretes have very high porosity so water may infiltrate through them very rapidly. As the water moves laterally in, or above, the kaolinised zone below, it can cause high seepage pressures and produce natural pipes which may give rise to springs on slopes beneath a dissected plateau surface. The slopes beneath an exposed ferricrete are, therefore, frequently remodelled by rilling and seepage in the weak materials of the kaolinised zone, and by the undercutting of the resistant cap (Fig. 3.5). Undercut ferricrete and

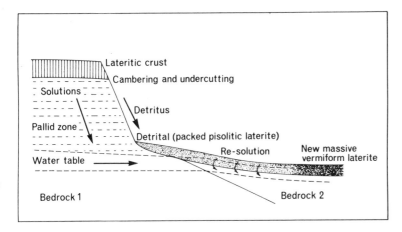

Fig. 3.5 Features of a laterite profile showing how a detrital ferricrete may form from the debris of an *in situ* crust (after Goudie, 1973).

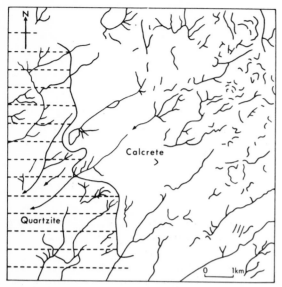

Fig. 3.6 Drainage pattern near Kalkrand, Namibia (after Goudie, 1973). As drawn by the author from air photographs.

silcrete caps may sag or collapse when caves are eroded in the pallid zone beneath them, and a complex landscape of collapse depressions and natural arches may form.

Ferricretes, silcretes, and calcretes which may be originally formed in hollows can produce a form of inverted relief when the less resistant rocks around them are eroded away and the crust is left capping a residual hill.

Benched or terrace landscapes may develop where solutions containing iron move laterally towards a seepage zone, hollow, or terrace edge, and into a zone of wetting and drying in which the iron is precipitated. A new massive ferricrete may be formed there. Slope debris may form cemented ferricrete taluses or, more commonly, recemented footslope, pediment, or detrital ferricretes (Fig. 3.5). Pavements of recemented ferricrete fragments are relatively common on valley floors and lower slopes below ferricrete-capped hills.

Calcretes may produce a very distinctive landscape when resolution of the lime opens joints and forms pipes through which surface water can drain. In parts of southern Africa areas of discontinuous surface drainage are features of calcrete outcrops, while areas with impermeable rocks nearby have a complete drainage network (Fig. 3.6).

Duricrusts as resources

It has already been pointed out that alcretes are the major sources of aluminium ores. Ferricretes have been used as sources of iron in East and West Africa as well as in India. Ferricretes and calcretes are widely used for road-making and both have been used for building materials: the famous temple of Angkor Wat in Southeast Asia is made of ferricrete. Most duricrusts, however, have soils of low fertility and hinder agriculture by their resistance to cultivation.

Further reading

Weathering processes are discussed in books by Keller (1957), Loughnan (1969), Carroll (1970), Reiche (1950), and in many texts on pedology and soil science. Ollier (1969) has emphasised the geomorphological aspects of weathering. Duricrusts, especially those of calcrete, are described by Goudie (1973) and Reeves (1976), and ferricretes by a number of authors including Maignien (1966) and McFarlane (1976). Langford-Smith (1978) has provided a study of silcrete in Australia.

4

Strength and behaviour of rock and soil

The term *strength* is used in two senses in geomorphology: (1) it may be used to denote the ability of rock or soil to resist abrasion — a property which is controlled by the hardness of constituent minerals and the strength of intergranular bonds; or (2) it may be used for the ability of material to resist deformation by tensile, shear, or compressive stresses. In studies of hillslopes it is the second type of strength which is of interest.

Tensile stress

When a force is applied to a solid body internal stresses are set up in the material of the solid. If, for example, one end of a rod of steel is placed in a vice and a tensile stress, or stretching force, is applied to the other end there is a tendency for the rod to be pulled apart. To prevent the rod breaking there is an internal stress acting in a direction opposite to the tensile stress. Stress has the dimensions of force per unit area, and is measured in newtons per square

metre, or pascals ($1\text{N/m}^2 = 1\text{Pa}$), so the tensile stress is equal to the force applied divided by the cross-sectional area of the rod.

As a tensile stress is applied the rod may stretch and thin: in this case the rod has suffered deformation or strain (Fig. 4.1).

Tensile strain is indicated by a ratio:

$$\frac{\text{change in length}}{\text{original length}} \text{ or } \frac{\delta l}{l}.$$

Strain is thus dimensionless; it is symbolised by epsilon (ϵ). The rod fractures when the tensile stress exceeds the tensile strength of the rod.

Shear stress

Shear stress, symbolised by tau (τ), is a force acting so that a body deforms by one part sliding over the other: thus shear stress changes the shape of the body but usually has only a minor effect on its volume (Fig. 4.2). A shear stress can be applied either by two forces acting not in line or by compression (as in Fig. 4.3).

Compressive stress

Rock or soil, beneath an overburden, has forces acting on it which are proportional to the weight of the overburden and which tend to compress it. True

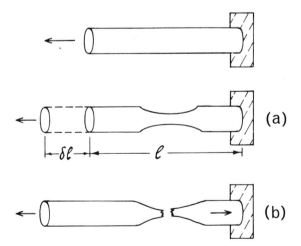

Fig. 4.1 (a) Strain in a rod produced by a tensile force;
(b) Strain followed by brittle fracture of a rod.

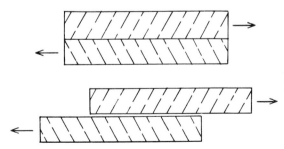

Fig. 4.2 Shear of a slab produced by forces not in line.

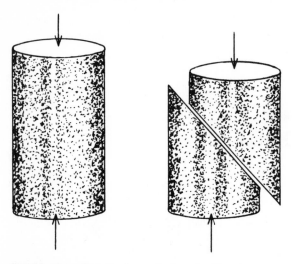

Fig. 4.3 Shear failure of a cylinder of soil under compression.

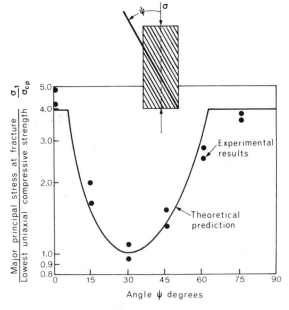

Fig. 4.4 The relationship between orientation of cleavage and the compressive strength of slate (after Roberts, 1977, from an original figure by E. Hoek).

compressive failure in a rock or soil occurs through internal collapse of the structure: in most rocks and soils this is accompanied by a decrease in the void space, and may involve fracture of mineral grains and movement along grain and crystal boundaries, but in some densely compacted soils compressive failure may cause an increase in void space. Sedimentary rocks with high void space tend to be weaker in compression than fine-grained igneous or metamorphic rocks.

In studies of rock strength uniaxial, or unconfined, compressive testing, in which a stress is applied in one plane only, is commonly used. Rocks containing planes of weakness usually fail along these when the angle between the applied stress and the plane of weakness lies between about 15 and 60°. Within this range the rock has markedly lowered strength (Fig. 4.4). Both rocks and soil can withstand greater compressive stresses than shear stresses.

Strength

The strength of a material, in a mechanical sense, is its ability to resist deformation and fracture without large-scale failure. No single criterion defines rock strength: engineers normally use unconfined compressive strength as an index because it is easily measured and they are often concerned with how material will resist crushing. Geomorphologists are usually more concerned with failure in direct shear and therefore regard strength as the resistance to shear stress: at failure shear stress is equal to shear strength.

Strength of rocks and soil

Rock and soil are not uniform solids but aggregations of solids with substantial voids which are filled, or partly filled, with water and air. The strength (*s*) of a rock or soil mass therefore depends partly upon the strength of the minerals composing the aggregates, but also upon the forces holding the aggregates together.

The shearing resistance of rock and soil depends upon many factors. A complete equation might take the form:

$$\tau = f(e, \phi, C, \sigma', c', H, T, \epsilon, \dot{\epsilon}, S \ldots)$$

in which τ is the shearing resistance

  e is the void ratio

  ϕ is the frictional property of the material

  C is the composition

  σ' is the effective normal load holding materials in contact

  c' is the effective cohesion of the material

  H is the stress history

  T is the temperature

  ϵ is the strain

  $\dot{\epsilon}$ is the strain rate

  S is the structure of the material

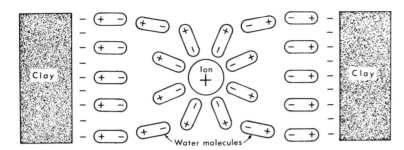

Fig. 4.5 Cohesion of clay particles induced by oriented water dipoles and a positively charged ion.

Many of these components are not independent, however, and many cannot be evaluated quantitatively. Consequently the two components which can be evaluated — cohesion and friction — are measured in conditions in which water contents, loads, rates of loading, confining pressures, and stress history can be controlled.

Cohesive forces are derived largely from cementing materials which bind rock and soil particles together. Frictional forces are derived from several sources including the resistance of grains to sliding past each other; resistance to grain crushing; resistance to rearrangement of grains; and resistance to volume change in rock and soil — a phenomenon known as dilatancy.

Cohesion

Cohesion is caused by chemical cementing of rock and soil particles and it is not controlled by compressive forces holding particles together. Chemical bonding between particles may be produced by carbonate, silica, alumina, and iron oxide cements as well as by organic compounds. Cohesive strengths as high as several kN/m^2 can develop from cementation of soils. The bonds are primarily molecular and ionic with particularly strong bonds being the result of sharing of ions between crystal units.

Between clay-sized particles, electrostatic and electromagnetic forces operate and between particles smaller than 1 μm they can produce stresses as large as 7 kN/m^2 when the particles are closer than about 30 nm. Electromagnetic bonding is aided by the presence of thin films (<0.5 μm) of ferromagnesian molecules on clay crystals. Electrostatic bonding also occurs because clay surfaces have a negative charge and a cation between them forms a direct link. Similar cohesive effects are produced by water molecules (Fig. 4.5). Because of its structure the water molecule is polar (i.e. one 'end' of the molecule is positively charged and the other is negatively charged). Water molecules can, therefore, be oriented

under the action of electrical charges such as those which exist on clay particles. The clay crystals are linked through the adsorbed water films and the positively charged ion. The strength of the electrical force giving cohesion is inversely proportional to the square of the distance of separation, so clays in a slurry or mudflow lose some of their cohesive strength because of separation. Cohesion between clay crystals is strongest where the crystals have edge-to-edge contacts.

Apparent cohesion

Apparent cohesion, as distinct from the true cohesion of cementation, is produced by capillary stresses occurring as surface tension in water films between particles, and as interlocking of particles at a microscopic level as a result of surface roughness. Such apparent cohesion is affected by the size of rock and soil particles, their shape and mineralogy, the amount of water present, and the state of packing of particles. In Fig. 4.6 are shown soil particles with water between them. In unsaturated soils attractive forces result from the capillary tension of adsorbed water and the attractive force is inversely proportional to the radius of curvature of the water surface between the soil particles (r). The smaller the radius the greater is the capillary tension and the greater is the apparent cohesion. In a saturated soil surface tension is completely eliminated and there is no strength available from apparent cohesion. Thus loose packing, large grain sizes, high moisture contents, and low wettability of particles all contribute to low cohesion.

Apparent cohesion is also produced by mechanical interlocking of surfaces of particles (Fig. 4.7). In shear, microscopic protrusions will have to fail before a macroscopic shear surface can develop.

Friction

The basic control on the strength of soil and of most rocks is the frictional resistance between

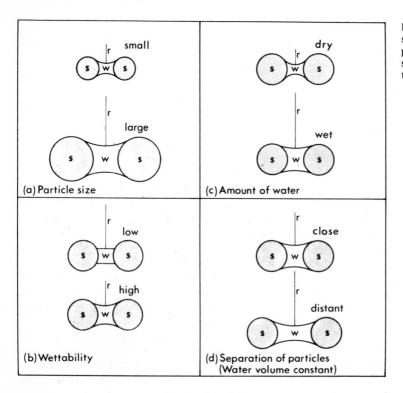

Fig. 4.6 The influence of soil particle size, wettability, water content (*W*) and particle packing on tensional force. The smaller the value of (*r*) the greater is the force.

mineral particles in contact. Frictional strength is thus directly proportional to the normal force holding grain surfaces in contact and it is influenced by the number of point contacts in a volume of rock or soil (there are probably about five million point contacts in 1 cm³ of a fine sand), by the arrangement, size, shape, and resistance to crushing of grains as well as by the voids and by dilatancy.

Where the packing of particles is open and particles are of uniform size the points of contact are relatively few and strength is low. Closer packing increases contact between grains, as does variability in grain sizes and more angular grain shapes (Fig. 4.8). Disturbance by shearing forces may require that closely packed grains have to ride up over other grains before shear failure can occur. This necessitates an initial decrease in volume before a new loosely packed state can develop (Fig. 4.9). The less the normal stress the greater is such dilatancy. In shear,

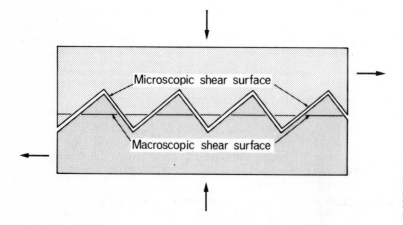

Fig. 4.7 Apparent cohesion of two surfaces resulting from interlocking. Failure can occur only by shearing through the protuberances.

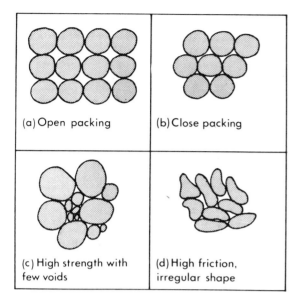

(a) Open packing

(b) Close packing

(c) High strength with few voids

(d) High friction, irregular shape

Fig. 4.8 Types of packing in idealised soils.

therefore, frictional resistance in a soil may be initially high before a lower resistance to continued shear can develop. This sequence is usually described as high initial peak strength followed by a lower residual strength of the remoulded material (note that 'strength' is equal to 'resistance' at the moment of failure, and the two terms can sometimes be used interchangeably).

Even polished surfaces have a roughness which induces frictional resistance. When two surfaces are brought into contact they will be supported initially on the summits of the highest asperities. As the load increases beyond a certain value the asperities may

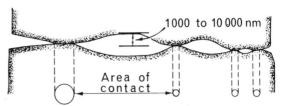

Fig. 4.10 Frictional contact, even along smooth surfaces, occurs at asperities which are of limited area.

fail by brittle fracture, first decreasing resistance to shear and then increasing the area of contact and so the frictional resistance, or they may deform plastically. For quartz the load to produce plastic deformation, or creep, is about 10 GN/m^2 and for diamond it is about 100 GN/m^2. Creep in the materials at a surface of contact increases the area of contact with time, and thus increases resistance to sliding.

At low loads materials will deform elastically, but as loads rise so resistance increases. The behaviour in shear is also influenced by the relative hardness of the materials in contact, for a hard surface will have asperities which can plough into the surface of a softer material, cutting grooves into it, and thus increasing the area of contact.

The roughness of a surface is the expression of a succession of asperities and depressions (Fig. 4.10). Highly polished metal surfaces may have a depth of depressions ranging from 1000 to 10 000 nm. Quartz surfaces of mirror smoothness have depressions 50 000 to 500 000 nm deep. The cleavage surfaces of mica may have depression depths of 100 to 10 000 nm. In soils and rocks, surfaces are seldom of mirror smoothness so frictional resistance to sliding

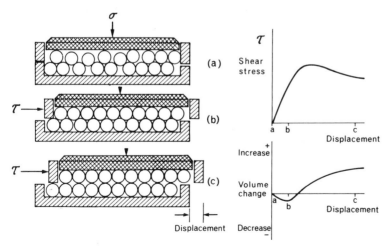

Fig. 4.9 During a shear box test a sample may initially consolidate and then exhibit dilatancy as grains rise over other grains during shearing. This effect is evident from the graphical plot of displacement against volume change in the sample during the shear test.

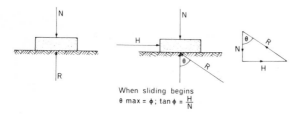

Fig. 4.11 Derivation of the coefficient of sliding friction.

is correspondingly high. The mineralogy of particles clearly influences surface roughness as does the history of weathering and transport (Horn and Deere, 1962).

Friction angle

Where a block of rock lies upon a horizontal surface the weight of the block (N) generates an equal and opposite reaction (R). N and R together form a compressive force normal to the plane of contact and the block is immobile (Fig. 4.11).

If a small horizontal force H (insufficient to move the block) is applied to the rock the reaction R will no longer be normal to the plane of contact. It adjusts in magnitude and direction to equal the resultant N and H. The triangle of forces represents, in magnitude and direction, the relationships between N, H, R, and the angle θ.

The normal force $N = R\cos\theta$, and the force H cuts across the plane of contact so that $H = R\sin\theta$. If the shear force H is increased until the block is just about to slide R will increase, and so will the angle θ. At the moment when sliding begins the frictional contact, holding the block and the surface stable along the plane of contact, will be broken and θ will have attained its maximum possible value, on that surface.

That maximum value is ϕ, the angle of plane static friction or angle of shearing resistance, and $\tan\phi = H/N$ and is the *coefficient of plane static friction*. The resultant stress acting along a plane is resolved into a stress, σ, acting normal to the plane of contact and a shear stress, τ, acting along the plane. If the plane is a plane of fracture, in a soil or rock material, movement along that plane in response to the shearing stress will be dependent upon the angle of internal friction of the material and

$$\tan\phi = \tau/\sigma.$$

The friction angle for pure quartz is 26 to 30° and for many clay particles it is close to 13°, but such values have little direct significance for determining the friction angles of soils and rocks because of the great heterogeneity of these materials. The volume of voids and particle sizes have a greater control on the friction angle of soils than does mineralogy (Fig. 4.12).

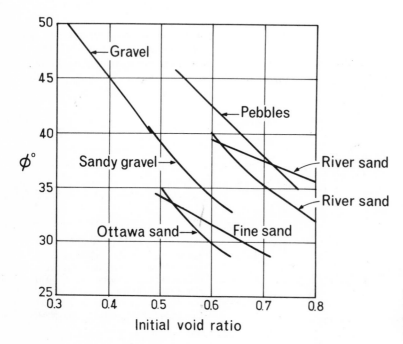

Fig. 4.12 The effect of void ratio on the angle of internal friction (ϕ) for various noncohesive materials (after Lambe and Whitman, 1979).

Water acts as an antilubricant when it is applied to the surfaces of minerals with massive crystal structures — such as quartz and calcite — but this effect is reduced as mineral surfaces become rougher and it is insignificant within soils. Water and clay particles separate large grains and behave like lubricants on the surfaces of platy minerals and reduce friction angles, but they also add apparent cohesion to a soil.

The friction angle, then, contains contributions from several sources; of these interparticle resistance to sliding probably accounts for at least half of the peak strength and much of the residual strength of most soils. The value of the friction angle decreases with increasing plasticity and water content at which soil flows (see the last section in this chapter).

The Coulomb equation

The importance of cohesion and friction in determining the strength of materials was recognised and expressed, as early as 1776, by the French engineer Coulomb: the total resistance to shear or strength at failure (s) is given by the normal stress (σ) multiplied by the coefficient of friction plus the cohesion (c) of the material or

$$s = c + \sigma . \tan\phi$$

for materials without cohesion the Coulomb equation is

$$s = \sigma . \tan\phi$$

Water and shear strength

In a perfectly dry soil or rock there is no apparent cohesion (as distinct from the true cohesion of cementation) caused by surface tension and the soil fabric is supported entirely on the point contacts of the constituent particles (Fig. 4.13). The voids in the soil are filled with air and the pressures in the voids are atmospheric; pore water pressures are then zero. In a moist soil the particles have an apparent cohesion caused by surface tension and are effectively under a suction so pore water pressures are said to be negative. In a fully saturated soil apparent cohesion is lost because there are no surface tension forces and part of the normal stress of the overburden is transferred from the soil fabric to the soil water. The transfer of load from soil to water is equivalent to a buoyancy or upthrust effect in which pore water pressures are positive. The effective stress then equals the differences between the total ground stress transmitted through the interparticle contacts and the stress supported by the pore water.

The value of the pore water pressure is proportional to the height of the free water column in the soil — known as the hydrostatic head when the water is at rest (i.e. static). This head is measured by inserting tubes, called piezometers, into the soil (Fig. 4.14), and for conditions in which lateral flow or seepage can occur is known as the piezometric head or pressure head.

A rise in pore water pressure decreases soil strength because although confined water can withstand a compressive stress it cannot withstand a shear stress; further the buoyancy effect counteracts the normal stress in proportion to the pore water pressure. Thus for saturated soils the Coulomb equation has to be written as:

$$s = c' + (\sigma - u) . \tan\phi'$$

where s is the shear strength at any point in the soil;
  c' is the effective cohesion, as reduced by loss of surface tension;
  σ is the normal stress imposed by the weight of solids and water above the point;
  u is the pore water pressure derived from the unit weight of water and the piezometric head ($\gamma_w . z_1$);
  ϕ' is the angle of friction with respect to effective stresses.
Alternatively the equation may be written as:

$$s = c' + \sigma' . \tan\phi'$$

where σ' is the effective normal stress ($\sigma' = \sigma - u$). The symbols c and ϕ thus refer to total stresses and c' and ϕ' to effective stresses. Effective stresses are thus modified by pore water pressure and conditions of loading or testing and are not fundamental properties of the material.

The above equation is usually related to soil but it can also apply to porous and permeable rock. Rock with low porosity, low permeability, and therefore low water content, will be less affected by saturation. Nearly all rocks which are not deeply buried, however, have numerous clefts in the form of joints, bedding planes, and other fissures. High water pressures in clefts can reduce effective stresses substantially so cleft water pressures are of great importance to rock strength.

Measurement of strength

Laboratory tests

Most tests of rock and soil samples are carried out in laboratories where carefully controlled conditions

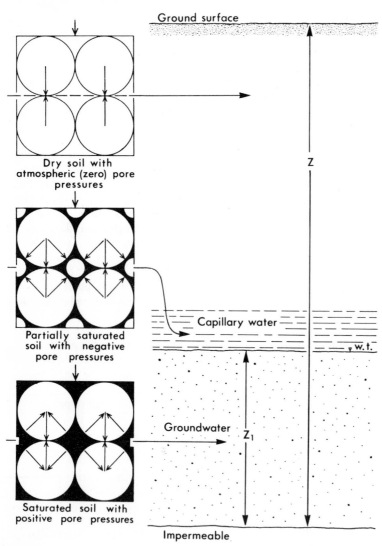

Ground surface

Dry soil with
atmospheric (zero) pore
pressures

Partially saturated
soil with negative
pore pressures

Saturated soil with
positive pore pressures

Z

Capillary water

w.t.

Groundwater Z_1

Impermeable

Fig. 4.13 Pore water pressures depend upon the water content of soil voids.

permit repeated and comparable tests of many samples. For studies of both soil and rock it is usually important to collect cores or carefully cut specimens whose orientation in the field, with respect to shearing forces, is known.

For tests of simple shear, samples of rock are placed in shear boxes in which shear forces are generated by a direct stress (Fig. 4.15a) or by a compressive stress on an angled specimen (4.15b). Alternatively a direct compressive stress (4.15c) may test the unconfined strength of rock. For many rocks direct shear strengths are commonly twice as great as tensile strengths, and compressive strengths four to ten times as great as direct shear strengths. The strongest rocks have high density and low porosity (Tables 4.1 and 4.2).

Soil shear box tests

For measurements of the strength of soils similar equipment is used. The shear box for soils consists of a square box, without a top or a bottom, split horizontally at the level of the centre of the soil sample. The sample is held between metal grilles and porous stones (Fig. 4.16). A gradually increasing horizontal (or shearing) force is applied to the lower part of the box until the sample fails in shear. The shear load at failure is divided by the cross-sectional area of the sample to give the shearing stress. For determining the shearing resistance under a normal stress a vertical load is applied to the sample by means of dead weights (Plate 4.1).

So that the shearing stress may be applied at a constant rate of strain the lower half of the shear

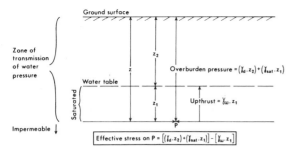

Fig. 4.14 Stresses in a soil resulting from the weight of the dry soil (γ_d) and of saturated soil (γ_{sat}) above point P, and from the buoyancy effect of pore water. If, for example, $\gamma_d = 12$ kN/m³; $\gamma_{sat} = 17$ kN/m³, $\gamma_w = 9.8$ kN/m³, $z = 10$ m, then: if the soil is perfectly dry the total stress on $P = 120$ kN/m², if it is fully saturated the effective stress is 72 kN/m², and if $z_1 = 6$ m then the effective stress on $P = 91.2$ kN/m². The effect of pore water thus increases with depth as in free-standing water. Capillary water is held suspended above the water table and thus does not alter the values of the pore pressures below the water table. The effective stress, however, is increased by the weight of water held within the capillary zone.

box is mounted on rollers and is pushed forward at a uniform rate by a gear-operated piston. The upper half of the box bears against a steel proving ring, the deformation of which is shown on the dial gauge and indicates the shearing stress. Change in volume of the soil sample during consolidation and shearing is measured by a dial gauge coupled to the normal load. Water escapes or enters the sample through the porous stones and the perforated metal grilles.

The horizontal stress is applied slowly, often at rates of about 0.6 mm per minute in a quick test, but at rates as low as 0.005 mm/minute if it is desired to prevent the build up of pore water pressures in the sample during testing. Failure of the sample is indicated by a sudden drop, or levelling off, of the proving-ring dial reading which records the reaction to a shear stress, that is, when the sample shears its resistance to shear stress may drop rapidly.

Table 4.1 Typical Rock Properties

	Strength (MN/m²)			Bulk density (Mg/m³)	Porosity (%)
	Compressive	Tensile	Shear		
Granite	100–250	7–25	14–50	2.6–2.9	0.5–1.5
Basalt	150–300	10–30	20–60	2.8–2.9	0.1–1.0
Gneiss	50–200	5–20	–	2.8–3.0	0.5–1.5
Slate	100–200	7–20	15–30	2.6–2.7	0.1–0.5
Marble	100–250	7–20	–	2.6–2.7	0.5–2.0
Shale	100–200	7–20	15–30	2.0–2.4	10–30
Sandstone	20–170	4–25	8–40	2.0–2.6	5–25

Note: density of water = 1 Mg/m³
Source: data from Attewell and Farmer, 1976.

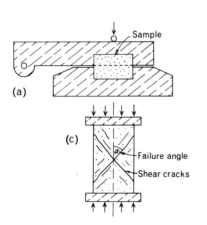

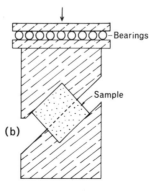

Fig. 4.15 (a, b) Shear strength testing of a cylindrical rock sample; (c) compressive testing.

Table 4.2 Typical Soil and Rock Properties

Description			Unit weight (Saturated/dry) kN/m³	Friction angle (1) degrees	Cohesion kPa
Type		Material			
Cohesionless	Sand	Loose sand, uniform grain size	19/14	28–34	
		Dense sand, uniform grain size	21/17	32–40	
		Loose sand, mixed grain size	20/16	34–40	
		Dense sand, mixed grain size	21/18	38–46	
	Gravel	Gravel, uniform grain size	22/20	34–37	
		Sand and gravel, mixed grain size	19/17	48–45	
	Compacted broken rock	Basalt	22/17	40–50	
		Chalk	13/10	30–40	
		Granite	20/17	45–50	
		Limestone	19/16	35–40	
		Sandstone	17/13	35–45	
		Shale	20/16	30–35	
Cohesive	Clay	Soft bentonite	13/6	7–13	10–20
		Very soft organic clay	14/6	12–16	10–30
		Soft, slightly organic clay	16/10	22–27	20–50
		Soft glacial clay	17/12	27–32	30–70
		Stiff glacial clay	20/17	30–32	70–150
		Glacial till, mixed grain size	23/20	32–35	150–250
	Rock	Hard igneous rocks – granite, basalt, porphyry	(2) 25 to 30	35–45	35000–55000
		Metamorphic rocks – quartzite, gneiss, slate	25 to 28	30–40	20000–40000
		Hard sedimentary rocks – limestone, dolomite, sandstone	23 to 28	35–45	10000–30000
		Soft sedimentary rock – sandstone, coal, chalk, shale	17 to 23	25–35	1000–20000

Notes: (1) Higher friction angles in cohesionless materials occur at low confining or normal stresses.
(2) For intact rock, the unit weight of the material does not vary significantly between saturated and dry states with the exception of some materials such as porous sandstones.
Source: data from Hoek and Bray (1977).

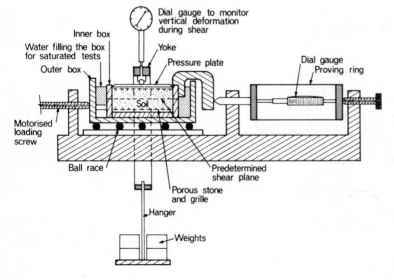

Fig. 4.16 Illustration of the principle of a shear box.

A series of tests are made for different values of normal stress and the results are plotted graphically with the shear stress on the vertical scale and the normal stress on the horizontal scale. The value of the effective angle of internal friction (ϕ') is determined from the slope of the line through the plotted points, and the value of effective cohesion (c') is the displacement of the line above the zero point (Fig. 4.17).

For a non-cohesive soil the plotted line passes through the origin of the graph and the Coulomb equation thus becomes:

$$s = \sigma' . \tan\phi' .$$

During a shear test many frictional materials exhibit dilatancy. Densely packed sands, silts and over-consolidated clays exhibit peak resistance to shear (ϕ'_p) and then decline to a residual or ultimate strength (ϕ'_r). Loosely packed sand and weathered clays by contrast can deform without the grains riding over one another and although peak strength is achieved there is no decrease in strength with further shearing (Fig. 4.19). The behaviour of clays in shear box tests is also related to the form of the clay crystals with plate-like crystals being readily reoriented parallel to the shear plane (Plate 4.2). Thus platy kaolinite type clays may exhibit greater dilatancy than, for example, allophanic clays with irregular crystal forms.

The shear box test is simple in principle but is open to a number of objections:

(1) The size of shear box normally used for clays and sands is 60 mm square and the sample is 20 mm thick; a larger size, 300 mm square, may be used for gravelly soils. Many soils, however, contain large clasts, plant roots, and structural features which provide major controls on soil shear strength in field situations. For engineering purposes organic matter, soil structures, and large clasts are usually irrelevant as the surface soil is removed before construction takes place. For geomorphic purposes, however, the characteristics of superficial soil materials are of fundamental importance. Tests of small samples in laboratory shear boxes may thus fail to represent natural soil strength which can only be determined in the field using a large shear box and undisturbed soil samples (see below).

(2) Another objection is that the stress distribution across a shear box test sample is complex, and the value of the shearing resistance obtained by dividing the shearing force by the area is only approximate.

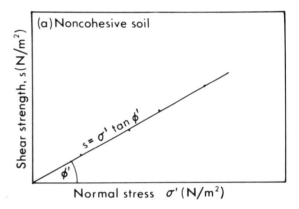

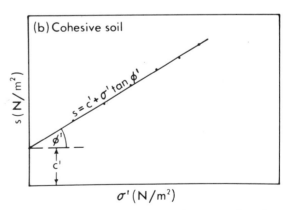

Fig. 4.17 Plots of effective normal stress against shear strength at failure for two soils. Note that the intervals of the horizontal and vertical scales of the graphs must be equal.

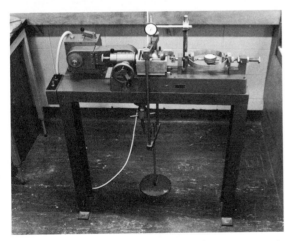

4.1 A shear box for soil testing. The normal force is applied by a load hanging below the box.

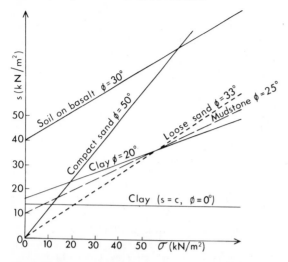

Fig. 4.18 Lines describing the Coulomb equation for a variety of soils. For clays with $\phi = 0$ the strength parameters are usually expressed as c_u and ϕ_u indicating the undrained test from which these parameters are determined.

(3) For many studies the peak strength of soil is required. This is the case where a 'first-time' landslide is being studied. But where soils on slopes have been repeatedly subjected to landsliding, or shearing of soil, the soil has reached a strength close to that of the residual value. Residual strengths are often difficult to determine accurately, even when the shear box is reversed so that the upper part of the sheared sample travels back over the lower half. Reversal of the direction of shear does not, of course, simulate natural conditions so an attempt has been made to overcome this problem by devising a ring shear apparatus.

In a ring shear test an annular ring-shaped sample (Fig. 4.20) is subjected to a constant normal stress. The sample is confined laterally and horizontally by upper and lower confining rings. The lower half of the sample is carried on a rotating table driven by a worm gear. The upper half of the sample reacts via a torque arm against a pair of fixed proving rings that measure the shear force (Plate 4.3).

The two main advantages of the ring shear test are that (1) there is no change in the cross-section of the shear plane as the test proceeds, and (2) the sample can be sheared through an uninterrupted displacement of any magnitude (Bishop *et al.*, 1971).

Triaxial tests

Rocks or very cohesive soils may be tested by simple or uniaxial compression, but granular soils derive much of their strength from lateral confinement

and a compressive test must be carried out so that this confining pressure is simulated. This is done using a triaxial cell.

The three axes (hence 'triaxial') along which stresses are applied are illustrated in Fig. 4.21. The major principal stress (σ_1) is transmitted in a vertical direction and the minor stresses (σ_2 and σ_3) are mutually at right angles with the major stress and with each other. Within the soil stresses operate in all three directions and all stresses can be resolved in terms of these three components (Plates 4.4 and 4.5). In a triaxial test the major stress is applied vertically through a piston and the lateral stresses ($\sigma_2 = \sigma_3$) are applied through the water which confines the soil sample within a Perspex cell (Fig. 4.22).

The pressure within the cell is raised to the desired level by means of a pump, and vertical loading is applied by moving the piston vertically at a constant rate. The load is measured by a steel proving ring and dial gauge.

Clay samples are trimmed to the correct size and, where the sample is to be consolidated before testing, metal caps and bases are used with porous plates to permit drainage of the sample during consolidation. Where pore-pressure measurements are to be made porous ends are used and the outlet is connected to the pore-pressure measuring apparatus. For quick undrained tests impervious endpieces are used.

The sample is sealed off from the pressure water surrounding it by enclosure in a thin rubber membrane. Loose granular soil materials are filled into the rubber tube by supporting the membrane in a metal former. The sample is placed in the cell and then flooded with water through the porous base. A vacuum is applied to remove all air and the metal support is removed. The Perspex cylinder is placed in position and the lateral pressure is applied.

The load shown on the proving ring dial gauge indicates the principal stress difference, $\sigma_1 - \sigma_3$, or

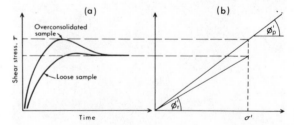

Fig. 4.19 (a) Direct shear test results for a dense and a loose soil, and in (b) the plot of several tests at various effective normal stresses.

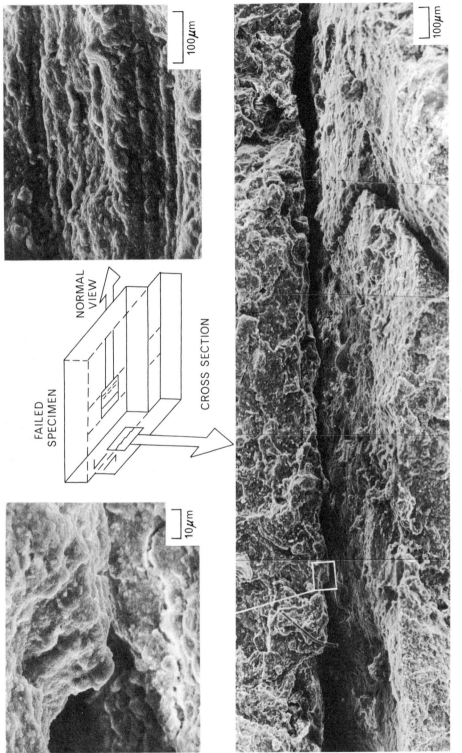

4.2 Scanning electron micrographs of a sample which has been sheared in a shear box. The orientation of the clay particles at the shear plane is shown (micrograph by N. W. Rogers).

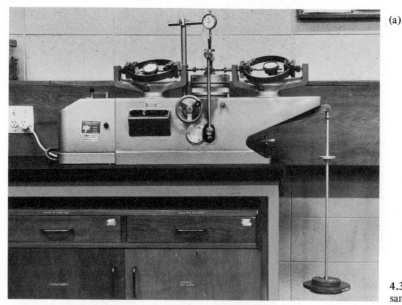

(a)

4.3 (a) A ring shear apparatus; (b) the sample holder.

deviator stress. At least three tests are normally carried out at different lateral pressures, and for for each test the total axial pressure at failure σ_1, and the lateral pressure σ_3, are plotted as a Mohr circle construction (Fig. 4.23).

Values of major (σ_1) and minor stresses are plotted on the abcissa and a semicircle is drawn through them. A tangent to these semicircles, which is known as a failure envelope, is drawn and extended to the ordinate. The intercept here gives

a value for the cohesion and the gradient of the line is the angle of internal friction (Fig. 4.24). The strength of the soil can thus be seen to depend on the confining stress through its influence on friction.

The value of the triaxial test compared with shear box tests is that pore pressure changes can be measured. This factor is of considerable importance in the study of landslides in cohesive soils (Bishop and Bjerrum, 1960).

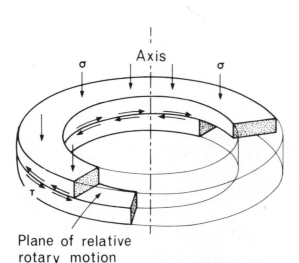

Fig. 4.20 The principle of ring shear (after Bishop *et al.*, 1971).

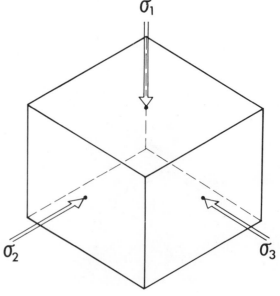

Fig. 4.21 The stresses on a cube.

(b)

Field measurements of soil strength

Where it is impossible, or difficult, to obtain soil samples for laboratory tests, or large numbers of readings are required, an indication of shear strength can be obtained by *in situ* testing using a shear vane.

A shear vane has four thin rectangular blades set at right angles to each other. Each blade is generally twice as deep as it is wide. The vane is pushed into the soil and rotated slowly (usually at about one revolution/two minutes). The size of the blades used depends upon the stiffness of the soil and the size of the particles within it. Large vanes are used for weak soils.

Most hand-held shear vanes have a dial gauge which records the resistance to a turning motion, i.e. the torque, directly in N/m^2.

Shear vanes may be used successfully only in fine-grained soils without roots, clasts, or strongly developed structures. Soil strength is closely controlled by soil moisture, so soil samples should be taken for determination of moisture content, or all measurements made at a nearly uniform moisture content such as field capacity (the state at which the soil has freely drained under gravity, but retained water in the micropores). As shear vanes indicate total stress they cannot be used to derive independently the values of c' and ϕ'.

Penetrometers have sometimes been used to derive an indication of soil strength. A penetrometer indicates the force needed to push or drive a cone or cylinder of known dimensions into the soil. It is a useful indicator of bulk density and for engineering purposes, of bearing capacity, but it does not always measure shear strength reliably and it is at least as susceptible as the shear vane to roots, clasts, soil

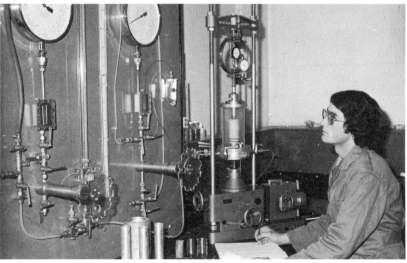

4.4 A triaxial cell and triaxial machine.

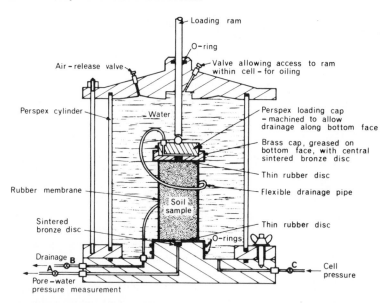

Fig. 4.22 A triaxial cell. The major principal stress (σ_1) is applied as a load and the minor stress (σ_3) through water in the cell.

structure, and moisture variations.

A field shear box may be used to determine the strength of very large *in situ* soil samples, in which it is wished to preserve the natural fissures which result from soil structure, and for studying the contribution of plant roots to soil strength. In field shear tests a steel box, either open at the base and sides, or at the base only, is placed on a monolith of soil which is left undisturbed after the surrounding soil has been cut away. A normal load is placed on the box and a shear force is applied by a winch. The shear force is measured by a proving ring or spring balance and the displacement, as the shear force is applied, by a scale. The monolith eventually

shears along its base and sides (Plate 4.7). This type of testing is particularly useful in studies of soil erosion by landsliding and for comparing the strength of soils under different kinds of vegetation (Chandler, Parker, and Selby, 1981). Field tests in shallow residual soils give values of effective strength because of rapid soil drainage, through macropores.

A similar type of large field shear box can be used on jointed rock with the shear force being applied through hydraulic jacks, but as a single test on a large sample (1 m³) may cost US $10 000 such testing is rarely undertaken.

Field measurements of rock intact strength

When discussing rock strength a distinction has to be made between the strength of a sample of intact rock and the strength of a rock mass inclusive of the joints or other planes of weakness.

4.5 A cylindrical soil sample at its original size and after shear and barrel failures in triaxial tests.

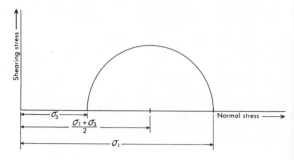

Fig. 4.23 Plotting a Mohr circle from one triaxial test.

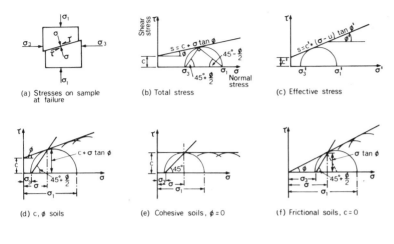

Fig. 4.24 Mohr circle constructions and characteristics.

(a) Stresses on sample at failure

(b) Total stress

(c) Effective stress

(d) c, φ soils

(e) Cohesive soils, φ = 0

(f) Frictional soils, c = 0

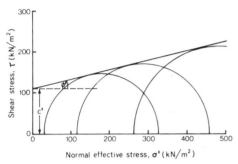

(g) Method of construction of Mohr – Coulomb failure line

Strength of intact rock

The most generally accepted measure of the strength of intact rock is the uniaxial (unconfined) compressive strength. Measurement of this parameter is simple in a laboratory but it requires precisely cut rock cores. It is, therefore, impossible to use this measure where drilling machinery is not available

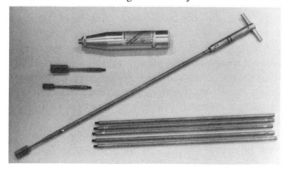

4.6 Field shear vanes with extension rods. The torque is measured on the ring in the handle and different size vanes permit measurements of strength in soft to stiff soils. The bottle-shaped instrument is a Schmidt hammer.

and in localities from which it is impossible to recover or transport drill cores or collect large samples of rock from which cores may be cut.

Two field tests of rock strength which may be correlated with unconfined compressive strength are the point-load strength test, and the Schmidt hammer test.

The point-load test was developed in Russia to provide a rapid strength test of irregularly shaped rock specimens (Protodyakonov, 1960, 1969). The International Society for Rock Mechanics (1973) has subsequently incorporated the Protodyakonov test as a standard technique. In the test specimens up to 100 mm in diameter are fractured between two cone-shaped platens (Plate 4.8). The maximum stress (I_s), occurring at the centre of the specimen, may be related to the applied load (P) and the distance between platens (D) by an expression with the form: $I_s = k.P/D^2$. The constant k is found to assume values which are dependent upon the geometry of the specimen. Correction charts for specimens of various sizes are provided by Broch

4.7 A field shear box. The normal force is applied through lead weights on top of the box and the shearing force is applied through the winch which is anchored in the ground. The shearing force is measured by the proving ring and dial gauge.

4.8 A point load strength tester.

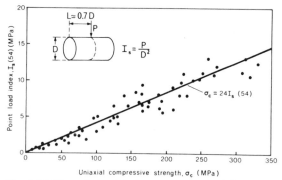

Fig. 4.25 The relationship between point load index and uniaxial compressive strength for 54 mm diameter cores (data from D'Andrea *et al.*, 1965; Broch and Franklin, 1972; Bieniawski, 1975).

and Franklin (1972) and strength indices are usually expressed as for a rock cylinder of either 50 or 54 mm diameter − I_s (54) − (Bieniawski, 1975).

Point-load strength tests have been carried out on a variety of rock types ranging from the strongest rocks to weathered materials which have the characteristics of strong soils. The irregularity of sample sizes causes scatter in strength data, but this can be compensated for by testing a large number of samples (15–20) and by trimming irregularities off samples with a hammer. If size correction curves are used the error is unlikely to exceed 15 per cent and in most cases should be much smaller. Anisotropies in materials can be studied by orienting samples so that they are fractured across or along planes of weakness and, in many cases, mean values may be used in a classification of rock mass strength.

Studies by Broch and Franklin (1972) indicate that there is similar scatter in the data from point-load tests and in data from unconfined compressive tests and that both tests are very dependent upon rock moisture content. Sandstones often undergo a strength reduction of 20–30 per cent as water content increases from a dry to a saturated state and even granite may lose more than 13 per cent of its strength. Field studies should normally be made at 'natural' moisture content even though this involves a different strength rating in an arid and a humid climate.

The point-load strength test is easily carried out with equipment which can be carried in a vehicle or, if necessary, on a back pack frame. Its results are highly correlated with unconfined compressive strength tests (Fig. 4.25 and 4.26): published correlation coefficients range from 0.88 to 0.95 (D'Andrea *et al.*, 1965; Irfan and Dearman, 1978).

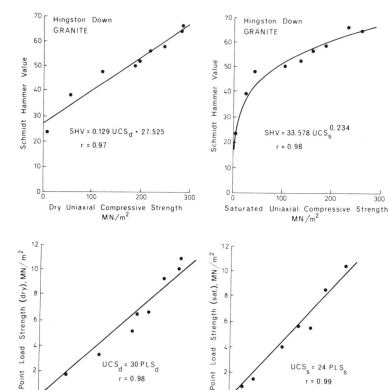

Fig. 4.26 Relationships, for Dartmoor granite, between Schmidt hammer and point load values against uniaxial compressive strength (after Irfan and Dearman, 1978).

The Schmidt hammer test was devised in 1948 by E. Schmidt for carrying out *in situ* non-destructive tests on concrete. The test hammer measures the distance of rebound of a known mass impacting on a rock surface. Because elastic recovery of the rock surface depends upon the hardness of the surface, and hardness is related to mechanical strength, the distance of rebound is a relative measure of surface hardness or strength (see Plate 4.6).

Three types of Schmidt hammer have value in geomorphic studies. The 'P' type is a pendulum hammer intended for testing materials of low hardness with compressive strengths of less than 70kPa (kN/m^2). The 'L' type hammer is spring loaded and has a small hammer head which makes it most useful for studying variations of hardness across a rock surface at intervals of a few millimetres. The 'N' type hammer is perhaps the most commonly used for testing concrete and rock. It is theoretically capable of testing rocks with compressive strengths in the range of 20 to about 250 MPa, but it is rather unreliable in weak rocks with strengths below about 30 MPa.

The 'N' type hammer has a scale from 10 to 100. Rocks like chalk have a rebound number (R) of 10 to 20 and at the other end of the scale basalts, gabbros, and quartzites have values of 55 or higher (see Table 4.3).

The advantages of the hammer are that it is light, weighing only 2–3 kg, very easy to carry, easy to use, and relatively cheap. Large numbers of tests may be made in a short time and data collected from a variety of rock surfaces. Weathering rinds, case hardening, individual large clasts, and rock matrices may be readily tested.

Among the disadvantages of the Schmidt hammer are that it is extremely sensitive to discontinuities in a rock, even hair-line fractures may lower readings by 10 points, hence fissile, closely foliated, and laminated rocks cannot be assessed with this instrument unless samples are clamped in a vice. It is also very sensitive to water content, especially of weak rocks. To eliminate as much variability as possible test impact sites should be more than 60 mm from an edge or joint; surfaces should be flat and free from flakes or dirt; and the hammer must be moved to a fresh site for each test. The number of impacts on each sample area of about 2 m² should be 20 to 50 with the larger number being undertaken if the rock is variable. The most reliable results are obtained if the lower 20 per cent of impact readings are ignored and measurements are continued until the deviation from the mean value of the remainder does not exceed ±3 points. The *R* value should be corrected for the inclination of the rock face by using either the manufacturer's chart or the table given by Day and Goudie (1977).

Each Schmidt hammer has a slightly different

rebound so the instrument number should be recorded and all instruments regularly calibrated against a test anvil.

On each test hammer there is a printed calibration curve giving a plot of unconfined compressive strength against the rebound number (*R*). This is unfortunate as the correlation is not universal, being affected by both the unit weight of the rock and its moisture content. A more reliable correlation chart which relates *R*, hammer orientation, rock unit weight, dispersion of strength, and unconfined compressive strength, has been provided by Deere and Miller (1966) (Fig. 4.27). For many purposes,

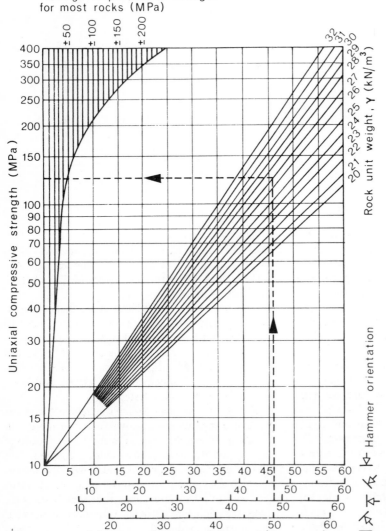

Fig. 4.27 Relationship between Schmidt hammer reading and the uniaxial compressive strength of rock for the 'L' type hammer (modified from Hoek and Bray, 1977, after an original figure by Deere and Miller, 1966).

however, it is unnecessary to convert R values to strength values as the R values alone can be used as indices (Table 4.3).

In the absence of any of the three instrumental methods of determining intact rock and soil strength an approximate indication may be obtained by a subjective field assessment. In Table 4.3 criteria for such an assessment are provided together with related values derived from the recommended strength tests. These values should not be taken as

exact equivalents as they are dependent upon such properties as rock unit weight, moisture content, and anisotropies. The terms 'strong' and 'weak' are used here to apply to intact strength alone and not to mass strength.

Rock mass strength

The strength, or resistance, of a rock mass to erosion is controlled by far more characteristics of the rock

Table 4.3 Approximate Strength Classification of Cohesive Soil and of Rock

Description	Uniaxial Compressive Strength, MPa	Point-Load Strength I_s (50), MPa	Schmidt Hammer, N-type, 'R'	Characteristic rocks
Very soft soil – easily moulded with fingers, shows distinct heel marks.	<0.04			
Soft soil – moulds with strong pressure from fingers, shows faint heel marks.	0.04–0.08			
Firm soil – very difficult to mould with fingers, indented with finger nail, difficult to cut with hand spade.	0.08–0.15			
Stiff soil – cannot be moulded with fingers, cannot be cut with hand spade, requires hand picking for excavation.	0.15–0.60			
Very stiff soil – very tough, difficult to move with hand pick, pneumatic spade required for excavation.	0.6–1.0	0.02–0.04		
Very weak rock – crumbles under sharp blows with geological pick point, can be cut with pocket knife.	1–25	0.04–1.0	10–35	Weathered and weakly compacted sedimentary rocks – chalk, rock salt
Weak rock – shallow cuts or scraping with pocket knife with difficulty, pick point indents deeply with firm blow.	25–50	1.0–1.5	35–40	Weakly cemented sedimentary rocks – coal, siltstone, also schist
Moderately strong rock – knife cannot be used to scrape or peel surface, shallow indentations under firm blow from pick point.	50–100	1.5–4.0	40–50	Competent sedimentary rocks – sandstone, shale, slate
Strong rock – hand-held sample breaks with one firm blow from hammer end of geological pick.	100–200	4.0–10.0	50–60	Competent igneous and metamorphic rocks – marble, granite, gneiss
Very strong rock – requires many blows from geological pick to break intact sample	>200	>10	>60	Dense fine-grained igneous and metamorphic rocks – quartzite, dolerite, gabbro, basalt

Source: Deere and Miller (1966), Piteau (1971), Robertson (1971), Broch and Franklin (1972), and Hoek and Bray (1977).

than intact strength alone. This is evident from a consideration of the ease with which a highly shattered rock may be eroded though each rock particle may have high intact strength as measured in a compressive or point-load test. The problem of determining total mass strength was first overcome by engineers engaged in tunnelling, mining, quarrying, and with designing foundations for structures in rock, by incorporating into one classification those parameters which contribute to rock mass strength and weakness.

Classification parameters

Numerous classifications of rock mass strength have been proposed for engineering purposes (Müller, 1958; Pacher, 1958; Deere and Miller, 1966; Piteau, 1971, 1973; Robertson, 1971; Wickham *et al.*, 1972; Bieniawski, 1973). Most of the commonly used classifications include several of the following parameters:

(1) strength of intact rock;
(2) state of weathering of the rock;
(3) the spacing of joints, bedding planes, faults, foliations, or other partings within the rock mass;
(4) orientation of the partings with respect to a cut slope;
(5) width of the joints, bedding plane, or other partings;
(6) lateral or vertical continuity of the partings;
(7) gouge or infilling material in the partings;
(8) movement of water within or out of the rock mass.

These eight parameters have been incorporated into one classification and assessed for field studies in geomorphology (Selby, 1980). A further three parameters are used in some engineering classifications:

(9) residual stresses within the rock;
(10) the angle of internal friction along partings;
(11) waviness and roughness along partings.

Because residual stresses are usually greatly reduced or eliminated by the development of open joints near the ground surface they may be disregarded in a geomorphic study. Neither the friction angle nor the roughness of rock surfaces along joints is readily measured in the field. Friction angles can only be measured by field shear strength tests involving heavy and expensive equipment or by an empirical method which needs further field assessment (Barton and Choubey, 1977), and roughness can only be

measured along adits cut into a rock face. Failure to take into account these two parameters may reduce the precision of a classification, but engineering experience suggests that this is not a serious deficiency. Some slight recognition of the importance of joint roughness has been incorporated into the orientation of partings parameter.

Most classifications for mining purposes indicate that it is possible to divide the values of each parameter into five classes and to apply a numerical rating to each. This practice has been followed in the geomorphic classification.

Intact strength

Rock intact strength is measured in the field using either a Schmidt hammer or point-load strength tester. Care is taken to ensure that the strength of unweathered rock is assessed.

Weathering

The state of weathering of a rock mass has a major influence upon its strength, permeability, and ease with which its material may be deformed (Figs. 4.28 and 4.29). Permeability and porosity may increase and strength decrease by more than two orders of magnitude during the conversion of unaltered rock to residual soil. It is, however, frequently difficult to quantify this change because weathering proceeds preferentially along rock fissures and zones of weathering may be extremely

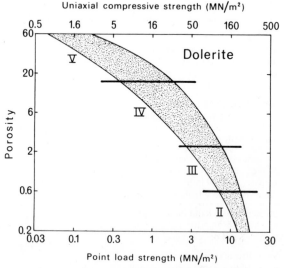

Fig. 4.28 The relationship between porosity and strength in a weathering profile. The horizons are indicated by Roman numerals (data from Dearman, 1974).

Table 4.4 A Scale of Mass Weathering Grades

Grade	Class	Description
VI	Residual Soil	A pedological soil containing characteristic horizons and no sign of original rock fabric.
V	Completely weathered	Rock is discoloured and changed to a soil but some original rock fabric and texture is largely preserved. Some corestones or corestone ghosts may be present.
IV	Highly weathered	Rock is discoloured throughout; discontinuities may be open and have discoloured surfaces and the fabric of the rock near to the discontinuities may be altered so that up to one half of the rock mass is decomposed and disintegrated to a stage in which it can be excavated with a geological hammer. Corestones may be present but not generally interlocked.
III	Moderately weathered	Rock is discoloured throughout most of its mass, but less than half of the rock mass is decomposed and disintegrated. Alteration has penetrated along discontinuities which may be zones of weakly cemented alteration products or soil. Core stones are fitting.
II	Slightly weathered	Rock may be slightly discoloured, particularly adjacent to discontinuities which may be open and will have slightly discoloured surfaces; intact rock is not noticeably weaker than fresh rock.
I	Unweathered fresh rock	Parent rock showing no discolouration, loss of strength or any other weathering effects.

Source: modified from Dearman, 1974, 1976.

irregular. For field classification several schemes have been devised for the description of granitic rock and these can be generalised to be applicable to a wide range of rock types. Many schemes recognise six grades of weathering ranging from fresh unweathered rock to a residual soil. The three main criteria for classification of weathering grades are: degree of rock discolouration; ratio of rock to soil; and presence or absence of original rock texture. Descriptions of these grades are given in Fig. 2.3 and Table 4.4. It should be noted that grade VI — the surface soil — lies outside any consideration of rock strength. Major papers on this subject include those by Dearman (1974, 1976),

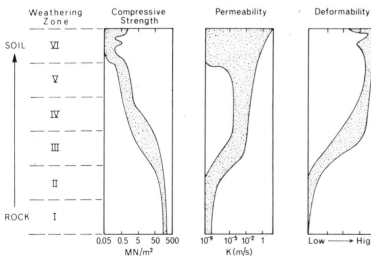

Fig. 4.29 An idealised weathering profile with some characteristic properties (based upon Dearman, 1974).

Baynes *et al.* (1978), Moye (1955) and Ruxton and Berry (1957).

Spacing of partings within a rock mass

Partings, or discontinuities, in a rock mass have a number of possible origins or forms:

(1) bedding: a regular layering in sedimentary rocks which marks the boundaries of small lithological units or beds;

(2) joint: a fracture in the rock mass along which there has been no identifiable displacement;

(3) shear: a surface which has been sheared but with no recognisable displacement; shearing can be recognised by slickensides, by polishing of, or by striations on, a rock surface;

(4) cleavage: a closely spaced parallel surface of fissility;

(5) contact: a surface between two rock types, one or both of which is not sedimentary;

(6) fault: surface of shear recognisable by the displacement of another surface that crosses it;

(7) gneissosity: surface parallel to lithological layering in metamorphic rocks;

(8) schistosity: surface of easy splitting in a metamorphic rock, defined by the preferred orientation of metamorphic minerals;

(9) unconformity: an eroded surface overlain by sedimentary rock;

(10) vein: a fracture, in rock, with a thin filling of intruded material.

For the sake of simplicity all partings will now be referred to as joints as their mechanical properties, not their origin, control their influence on rock strength.

All intact rocks have cohesive and frictional strength. As joints in rocks open cohesion is lost and frictional strength is reduced, although the strength along a joint in which rock grains are interlocking may still be high. Consequently the effective strength of a rock mass containing only microfissures may be nearly as great as that of an unfissured rock.

It has been shown by Hoskins *et al.* (1967) that in low stress situations, such as exist close to the ground surface, the strength along a joint may be described by a Mohr envelope in a straight line form, and that the frictional strength is independent of the area of contact and directly dependent upon the normal load. They also discovered that cohesion values are appreciably large for clean closed joints and often lie between 350 and 1400 kN/m^2 (Fig. 4.30). This phenomenon may be explained either by cohesive interlocking of rock grains on either side of a 'hairline' joint or, as Patton (1966) suggested, by showing that the Mohr envelope is in fact curved at very low normal stresses and cohesion is consequently negligible. Since cohesion is destroyed even after very small movements, and there is no way of measuring it in the field, it is always assumed to be zero. The possibility that it does exist, however, suggests that very tightly closed hairline joints should not normally be included in an assessment of joint spacing.

The more closely spaced are joints the weaker is the rock mass, and the greater is the opportunity for water pressures and weathering processes to weaken it further. Any joint spacing classification thus has a simple logic behind it. The classification

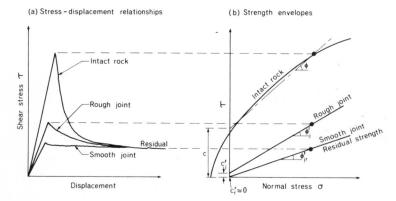

(a) Stress – displacement relationships   (b) Strength envelopes

Fig. 4.30 Cohesion and friction angles for joints and intact rock (after Lo and Lee, 1975).

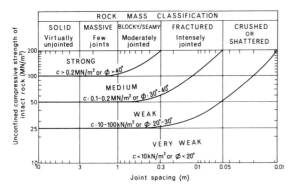

ROCK MASS CLASSIFICATION				
SOLID	MASSIVE	BLOCKY/SEAMY	FRACTURED	CRUSHED OR SHATTERED
Virtually unjointed	Few joints	Moderately jointed	Intensely jointed	

Fig. 4.31 Strength classification of jointed rock masses (developed by Bieniawski, 1973, from an original classification by Müller, 1963).

of Deere (1968) has been widely used by engineering geologists and is adopted here as it provides a convenient five-class scale (Fig. 4.31). Müller (1963) and Bieniawski (1973) have demonstrated how the spacing of joints modifies rock mass strength in comparison with intact rock material. Thus a rock with high intact strength (100 to 200 MPa) but intensely jointed, with joints 50 mm or less apart, has a weak rock mass.

On many sedimentary rocks weathering processes produce a close fissuring of the outcrop surface. This pattern may obscure the more widely spaced true joint pattern. Superficial fissuring does not reduce mass strength, even though it may aid the production of talus material, and it is frequently essential to remove the shattered rock debris before measurements can be made of true joint spacing, orientation, and width (Plate 4.9).

Orientation of joints

The significance of joint orientations may be gauged from Terzaghi's (1962) comment that the stability of a slope formed on chemically intact rock, with uniaxial compressive strength greater than about 30 MPa, depends primarily on the orientation of the major continuous joints with respect to the hillslope.

4.9 An outcrop of sandstone showing the clean face and widely spaced joints exposed by a recent block fall, and the superficial spalling and blocky disintegration which gives the appearance of more closely spaced fissures. The strength of the rock mass is related to the spacings of the main joints, not to the weathering features.

The theoretical critical hillslope angle for stability is, in the absence of cohesion across the joint, controlled by the friction angle along the joint ($\phi_j{}^\circ$) in the direction of potential shear. Where the major continuous joints have a random orientation the critical hillslope angle for stability on igneous rocks is commonly about 70° because of the high friction angle of interlocking joint blocks. In sedimentary rocks the major joints are usually along bedding planes and hence essentially planar. The friction angle in planar joints is commonly close to 30° (Table 4.5) and if bedding planes dip towards a valley at an angle smaller than $\phi_j{}^\circ$ the theoretical critical hillslope angle (for stability) is 90°. For bedding planes dipping towards a valley at angles greater than $\phi_j{}^\circ$ the critical hillslope angle is equal to the angle of dip of the bedding planes (these relationships are explored in more detail in Chapter 7).

The theoretical condition seldom applies exactly to a field situation because the water content of a joint varies and produces a range of values of ϕ_j as well as upthrust, and hence decreased stability, from cleft water pressures. Joint roughness also varies; joint blocks may be cemented together or interlocked and hence stability increased. It has been shown experimentally by Barton (1973) that rough joints have a friction angle which is greater than that of a smooth joint in intact rock in proportion to the amplitude of the asperities along the joint.

It is rare for outcrops to provide sufficient exposure for a full evaluation of joint roughness, so this parameter is included in the rock mass classification to only a very limited extent.

It must be recognised that high water pressures, very smooth joints — such as cleavages in slates — and cementation may produce considerable variation from theory, but some guide to orientation significance may be obtained from Table 4.6. In this system horizontal joints are taken as an intermediate case with greater strength being attributed to joints dipping into the slope, and low strength ratings being adopted for increased dips out of the slope. Vertical jointing is classed with horizontal joints for rocks with high compressive strength, but it is regarded as unfavourable in low compressive strength rocks which may fail by buckling. Random orientation of joints is regarded as favourable in hard rocks with rough joints produced by tensile failure, but fair or unfavourable in shattered rock in which shear failure may occur along multiple intersecting smooth joints, foliation, or cleavage. Table 4.6 is an approximate guide only and anyone using the classification should be primarily guided by local field evidence, particularly of slope failures. It should also be noted that in rocks with strong planar joints dipping steeply into a slope, failure is likely to occur in cross joints rather than along the continuous joints, and the cross-joint angle should then be considered as the control (Plate 4.10).

The accuracy with which dip measurements, on limited outcrops, can be made is indicated by studies of the accuracy of joint surveys in tunnels and open pit mines (Piteau, 1973). The estimated average maximum error is ±5°. For planar joints this may be improved to ±1° where dips are greater than 70°, and ±3° where inclinations are between 30 and 70°.

Width of joints

Separation is important to mass strength because it largely controls the frictional strength along the joint as well as the flow of water and the rate of weathering of the wall rock. Widely open joints have no inherent strength and their resistance to shear is only that of the joint infill. Where the infill is of clay then the stability of a rock mass with joints

Table 4.5   Roughness Classification and Friction Angles (ϕ_j°) for Joints

Roughness	Joint surface features	ϕ_j° at low normal stresses (i.e. 350 to 2000 kPa)	ϕ_j° at high normal stresses
1	with clay infill	infill controls	infill controls
2	smoothest natural joints	31–40	29.5
3	moderately rough	38–47	32.5
4	very rough with steps	40–50	36
5	extremely rough	>50	42

Source: after Robertson, 1971.

4.10 Joints in dolerite. The main joints dip into the cliff but the control on cliff stability is exercised by cross-joints nearly parallel to the cliff face.

dipping out of the hillslopes may be little more than that of the clay.

Tightly closed joints may have high frictional and some cohesive strength, while those with widths between the two extremes have strengths dependent upon the joint roughness and contact between the wall rocks.

In engineering classifications joint width is usually judged according to that likely to be encountered in the high stress conditions of newly opened tunnels

Table 4.6 Strength Classification for Joint Orientations

	Mode of joint formation	
	Tensile (rough)	Shear (smooth)
Very unfavourable	Joints dip out of the slope: planar joints 30–80°; random joints > 70°.	Joints dip out of the slope: planar joints > 20°; random joints > 30°.
Unfavourable	Joints dip out of the slope: planar joints 10–30°; random joints 10–70°.	Joints dip out of the slope: planar joints 10–20°; random joints 10–30°.
Fair	Horizontal to 10° dip out of the slope. Nearly vertical (80–90°) in hard rocks with planar joints.	Horizontal to 10° dip out of the slope.
Favourable	Joints dip from horizontal to 30° into the slope: cross joints not always interlocked.	
Very Favourable	Joints dip at more than 30° into the slope: cross joints are weakly developed and interlocking.	

4.11 The dominant joints controlling the cliff slope are stress release joints nearly parallel to the face. The horizontal bedding joints are so tightly closed that they have little or no influence on mass strength.

or mines. In geomorphic situations joints exposed at the ground surface may be several centimetres open. Furthermore such joints can be further opened by weathering and erosion. Consequently the width which has to be considered is that of the joint at a depth of 10 or more centimetres into the outcrop. Another problem is that 'hairline' fractures may have little effect in reducing strength and may have to be ignored (Plate 4.11). Judgement has to be exercised in assessing width and a conservative view is advocated in interpreting the classes of Table 4.7.

Continuity and infill of joints

The significance of continuity of joints is illustrated by the observation of Orr (1974) that where relative continuities of joints are less than 60 per cent then shear along their surfaces is very improbable. Terzaghi (1962) analysed the significance of the cohesive strength of rock in cliff slope stability. He suggested that the effective cohesion within a jointed rock mass is given by:

$$c_e = c \, \frac{(A - A_j)}{A}$$

where c_e is the effective cohesion of the rock mass;
  c is the intrinsic cohesion of intact joint blocks;
  A is the total area of the shear plane;
  A_j is the total joint area within the shear plane.

  Thus the larger the bridges of intact rock between joints the greater is the effective cohesion (i.e. the smaller is A_j). Continuous joints thus reduce cohesive strength, provide zones of shear, and permit the circulation of water.

  Joint infill (often called gouge by mining engineers) may be composed of clay-rich soil. Where the infill contains swelling clays, such as montmorillonite, swelling pressures may be exerted, or the gouge may prevent frictional contact of the rock walls of the joint (Fig. 4.32). Because of the difficulty of measuring fill in inaccessible joints which

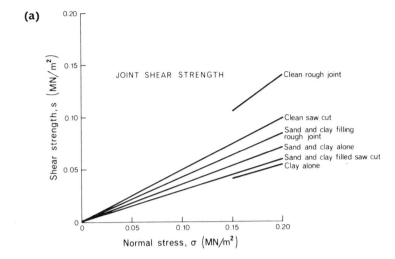

(a)

JOINT SHEAR STRENGTH

Clean rough joint

Clean saw cut
Sand and clay filling rough joint
Sand and clay alone
Sand and clay filled saw cut
Clay alone

Shear strength, s (MN/m^2)

Normal stress, σ (MN/m^2)

Fig. 4.32 (a) Direct shear test results for joint strengths with varying roughness and fill of the joint (data from Kutter and Rautenberg, 1979); (b) Influence of joint filling thickness on the shear strength of an idealised sawtooth joint; as the filling increases in thickness strength along the joint reduces towards the strength of the fill alone (after Goodman, 1970).

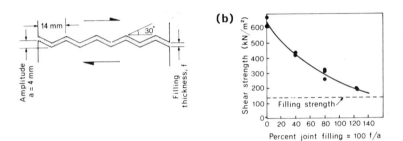

(b)

14 mm   30°

Amplitude a = 4 mm

Filling thickness, f

Shear strength (kN/m^2)

Filling strength

Percent joint filling = 100 f/a

are seldom exposed in three dimensions an entirely qualitative classification is used (Table 4.7).

Ground water

The flow of water out of a rock face is probably of greater significance in reducing the stability of rock exposures in mines and tunnels than in landforms. However, seepage forces develop along joints and the incidence of rockfalls from cliffs in mountains is frequently correlated with high water tables and rainfall maxima.

Subsurface water promotes instability in at least four ways:

(1) pore water in joint filling materials promotes weathering and solution and thus reduces the cohesive and frictional strength of the infill;

(2) cleft-water pressures along joint surfaces produce uplift forces along potential failure planes, thereby reducing frictional resistance and effective normal stresses;

(3) water in the pores of intact rock promotes weathering and solution and decreases rock compressive strength, especially where confining stresses have been reduced;

(4) mudstones, shales, and argillites may disintegrate by slaking and be converted to slurries.

The extreme effects of water upon shale slopes, in the De Beers diamond mine near Kimberley, were noted by Jennings (1971) who reported that they will stand at angles greater than 45° when drained, but undrained slopes in the same material will stand at only 18-29°. A similar situation occurs in cohesionless sand which will support a hillslope of about 30° when it is dry, but of only half that when the water table has risen to the slope surface.

A rock mass strength classification should, ideally, use pore- or cleft-water pressures within the rock mass at the least favourable season of the year. This is seldom, if ever, possible in a geomorphic study and instead it is recommended that an estimate be made of the volume of water flowing out of each 10 m² of rock face in the wettest season. It is recognised that in many situations it may be impossible to do better than express the flow as 'none, trace, slight, moderate, or great'.

Unified classification and rating of parameters

In a total classification of rock mass strength all of the parameters have to be combined in a graded scale. For this purpose five classes are used: this number of classes can be readily recognised in the field and is commonly used in engineering classifications. Class 1 denotes a very strong rock mass and class 5 a weak mass.

The parameters which are assessed in the field provide semi-quantitative data for the classification. The parameters do not necessarily conform to the same class. Thus at one exposure the intact rock strength may be class 2; the joints may be widely spaced (class 1), discontinuous (class 1), but the dip of the strata may be unfavourable (class 4); the same rock mass may be only slightly weathered (class 2), but have a moderate outflow of ground water (class 4).

Not all parameters are of equal importance in producing rock strength and it is thus necessary to assign a numerical weighting to each. The final rock mass strength class will then be the sum of the weighted values determined for the individual parameters at each site. Higher numbers reflect greater mass strength.

Intact rock strength is given a 20 per cent rating; separation of joints 30 per cent; joint orientations 20 per cent; joint width, continuity, and water flow collectively 20 per cent. Weathering is accorded only 10 per cent because much of the effect of weathering is subsumed in the other parameters: thus loss of strength and opening of joints at a rock face are largely weathering phenomena, and water movement and formation of joint infill both promote and are an effect of weathering.

The classification, with the ratings (r) for each parameter and each class is presented in Table 4.7. Field experience indicates that most of the ratings can be applied without much difficulty. The greatest problem arises where the spacing of joints along a face is variable and more than one class of spacing is present. This problem may be resolved either by dividing the exposure into several units and classifying each one separately, or by using an intermediate rating value between the set values of each class. The latter procedure is most relevant to the joint spacing parameter where the differences in rating between classes 2, 3, 4, and 5 are six or seven points.

In this classification the lowest possible score is 25, leaving open the possibility of adding soil into the lower range of a comprehensive strength scale.

Very rarely will very weak rock masses be exposed in the field for they will usually be vegetated in humid climates and in arid climates the absence of a high water outflow will prevent application of the lowest possible rating.

Field procedures

The only items of equipment needed for the application of the classification are: a tape measure, an inclinometer, and a strength tester — in most cases a Schmidt hammer is the most convenient device.

The classification can be applied to any rock mass with sufficient exposure for measurements to be made of the joints. The outcrop must be divided into a number of regions, each having similar characteristics; for example, the same joint spacing, dip, and lithology. The boundaries of a region will usually correspond with a prominent feature such as a distinct break of slope, a fault, dyke, or change in lithology.

Measurements of joint spacing are made in both a vertical and horizontal direction and the area of outcrop measured is usually about 10 m^2, although this has to be adjusted to suit field conditions. By using Table 4.7 the ratings for each parameter may be decided in the field but it is recommended that a detailed description and field sketch should also be made of each rock unit so that the operator can check on the consistency of his, or her, measurements.

The classification may be used to further an understanding of changes in slope profiles, the location of hills and depressions, the location of stacks, tors, inselbergs, roches moutonnées, or other features. Where rock exposure is adequate the rock mass classification rating can be incorporated into a geomorphic map. Examples of applications are given in Chapter 9.

Behaviour

Rheology is the study of behaviour of all materials under stress. There are four basic rheological behaviours — elastic, plastic, viscous, and fracture. To illustrate these behaviours imagine a ball of rubber, a ball of soft clay, a cube of honey, and a cubic crystal of common salt (halite). These balls and cubes are dropped from the same height on to a concrete floor:

(1) the rubber ball deforms momentarily as it strikes the floor, but then recovers its shape as it bounces — it behaves as an elastic substance;

Table 4.7 Geomorphic Rock Mass Strength Classification and Ratings

Parameter	1 Very Strong	2 Strong	3 Moderate	4 Weak	5 Very Weak
Intact rock strength (N-type Schmidt Hammer 'R')	100–60 r : 20	60–50 r : 18	50–40 r : 14	40–35 r : 10	35–10 r : 5
Weathering	unweathered r : 10	slightly weathered r : 9	moderately weathered r : 7	highly weathered r : 5	completely weathered r : 3
Spacing of joints	>3 m r : 30	3–1 m r : 28	1–0.3 m r : 21	300–50 mm r : 15	<50 mm r : 8
Joint orientations	Very favourable. Steep dips into slope, cross joints interlock r : 20	Favourable. Moderate dips into slope r : 18	Fair. Horizontal dips, or nearly vertical (hard rocks only) r : 14	Unfavourable. Moderate dips out of slope r : 9	Very unfavourable. Steep dips out of slope r : 5
Width of joints	<0.1 mm r : 7	0.1–1 mm r : 6	1–5 mm r : 5	5–20 mm r : 4	>20 mm r : 2
Continuity of joints	none continuous r : 7	few continuous r : 6	continuous, no infill r : 5	continuous, thin infill r : 4	continuous, thick infill r : 1
Outflow of groundwater	none r : 6	trace r : 5	slight <25l/min/10 m² r : 4	moderate 25–125l/min/ 10 m² r : 3	great >125l/min/10 m² r : 1
Total rating	100–91	90–71	70–51	50–26	<26

Source: Selby, 1980.

(2) the ball of clay sticks to the floor as a blob — it behaves as a plastic substance;

(3) the cube of honey slowly spreads out on the floor — it behaves as a viscous material;

(4) the halite shatters and fragments are scattered on the floor — it has behaved as a brittle substance.

In elastic behaviour all of the strain is recoverable, as when a spring is stretched, and the strain is proportional to the stress. Material behaving in this way is said to be a Hookean substance.

With a plastic material if the shear stress is less than the yield strength the body remains rigid, but if the shear stress equals the yield strength the plastic body deforms. This type of behaviour can be modelled by a weight resting on a flat surface. In order to move the weight the friction between it and the surface must be overcome. If the force applied is too small the weight does not move, once the force is sufficiently large the weight moves and the deformation is permanent. Such behaviour is said to follow a Saint Venant model.

Viscous substances behave in a manner which can be modelled by a dashpot — that is a tube filled with oil in which a fitting piston can be pulled back and forth. Pushing or pulling the piston causes the oil to move from one end of the tube to another by flowing rourd the edge of the piston. The rate at which the oil moves is proportional to the force applied to the piston. Material behaving like the oil is said to behave as a Newtonian substance.

Some substances do not behave as simple brittle, plastic, or elastic solids nor as viscous liquids, but have a complex behaviour which can best be described as combinations of the basic models. Rock at normal temperatures and pressures behaves as an elastic-brittle substance but soil is an elastic-plastic substance which at high water contents can behave as a viscous fluid. The particulate nature of soil, however, makes even this complex model inadequate to describe all soil behaviour. Ice is another complex

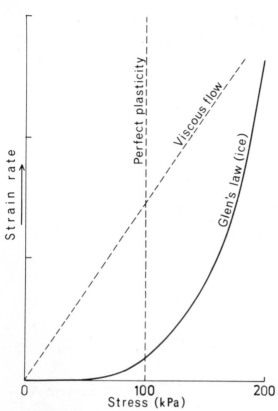

Fig. 4.33 Generalised relationships between stress and strain for various materials.

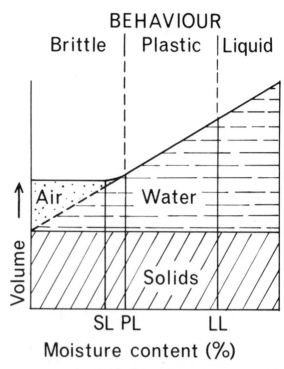

Fig. 4.34 Composition of soil ranging from air-filled voids to water-filled voids to a liquid, with the Atterberg limits.

material: it deforms according to what is known as Glen's Law — as the stress is increased beyond a critical amount the strain rate increases until a steady state is reached at which for a constant stress the strain rate is nearly constant (Fig. 4.33).

Properties of soils

The resistance of soils to erosion depends upon many factors. Among those factors which can be determined are: types of minerals and their amounts; types of adsorbed cations in soils; shape and size distribution of particles; water content; soil fabric; and a variety of external factors, such as climate and vegetation. The external factors will be discussed in Chapters 5 to 8. In this section some major compositional factors will be discussed.

In soils the most reactive components are the clay minerals. In general the greater the quantity of clay minerals in a soil the greater is the potential shrinkage and swelling, the lower is the permeability, the higher the plasticity, the higher the cohesion, and the lower the angle of friction. Consequently

the focus of attention in investigation of soil strength is often the clay content. This compares with the spacing of joints as a focus in rock investigations.

Atterberg limits

Soils contain varying proportions of solids, and voids which may be filled with gases or with water (Fig. 4.34). The proportions of these three components greatly affects the behaviour of soil under stress.

The behaviour of a soil — whether it behaves as a brittle solid, a plastic substance, or a viscous liquid — is largely dependent upon its water content. Soils have varying textural and mineral compositions so series of tests are commonly used to determine the water content at which behaviour changes from one state to another. These tests, known after their originator A. Atterberg, apply only to silts and clays and not to cohesionless coarser materials.

A cohesive soil with a very high water content would be a suspension of soil particles in water and if a stress were applied to it there would be a continual deformation. It would behave as a Newtonian substance.

4.12 Apparatus for determining Atterberg limits: a drop-cone penetrometer, a liquid limit dish with grooving tool, a thread of soil on a glass plate for determination of the plastic limit.

As the soil dried out a point would be reached at which it would exhibit a small resistance to shear and the soil would be behaving as a plastic substance with permanent deformation. The moisture content at which the liquid starts acting as a plastic solid is known as the liquid limit (LL) (Fig. 4.34).

As further moisture is driven from the soil resistance to larger shearing stresses occurs. Eventually the soil exhibits no plastic deformation and fails by brittle fracture. The limit between plastic and brittle failure is known as the plastic limit (PL).

If the drying process is prolonged after the plastic limit has been reached the soil will continue to decrease in volume until a certain moisture level has been reached. Below this moisture level, known as the shrinkage limit (SL), the soil will retain a constant volume even though further drying is possible.

Determination of the liquid limit is carried out by placing a sample in a hinged cup which rests on a base plate. A groove is cut across the surface of the sample, and the moisture content (per cent) at which 25 taps of the cup on the base cause the soil to flow and close the groove, is the liquid limit. The plastic limit is defined as the minimum moisture content at which the soil can be rolled into a thread 3 mm in diameter without breaking. Alternatively the liquid limit may be determined from the depth of penetration of a standard cone penetrometer head which is allowed to sink into a sample from a set height (Plate. 4.12).

The importance of the Atterberg limits is that they are greatly influenced by soil mineralogy. Some soils with high water contents will behave plastically while others at the same water content may behave as liquids or as brittle solids. Soils with clay minerals like kaolinite do not swell readily and they cannot adsorb much water because of their limited area of adsorption surfaces. Montmorillonite, by contrast, has very high swelling and shrinkage characteristics

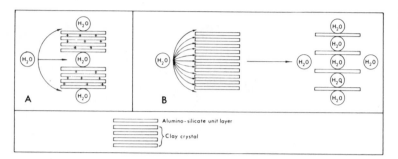

Fig. 4.35 The adsorption of water between the platelets of (A) illite and (B) montmorillonite crystals.

because, unlike kaolinite, it can adsorb water molecules between the crystal units (Figs. 4.35 and 4.36). Even a small quantity of montmorillonite can greatly increase the capacity of a soil to adsorb water and, as shear strength is largely controlled by water content, thus greatly influence soil strength.

Soils with montmorillonite clays therefore are liable to extremes of plastic and liquid behaviour; they will shrink on drying and swell on wetting thus causing the formation and closing of cracks. In extreme conditions of very high water content soils will flow. Such soils are potentially very unstable. The capacity of clays to exchange cations is directly related to the capacity of a clay mineral to adsorb water, so cation exchange capacity (CEC) is a useful index of water adsorption capacity.

Because of the variable controls on the liquid and plastic limits it is usual to express the consistency limits of the soil as indices — plasticity index, liquidity index, activity:

$$\text{plasticity index} = \text{liquid limit} - \text{plastic limit}$$

$$\text{liquidity index} = \frac{\text{water content} - \text{plastic limit}}{\text{plasticity index}}$$

$$\text{activity} = \frac{\text{plasticity index}}{\text{percentage clay fraction}}.$$

The plasticity index (PI) is the range of moisture content over which a soil is plastic. The higher the plasticity index the greater is the potential instability of that soil on a slope. The relationship between plasticity, clay content, and clay mineralogy of soil clays is indicated in Fig. 4.37. The exchangeable-cation composition of montmorillonite not only affects plasticity but also shear strength with

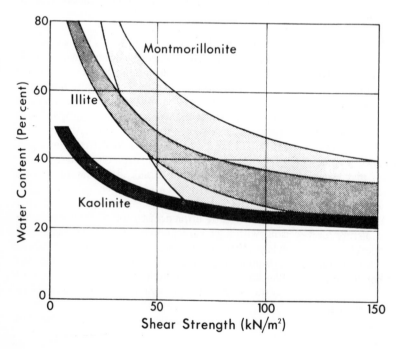

Fig. 4.36 The relationship between shear strength and water content for various clay minerals.

Table 4.8 Clay Mineral Properties

Mineral	Surface area (m²/g)	Plastic limit (%)	Liquid limit (%)	Shrinkage limit (%)	Volume change	Cation exchange capacity (CEC) (milliequivalents per 100 g)
Montmorillonite	800	50–100	100–900	8.5–15	high	80–150
Illite	80	35–60	60–120	15–17	medium	10–40
Kaolinite	15	25–40	30–110	25–29	low	3–15

Note: the great variability in plastic and liquid limits, and in CEC, is a result of the effect of different adsorbed cations.
Source: data from Grim, 1968.

increases in the following order: sodium < calcium < aluminium. The non-clay minerals quartz and feldspar do not develop plasticity with water, even when they are ground to sizes of less than 2 μm. Plasticity is thus due to characteristics of the clay minerals.

In Fig. 4.37 the relationship between the amount of clay-sized particles present and the plasticity index is a straight line passing through the origin of the graph. The slope of the graph line indicates the magnitude of the surface forces of the clay particles and is called the activity ratio. The active clays (those with a large activity ratio) exhibit plastic properties over a wide range of water content.

Activity of clays has been correlated with their geological history and with the proportion of their strength which is due to the cohesion of one particle to another. It has been suggested by Skempton (1953b) that clays with an activity of less than 0.75 should be termed inactive, those between 0.75 and 1.25 normal, and greater than 1.25 active. These divisions correspond roughly with the broad groups of silicate clays: kaolins are generally inactive, the normal clays are illitic, and the active clays are either organic colloids or montmorillonitic (Seed *et al.*, 1964).

The strength of soils, as indicated by the residual angle of internal friction, is closely related to the liquid limit, percentage of clay present in the soil, and to the plasticity index. Because the friction angle can only be determined from the relatively difficult and expensive ring shear test it is sometimes convenient to estimate it from published data (Figs. 4.38 and 4.39). The relationship of ϕ_r' with plasticity index is shown on an arithmetic scale which emphasises the differences between clay mineral species and a semi-logarithmic scale which is easier to use for estimations (Voight, 1973; Jamiolkowski and Pasqualini, 1976; Seyček, 1978).

It should be noted that Atterberg limits are always determined on remoulded samples, hence a natural soil or clay material may exhibit a different consistency from that determined in a laboratory test.

Sensitivity

Within a soil the clay particles may be aggregated into clusters in a manner which is controlled by the electrostatic forces on their surfaces. If clay minerals are being deposited in saline waters they are likely to form a honeycomb or a flocculent structure in which the clay platelets have an edge to edge or 'cardhouse' structure. This structure may be caused by negative charges on the flat mineral surfaces and positive charges on the edge of platelets where there are adsorbed cations. A cardhouse structure can thus be stable. If, however, adsorbed

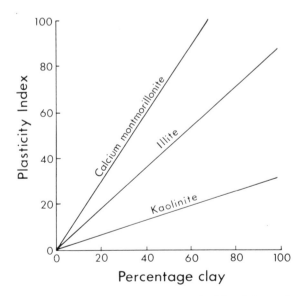

Fig. 4.37 The relationship between plasticity index and percentage of clay-sized particles in a soil (after Skempton, 1953b).

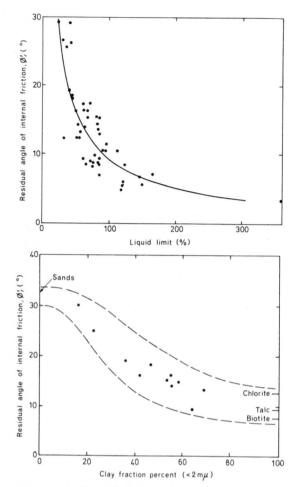

Fig. 4.38 The relationship between liquid limit and ϕ'_r for soils (data from Jamiolkowski and Pasqualini, 1976), and the relationship between percentage of clay in the soil and ϕ'_r (after Skempton, 1964).

cations are removed in percolating drainage water, or very high normal loads develop during deep burial of sediments, or if high shear stresses develop, the clay platelets may be forced to align parallel to each other (Fig. 4.40).

The effect of remoulding under a shear stress is characterised quantitatively by the term sensitivity; for samples at the same water content:

$$\text{sensitivity} = \frac{\text{undisturbed, undrained strength}}{\text{remoulded, undrained strength}}.$$

Soils which do not lose their structure under shear stress have a sensitivity of 1. Normally consolidated clays have sensitivity values ranging from 5 to 10,

but certain clays in Canada and Scandinavia have sensitivities as high as 100 or even 1000. Sensitivities higher than 16 are characteristic of quick clays. In Norway quick clays have formed where marine clays have been raised above sea level and then leached of their salt content. Their 'house of cards' structure becomes unstable and they are readily remoulded and large-scale landsliding may result from disturbances produced by heavy rain, earthquake shocks, or even pile-driving (Fig. 4.41). Once instabilities develop they may spread throughout an area by retrogressive failure (see Chapter 6).

Liquefaction

Not only clays but also sands can behave in a highly sensitive manner (although the term 'sensitivity' is confined to clay materials). A loosely packed sand owes its strength to point-to-point contacts and if

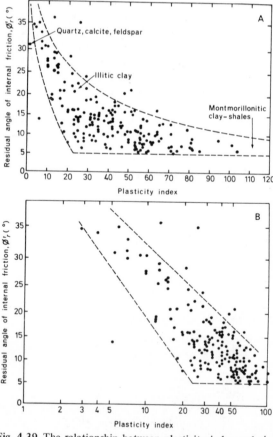

Fig. 4.39 The relationship between plasticity index and ϕ'_r for various soils shown on both arithmetic and semilogarithmic plots (compiled from many sources by Seyček, 1978).

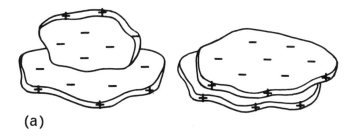

Fig. 4.40 (a) Types of bonds between platy clay particles; (b) the effect of consolidation on clay structure; (c) the effect of remoulding; (d) the effect of shearing.

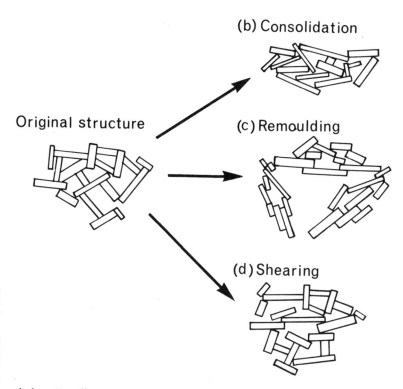

shaken it will compact into a denser, and stronger, form. If a loosely packed sand is saturated with water, compaction on shaking is retarded where water cannot drain away to accommodate the volume change. Consequently the relaxing sand skeleton transfers some of its intergranular effective stresses to the pore water and pore-water pressures increase. In extreme cases all of the effective stresses are transferred to the water and interparticle frictional strength becomes zero. In this state the sand has no significant shearing resistance and it deforms like a liquid.

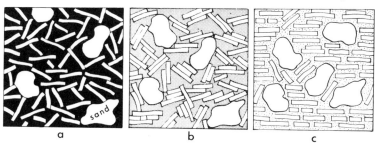

Fig. 4.41 Structural changes in a soil which has been leached and then failed as its structure has collapsed to a more compact form. The rectangular forms represent clay crystals: (a) flocculated structure; (b) leaching has caused some breakdown of structure; (c) collapsed structure.

Liquefaction usually results from shocks such as earthquake, pile-driving, or traffic vibrations. Alluvial sands, coastal sand dunes inundated by high sea levels, and badly compacted engineering structures are prone to liquefaction.

Further reading

The strength and properties of soils are discussed at length by Kezdi (1974) and by J. K. Mitchell (1976). The cohesive strength of clays is particularly well treated by Osipov (1975), and the effect of asperities in frictional strength by Scholz and Engelder (1976). Wesley (1977) has discussed the peak and residual strength of clays. The applications of soil mechanics are well described by Lambe and Whitman (1979) and in many other introductory texts such as Smith (1978). Soil strength testing is described by Skempton and Bishop (1950) and Vickers (1978). Rock strength is treated by Zalesskii (1964) and in many textbooks of rock mechanics such as Attewell and Farmer (1976) and Vutukuri *et al.* (1974). The rheology of rocks and soils is described by Johnson (1970), and the influence of mineralogy and pore fluids on residual strength of clays by Kenney (1967, 1977).

5

Water on hillslopes

Hillslopes with a soil cover undergo a great variety of weathering, soil formation, soil erosion, mass wasting, and deposition processes. The energy for these processes is that of gravity and received solar radiation and the agent for change is almost invariably that of water. The action of water and ways in which its action is modified by vegetation, soils, slope angle, and surface relief are thus the focal points of studies of slope processes.

Hillslopes in the hydrological cycle

Water reaches the ground by falling from clouds as rain, snow, hail, or by the condensation of dew and fog – these forms of water are collectively called precipitation. Some of the falling precipitation may directly hit soil or rock or fall into the streams, but in a humid region much hits the vegetation and is intercepted. Part of the intercepted water is evaporated back into the atmosphere but in prolonged or intensive rain some water falls between the plants or drips from the leaves to the ground, some flows down the plant stems, and some may be absorbed by the plants. Of the water which reaches the ground part may flow off directly into the streams but, except on frozen or bare rock surfaces, most of the surface water is infiltrated into the soil or is temporarily ponded in depressions on the surface. The water which infiltrates into the soil first fills the voids in the soil and, if the soil is very impermeable, any excess water must flow off the soil surface – as overland flow – but, if the properties of the soil profile permit, subsurface water may flow laterally as interflow or move down vertically to become part of the ground water. Both interflow and ground water may eventually reach the stream occupying valley floors. The water of streams flows into the oceans from where it may again be evaporated and then condensed into cloud droplets, and be precipitated on the land. This system of circulation of water between the atmosphere, the land, and the oceans is called the hydrological cycle. The main elements are shown pictorially and schematically in Figs. 5.1 and 5.2.

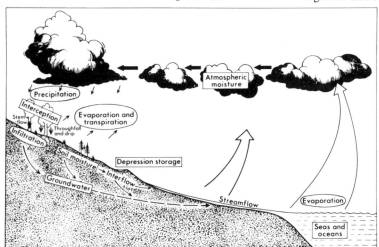

Fig. 5.1 A pictorial representation of the hydrological cycle.

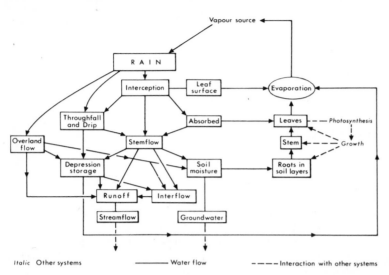

Fig. 5.2 A schematic representation of the links between components of the hydrological cycle.

Italic Other systems ———— Water flow – – – –Interaction with other systems

It must be emphasised that the following discussion is concerned with hillslopes and not with large river valleys in which terraces and floodplains are major collecting areas for water.

Interception

As rainfall is intercepted on its path to the soil the water may drip off the leaves of the intercepting tree or shrub, or it may become stemflow as it runs down the trunks of the plants. When interception occurs at the beginning of a storm the leaves are dry

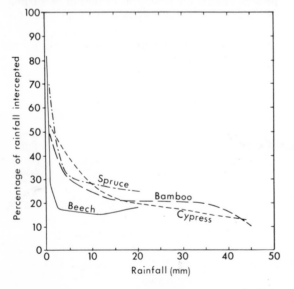

Fig. 5.3 Interception by various tree species (data from Lull, 1964).

and water may be directly absorbed or evaporated from the plant surface. Small depressions in the bark and the crotches of trees will also hold water. Thus there is an initial interception loss. The capacity of the plants to store water and lose it by evaporation declines with increasing duration of the storm, and under very prolonged rainfall the canopy may become saturated and the loss may decline to virtually nothing.

It is impossible to draw quantitative universally applicable conclusions about interception losses because they depend upon various meteorological factors: the duration, amount, intensity, and frequency of rainfall; windspeed; and the type of vegetation. The bulk of the experimental evidence indicates that interception losses are generally greater beneath evergreen than beneath deciduous trees. Winter and summer losses beneath evergreens are usually about the same but winter losses beneath deciduous trees are much lower (about 4-7 per cent) than the summer losses (10-15 per cent) (Lull, 1964).

A high initial loss with a decline to a much lower loss is normal for all reported studies (see Figs. 5.3 and 5.4). In New Zealand Aldridge and Jackson (1968) found that manuka intercepted 39 per cent of the total gross rainfall; the lost water was presumed to be accounted for by evaporation. Stemflow accounted for about 23 per cent of the gross rainfall and throughfall about 39 per cent. The rainfall measured at the ground surface increased with gross rainfall amount and intensity. Interception losses beneath grasses and crops are

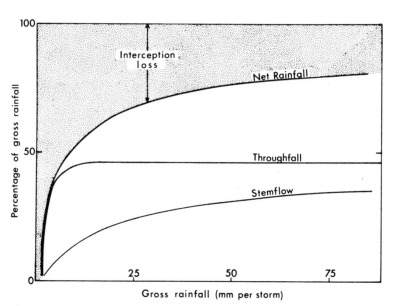

Fig. 5.4 Distribution of rainfall after it has reached the canopy of 5 m tall manuka (*Leptospermum scoparium*) (after Aldridge and Jackson, 1968).

usually in the range of 10 to 20 per cent and beneath forest 5 to 50 per cent.

Not all effects of vegetation on moisture balance are negative; in cloudy and foggy areas condensation on leaves and the resulting drip can account for a large proportion of the moisture supply; some so-called cloud forests get nearly all their moisture this way. Methods of measuring stemflow, throughfall, and interflow are shown in Plates 5.1 and 5.2.

Infiltration

Infiltration is the process by which water enters the surface horizon of the soil. Percolation down to the water table occurs beneath the surface horizons. Infiltration may be controlled by a variety of factors including: (a) the type of precipitation, especially its intensity; (b) surface soil compaction; (c) soil porosity, especially at the surface; (d) splashing of fines on a bare surface to form a crust which blocks soil pores; (e) depth of surface detention; (f) cracks; (g) slope; (h) freezing; (i) cultivation; (j) swelling clays which may block pores; (k) low soil pH (where the soil cation-exchange complex is dominated by H^+ ions soil particles disperse, but divalent cations such as Ca^{2+} encourage flocculation and soil structure development); (l) animal trampling; (m) vegetation; (n) litter; (o) soil organisms; and (p) soil moisture. The actual controls on infiltration into a particular soil depend upon the local operation of these factors in various combinations and magnitudes.

Pre-existing soil moisture is important, partly because some dry soils exhibit an initial resistance to wetting, but mostly because water added to a soil moves through the film of water adhering to the surface of soil particles. Only when the films have thickened so that they almost fill the larger voids can water move at the maximum rate for that soil. This condition may be expressed by saying that the initial, or dry, hydraulic conductivity is low and that saturated hydraulic conductivity is the maximum rate of water movement under gravity.

After initial wetting the surface horizons form a transmission zone above a wetting front. Existing soil water is displaced downwards out of the macropores by newly infiltrating water which moves as a pressure wave through the profile. In multilayered soils hydraulic conductivities vary so that saturated layers may form above each less permeable horizon and interflow may develop in the more permeable horizons.

A steady rate of infiltration is achieved when the entire profile is transmitting water at the maximum rate permitted by the least permeable horizon. This may occur in as little as 10 minutes after the onset of a storm or only after several hours. Excess water which cannot infiltrate is stored initially in surface depressions (depression storage), and once these are filled the excess spills downslope as overland flow. The changing rate of infiltration may be represented by a curve (Fig. 5.5).

In a soil which is very uniform in its properties,

5.1 Experimental plots with long troughs for catching throughfall and collars for catching stemflow. The segmented collar in the inset is for measuring fall and drip close to the trunk.

slope, vegetation, and land use, a single curve may give a reasonable representation of the infiltration occurring during a storm, but for a soil with variable properties a family of curves gives a better indication of infiltration (Fig. 5.6). Where the infiltration rate is highly variable from one soil plot to another a few centimetres away, the runoff from one plot of low infiltration capacity may well be absorbed by another plot of higher capacity lower down the slope. Surface runoff from one part of a slope does, therefore, not necessarily reach a stream by overland flow.

Infiltration is strongly influenced by the structure and texture of a soil as sands and gravels may be a million times more permeable than clays, but permeabilities alone do not control infiltration (Table 5.1). Land use practices and vegetation cover greatly modify the ability of a soil to absorb water. One pass of a tractor wheel, for example, has been known to reduce non-capillary pore space by half and infiltration rate by 80 per cent (Steinbrenner, 1955). Compaction by machinery or by trampling animals may be so great that infiltration into cultivated fields is markedly less than beneath nearby woodland. Wise land use is, therefore, aimed at reducing compaction, especially in critical areas in valley bottoms and near gullies where increased runoff could start rapid rill and gully erosion.

Reported infiltration capacities vary from 2 to 2500 mm/hour. The range of infiltration in relation to vegetation, slope, and pre-existing moisture content of a variable soil is indicated in Fig. 5.7.

5.2 Boxes inserted into the wall of a trench for collecting interflow water.

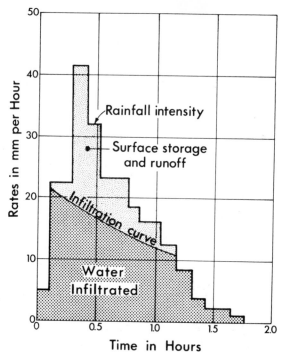

Fig. 5.5 Rainfall and infiltration during a storm. At low intensities of rainfall all the water infiltrates.

Sources of runoff

In most drainage basins between about 1 and 5 per cent of the storm water moving in the channel is derived from rainwater which falls directly into it; the greater part of the water is derived from the surrounding catchment. The routes and time taken by the water from the moment it hits the ground as rain until it reaches the channel depend upon the physical characteristics of the basin, particularly upon the slopes, soils, and vegetation. In a single drainage basin several routes may be recognised (Fig. 5.8).

The movement of water as a broad sheet, or miniature threads of water, over the surface of the ground is called overland flow or surface runoff. It occurs on slopes when rainfall intensity is so great that the infiltration capacity of the soil is exceeded (Horton, 1933). This type of flow is relatively common in environments in which there is a thin vegetation cover and soil mantle. It is important, therefore, in mountain country where steep slopes, bare rocks, and skeletal soils promote rapid runoff, and also in areas where frozen soil prevents infiltration. It is perhaps most common in semi-arid environments where a thin vegetation cover and soil mantle are characteristic, but is rare in most humid temperate environments except during melting of snow (Hewlett, 1961; Dunne, 1978b). Infiltration excess surface flow is now sometimes called 'Hortonian overland flow' to distinguish it from 'saturated overland flow' (see below).

Hortonian overland flow usually has a depth not exceeding about 10 mm but the thickness of water on a slope will vary depending on the shape of the slope, the variation in infiltration capacity of the soils, the position on the slopes, and the rainfall intensity. Once overland flow has commenced rain falls into the flowing water and is not available for infiltration, hence there is an increase in discharge downslope which results in an increase in both the depth and velocity of flow. Depth absorbs about two-thirds of the increase and velocity absorbs about one-third (Emmett, 1970). Typical velocities for Hortonian overland flow are about 200–300 m/h, so that in a typical rainfall of one hour water from all parts of a basin will reach a stream channel, where 200–300 m is a typical distance from the drainage divide to the stream.

In well-vegetated areas there is usually a humid climate without a severe drought season, and soil profiles are well developed. Except where prolonged intense rain occurs, most of the water reaching the soil surface will infiltrate into it because vegetation has the effect of promoting the formation of

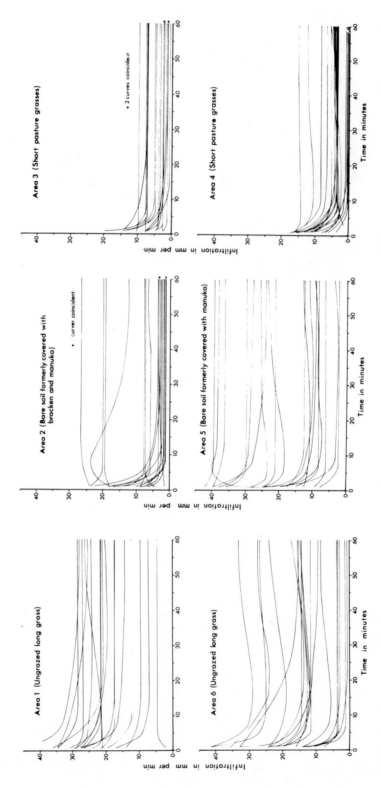

Fig. 5.6 Families of curves showing rates of infiltration into yellow-brown pumice soils beneath ungrazed long grass, short grazed grass, and a shrub vegetation of braken and manuka (from Selby, 1970).

Table 5.1 Coefficient of Permeability (k, m/s)

clay	$<1 \times 10^{-9}$	fine sands	1×10^{-7} to 1×10^{-5}
silts	1×10^{-9} to 1×10^{-7}	coarse sands	1×10^{-5} to 1×10^{-2}
		gravels	$>1 \times 10^{-2}$

granular soil structures which permit high infiltration rates and, in forests, of producing a porous and absorbent litter layer. Not all soil horizons are equally permeable and the movement of a wetting front down the profile may be impeded by a change of soil texture or structure. A barrier may occur, for example, beneath the litter layer, at a silica pan in the eluvial horizon, or at clay or iron-rich horizons (Fig. 5.9). Water reaching a barrier flows laterally downslope as interflow (also called throughflow) which moves at rates of a few cm/h. If rainfall is not intense nor prolonged all the water

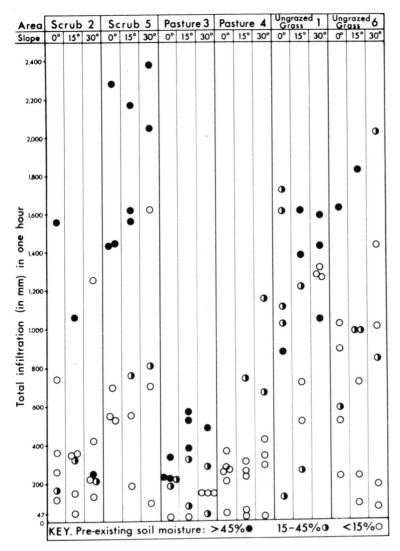

Fig. 5.7 Total infiltration in one hour into pumice soils with three classes of vegetation, three set angles of slope, and three classes of pre-existing soil moistures. The variability of infiltration is characteristic of soils with weakly developed profiles (from Selby, 1970).

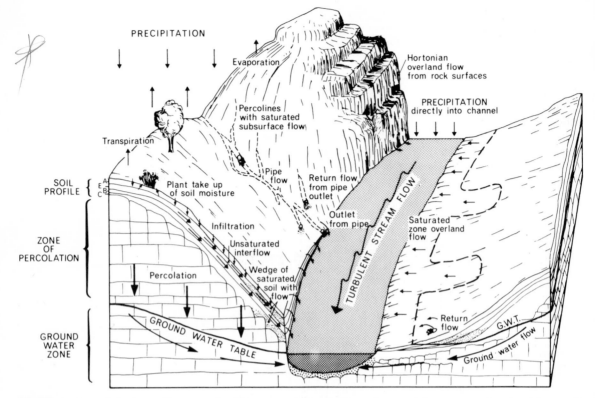

Fig. 5.8 A schematic landscape illustrating the different types of runoff from slopes and the sources and paths of the runoff water.

falling in a period may move as interflow so that the horizons in which it is moving will become more nearly saturated downslope. Such conditions have been described in temperate forests by Whipkey (1965).

Where the soil is porous and infiltrating water passes rapidly through the profile there is neither overland flow nor interflow. The newly percolating water is added to the ground water and the effect is to raise the water table, especially near the

Table 5.2 Flow Velocities of Selected Hydrologic Processes

Flow type	Velocity range (m/h)
Open channel	300–10 000
Overland flow	50–500
Pipe flow	50–500
Interflow	0.005–0.3
Ground water flow:	
jointed limestone	10–500
sandstone	0.001–10
shale	0.00 000 001–1

stream channel, and produce accelerated ground water flow. The addition of water to the stream may occur with little indication on the soil surface, as the increase occurs only by seepage from the channel walls. Where the water table is sufficiently high it may intersect the soil surface and a saturated zone forms which is fed by interflow and rising ground water as return flow. The saturated zone also receives precipitation falling into the surface water and transmits it directly and quickly into the stream. The area of a valley contributing saturated overland flow will gradually expand, with the area immediately alongside channels contributing first, then the bases of slopes and slope depressions. Such a situation is indicated in Fig. 5.10. In semi-arid zones, or areas with very permeable soils, perennial (i.e. permanent) streams carry water all the time, and after a period of rain the zone of intermittent flow becomes incorporated in the contributing network. Finally the finger-tip tributaries of the ephemeral channels contribute also (Fig. 5.11).

The nature of perennial, intermittent, and ephemeral flow is illustrated in Figs. 5.12 and 5.13.

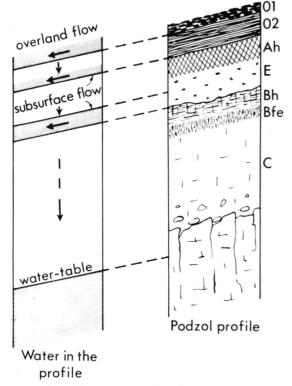

Fig. 5.9 Interflow in the profile of a podzol is controlled by the characteristics of the horizons.

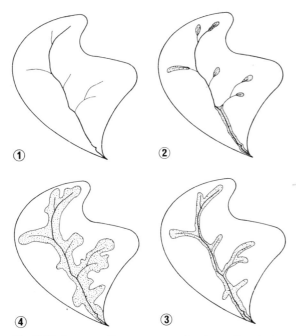

Fig. 5.10 An expanding source area for stream flow during a storm (dotted).

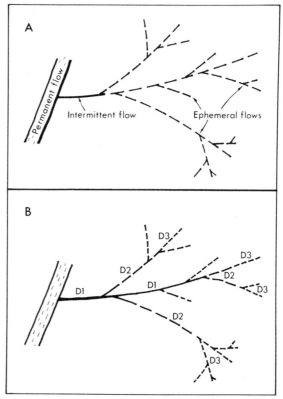

Fig. 5.11 Expanded channel network flow extends from the permanent channels progressively up the tributaries until in prolonged rain the entire channel network is contributing.

Where the water table always contributes water to the stream the flow is permanent or perennial. Where the water table falls below the stream during a dry season there will be a decline in flow along the channel, and so much loss may occur that flow ceases until the water table rises again to stream level — in such situations flow is intermittent. Ephemeral flow occurs where the water table is usually below the channel level, and it is therefore the result of ground water rise or additions from the slopes during and after storms.

Variable contributing area to stream flow

During a storm the area of a catchment which contributes water to a stream gradually extends and after the end of rainfall contracts. The concept of a variable contributing area was elaborated by Betson (1964) whose analysis of storm runoff from catchments in North Carolina suggested that only about 10 per cent of the catchment directly contributes to runoff; other studies suggest that the contributing area may range from 5 to 80 per cent

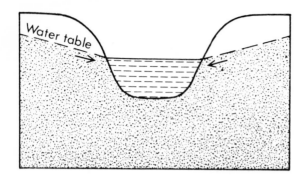

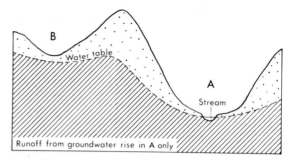

Fig. 5.13 The water table is below the level of channel B which will not contain a stream until the water table rises. A is a permanent stream.

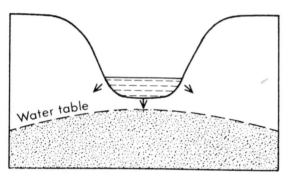

Fig. 5.12 Influent (top) and effluent (bottom) movement of water to and from a channel.

of the catchment. Hydrologists of the Tennessee Valley Authority (1964) developed the idea of a variable source area for stream flow under the title the 'Dynamic Watershed Concept' (Fig. 5.14); an alternative name is the 'partial area concept' (see Hewlett and Hibbert, 1967; Ragan, 1968; Dunne, 1978b). (Note: in American usage 'watershed' means the whole area of the drainage basin, or what

English usage calls the 'catchment'. English usage is to call the perimeter of the basin the watershed and Americans call this the water divide.)

Dynamic contributing area concepts are very important to an understanding of geomorphic processes in a valley. Between about 1940 and 1970 the idea that Hortonian overland flow, and resulting surface wash of soil particles, is a dominant process on hillslopes was paramount in much geomorphic thinking. Perpetuation of this idea was possible for four main reasons: it is supported (1) by research carried out in semi-arid areas; (2) by research on clay and shale slopes with thin soils and limited vegetation; (3) by studies of runoff on cultivated soils where intense rain causes splash, wash, and rilling; and (4) by the apparent conformity of conventional mathematical hydrograph analyses with the occurrence of overland flow. The last point has been the most persuasive for the rapid response of streams to rainfall seemed to be a direct confirmation of the existence of Hortonian overland flow. It was not until detailed measurements were made of discharge from various slope

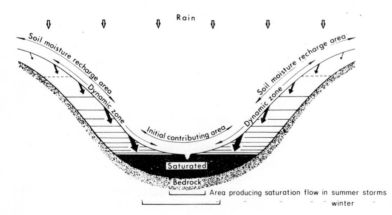

Fig. 5.14 The dynamic watershed runoff model with an indication of the area which may contribute runoff at different times of the year for a temperate climate with wet winters and drier summers with isolated storms.

segments that it was recognised that, under a full vegetation cover on thick moist soils, Hortonian overland flow is rare or non-existent and that which does occur reaches streams too late to make a significant contribution to the measured channel storm hydrograph (Dunne and Black, 1970a, b). This conclusion also implies that surface wash and soil transport on slopes are minimal except where soils are impermeable, thin, or unprotected by vegetation.

It is now recognised that initial hydrograph responses to storm rainfall are derived from rain falling directly into the channel and from the saturated contributing area immediately adjacent to the stream (Fig. 5.15). The importance of subsurface storm flow increases with the duration of the storm, contributing to delayed peaks in the hydrograph and to the recession. In deep, well-drained, soil on steep slopes subsurface flow may be the only source of storm flow as basal slopes drain rapidly into the streams and saturated zones cannot expand upslope. In catchments of low relief, and especially in those with concave basins and hollows, saturated overland flow may be very important with the contributing area steadily enlarging during the storm.

In spite of the generalisations offered above it has to be recognised that they are made from a limited number of research catchments in a few small areas of the humid temperate zone. The generalisations may not be as universally valid as is commonly thought. There is some evidence, for example, that although forested catchments in northern New Zealand respond to storms with partial area contributions from riparian zones, hillslopes with pasture grass vegetation respond differently in storms occurring in otherwise dry summer seasons. Summer storms may initially produce Hortonian overland flow from spurs because the dry hydraulic conductivity of the soils is low and infiltration is at first negligible; only after some hours will soils 'wet up' and infiltration occur at the rate permitted by soil saturated hydraulic conductivity. Concavities on the slopes may produce Hortonian overland flow if they also are initially dry, but by a combination of Hortonian overland flow, interflow, and return flow if parts of the depressions have moist soils. In dry summers hillslopes frequently produce greater initial storm flow responses than the riparian zones because the latter have low water tables, reduced hydraulic conductivities and have to become saturated before

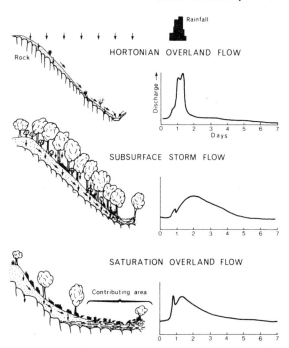

Fig. 5.15 Types of storm flow on hillslopes with stream hydrographs it generates. The initial rise in the hydrograph is produced by water falling directly into the channel. The shape of the recession curve indicates the storage and rate of drainage from the soil on the slopes.

contributing to stream flow with both subsurface storm flow, return flow, and direct precipitation into the saturated zone. Initial responses are from the spurs, followed by hillslope concavities and lastly from riparian areas if storms are of sufficient duration (R. A. Petch, personal communication). Wet season runoff from pasture-covered hillslopes follows the partial area responses as in Fig. 5.10. Only in rather uncommon circumstances can deep subterranean water contribute directly to storm flow. In well-jointed rocks, such as some limestones and basalts, open joints may reach the surface of the ground so that water can be transmitted very rapidly to the ground water store and cause a rise of the water table, or a flood in cave streams which connect with the surface drainage network.

The shape of a catchment has a major effect upon the hydrograph and also upon the geomorphic processes operating within it. Long straight slopes contribute water to channels mainly by subsurface flow and consequently they are primarily modified by solution and soil creep processes in the regolith: spurs by contrast, shed water rapidly (Fig. 5.16). Concavities produce convergence of subsurface and

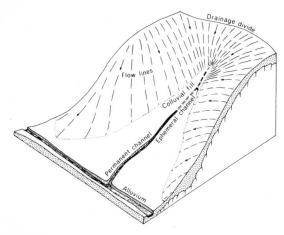

Fig. 5.16 Spacing of the flowlines on a hillslope indicates areas of convergence in valley heads, and divergence on spurs.

surface flow so promoting solution and mass movement processes, because of the high pore-water pressures and loss of soil strength as the soil gets closer to its plastic and then liquid limit. Entrainment of soil particles by overland flow is concentrated at the base of soil slopes adjacent to channels, in hollows and cavities, and in areas of thin or impermeable soils, especially where vegetation is disturbed. Because of the concentration of water movement in old landslide scars and other concavities, these often become zones of increased soil weathering and sites of further landsliding. There is thus a reinforcing effect, or feedback, in operation with spurs shedding water being least affected by erosion, and concavities collecting water and being most affected by erosion. Drainage network extension thus concentrates in hollows and valley heads.

Long side-slopes may be immune from landsliding until a critical shear stress is produced in a severe storm or earthquake (Kirkby and Chorley, 1967; Beven, 1978). Sediment supply to streams will thus be largely by bank collapse and by landsliding processes, and only in areas of limited vegetation cover will it be from surface wash.

Erosion by raindrops and running water on slopes

Erosion is an inclusive term for the detachment and removal of soil and rock by the action of running water, wind, waves, flowing ice, and mass movement. On hillslopes in most parts of the world the dominant processes are action by raindrops, running water, subsurface water, and mass wasting. The activity of waves, ice, or wind may be regarded as special cases restricted to particular environments.

Climate and geology are the most important influences on erosion with soil character and vegetation being dependent upon them and interrelated with each other (Fig. 5.17). The web of relationships between the factors which influence erosion is extremely complex. Vegetation, for example, is dependent upon climate, especially rainfall and temperature, and upon the soil which is derived from the weathered rock forming the topography. Vegetation in its turn influences the soil through the action of roots, take-up of nutrients, and provision of organic matter, and it protects the soil from erosion. The importance of this feedback is most obvious when the vegetation cover is inadequate to protect the soil, for eroded soil cannot support a close vegetation cover. The operation of the factors which influence erosion is most readily seen in their effect upon the disposition of storm

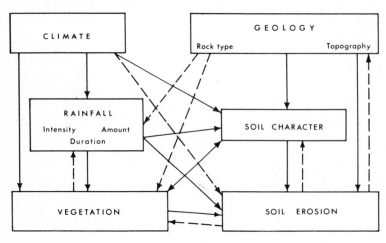

Fig. 5.17 The interrelationships between the main factors influencing soil erosion (after Morisawa, 1968).

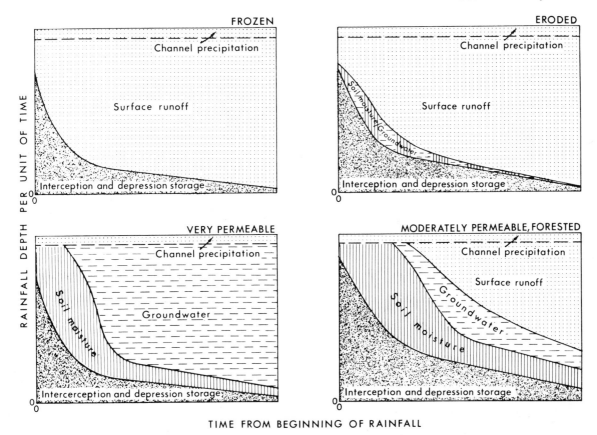

Fig. 5.18 Proportions of water in various parts of the drainage basin vary with time from the onset of rain and with the characteristics of the drainage basin. Interflow and saturated overland flow are included in soil moisture. The greater the Hortonian surface runoff the greater is the probability of soil erosion.

rainfall (Fig. 5.18). By comparison with the high runoff from an eroded catchment a well-vegetated catchment with a permeable soil will experience higher infiltration, lower surface runoff, and less surface erosion.

Erosion is a function of the eroding power (i.e. the erosivity) of raindrops, running water, and sliding or flowing earth masses, and the erodibility of the soil, or:

$$\text{Erosion} = f \text{ (Erosivity, Erodibility).}$$

Erosivity is the potential ability of a process to cause erosion, and for given soil and vegetation conditions one storm can be compared quantitatively with another and a numerical scale of values of erosivity can be created. Erodibility is the vulnerability of a soil to erosion and for given rainfall conditions one soil can be compared quantitatively with another, and a numerical scale of erodibility created. Erodibility of the soil can be divided into two parts –

first, the characteristics of the soil, such as its physical and chemical composition, and second, the manner of treatment of the soil beneath land use (i.e. cropping, forestry, or grazing, etc.) and management (i.e. fertiliser applications, cropping, harvesting, etc.). All these factors operate together and are expressed in the Universal Soil Loss Equation (Fig. 5.19). The equation is widely used in soil erosion studies in croplands with Hortonian overland flow, but it has not been generally applied to areas with a complete grass or tree cover and it does not apply to soils being eroded by mass wasting. It does, however, demonstrate the inter-relatedness of the various factors which influence the rate of soil erosion (Wischmeier and Smith, 1965; Foster *et al.*, 1977a, b).

Factors of erosion

The factors affecting soil erosion – climate, topography, rock type, vegetation, and soil character –

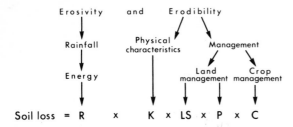

Fig. 5.19 Components of the Universal Soil Loss Equation. For details of its application see Hudson (1971) and Wischmeier (1976, 1977).

can be summarised into a descriptive equation:

$$E = f\,(C, T, R, V, S, \ldots [H], \ldots)$$

which has to be left open for additional factors, such as human interference [H], which may be significant at a particular site. This equation has seldom been expressed in quantified terms because of the extreme complexity of the variables, but a few attempts have been made to isolate some of the most significant factors and to express their relationship.

The human factor can exert its influence in a great variety of ways, especially by modifying the other factors, as when land is cleared of forest for pasture or cultivation; topography is modified by terraces or drainage ditches; and soils are changed by effects on vegetation, by cultivation, compaction by machinery, or application of fertilisers. Human disturbance has frequently disrupted the approximate balance between soil formation and soil erosion. It is, of course, possible for man to have the opposite influence and to protect soil from erosion by such activities as afforestation.

The major *climatic factors* which influence runoff and erosion are precipitation, temperature, and wind. Precipitation is by far the most important. Temperature affects runoff by contributing to changes in soil moisture between rains, it determines whether the precipitation will be in the form of rain or snow, and it changes the absorbtive properties of the soil for water by causing the soil to freeze. Ice in the soil, particularly needle ice, can be very effective in raising part of the surface of bare soil and thus making it more easily removed by runoff or wind. The wind effect includes the power to pick up and carry fine soil particles, the influence it exerts on the angle and impact of raindrops and, more rarely, its effect on vegetation, especially by windthrow of trees.

The erosion caused by rain is determined by the amount, intensity, and duration of the rainfall. A large total rainfall may not cause much erosion if the intensity is low, and likewise an intense rainfall of short duration may cause little erosion because the amount is small. Where both intensity and amount of rain are high, erosion is likely to be rapid. The erosive power of rain may, of course, be reduced to nothing if a complete vegetation cover prevents raindrops from hitting the soil.

In most temperate climates rainfall rate seldom exceeds 75 mm per hour, and then only in summer thunderstorms. In many tropical countries intensities of 150 mm per hour are experienced regularly. A maximum rate, sustained for only a few minutes, was recorded in Africa at 340 mm per hour (Hudson, 1971).

Raindrop sizes may be measured by photographing them or, more commonly, by using an absorbent paper with a dusting of very finely powdered water-soluble dye on its surface. When dry the dye is invisible, but on exposure to rain each raindrop makes a roughly circular stain which can be measured. Stain sizes can be calibrated with drops of known size produced in the laboratory.

The largest natural raindrops appear to be about 5 mm in diameter and drops bigger than about 4.6 mm diameter break up into smaller drops. Only rarely are very large drops (6 mm or more) produced, and then only by the collision of two smaller drops.

During a storm the rain is made up of drops of all sizes and the distribution of these sizes depends upon the type of rain. Low-intensity rain is made up of small drops, but high-intensity rainfalls have a greater proportion of medium and large drops (Fig. 5.20).

A body falling under the influence of gravity only will accelerate until the frictional resistance of the air is equal to the gravitational force, and it will then continue to fall at that speed. This is known as the terminal velocity and depends upon the size, density, and shape of the body. The terminal velocity of raindrops increases as the size increases; large drops with diameters of about 5 mm have a terminal velocity of about 9 m/s (Fig. 5.21).

In a series of experiments Laws (1941) found that the terminal velocity of raindrops in the open is affected by wind velocity and turbulence, but in spite of these effects most raindrops reach the ground at 95 per cent of their 'still air' terminal velocity. During rain wind may 'drive' raindrops, and the resulting vector may be greater than the 'still air' velocity.

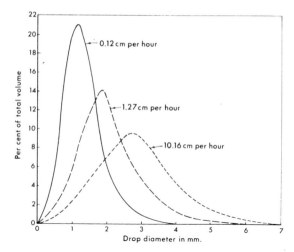

Fig. 5.20 A distribution of raindrop sizes in storms of given intensities (data from Laws and Parsons, 1943).

If the size of raindrops is known, and also their terminal velocity, it is possible to calculate the momentum of the falling rain, or its kinetic energy, by the summation of the values for individual raindrops. The kinetic energy of rain is closely related to the intensity. The results of a number of measurements of this relationship are shown in Fig. 5.22.

Raindrops have the effect of breaking down soil aggregates, splashing them into the air, causing turbulence in surface runoff, and carrying away soil particles. The kinetic energy of falling raindrops is far higher than that of the resulting runoff as is

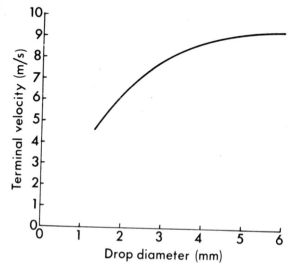

Fig. 5.21 The terminal velocities of raindrops of various sizes (data from Gunn and Kinzer, 1949).

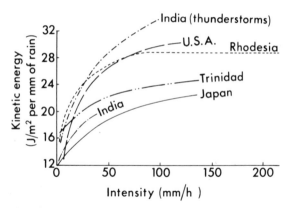

Fig. 5.22 Relationships, determined in a number of countries, between kinetic energy and rainfall intensity. Each curve extends to the highest intensity recorded.

shown by a simple example (Hudson, 1971): kinetic energy $= \frac{1}{2} \times$ mass $\times$ (velocity)^2; therefore, assuming that the falling rain has a mass R with a terminal velocity of 8 m/s its kinetic energy is $\frac{1}{2} \times R \times 8^2 = 32\,R$; assuming that 25 per cent of the rain becomes surface runoff the mass of runoff is $R/4$; assuming that the speed of surface runoff is 1 m/s then its kinetic energy is $\frac{1}{2} \times R/4 \times 1^2 = R/8$: therefore the rainfall has $32 \times 8 = 256$ times more kinetic energy than the surface runoff. Raindrop erosion is therefore a most potent process on exposed soil. The same conclusion may be reached from the observation that the main effect of raindrops is to detach soil particles from the aggregated mass, whereas the main effect of a thin film of surface runoff water is to transport the detached particles. Raindrop splash also has the effect of compacting the surface soil and sealing soil pores by depositing fine particles in them (Farres, 1978), thus reducing infiltration and increasing surface runoff.

The rate of detachment of soil particles depends closely on the energy of the rainfall and therefore upon its intensity. Experimental work by Wischmeier *et al.* (1958) showed that there is a strong statistical correlation between the soil eroded during a particular storm and the product of the kinetic energy of the storm and the 30-minute intensity. The 30-minute intensity is the greatest average intensity experienced in any 30-minute period during the storm. Wischmeier (1959) was thus able to propose that these factors could be used as an index of erosivity or an EI_{30} index. Development of this index has lead to the idea that there is a threshold value of intensity at which rain starts to become erosive, because low-intensity rainfall may be observed to cause little or no erosion. Studies in Africa indicate that rain falling with an intensity of less than 25 mm/h is not erosive. By ignoring the rainfall with less intensity than 25 mm/h it has been possible to produce a more reliable index of erosivity called the $KE{>}25$ index. A simplified calculation of this index is given below, using only rainfall intensities during a given storm which are derived from the chart of an automatic rain-gauge (after Hudson, 1971).

		0–25	25–50	50–75	>75	–
(1)	Rainfall intensity (mm/h)	0–25	25–50	50–75	>75	–
(2)	Rainfall amount (mm)	30	20	10	5	Total 65
(3)	Energy of rain (J/m² per mm of rain, derived from Fig. 5.22)	ignore	26	28	29	–
(4)	Total line 2 × line 3	ignore	520	280	145	Total 945

The erosive power of the storm is thus 945 J/m².

(Note: a joule (J) is a measure of the work done when the point of application of a force of one newton is displaced through a distance of one metre.)

Indices of erosivity are most useful where they can be used to predict the erosive effects of rainfall and so influence the design of land management and soil conservation plans. Rainfall erosivity can be mapped for any area for which there are sufficiently detailed rainfall records. Information can also be prepared on how erosivity varies during the year for any location. As an example (Hudson, 1971) two locations may be compared.

At location A 5 per cent of the rain is erosive, and the total annual rainfall is 750 mm, hence there is 37.5 mm of erosive rain.

At location B 40 per cent of the rain is erosive, and the total annual rainfall is 1500 mm, hence there is 600 mm of erosive rain.

Furthermore locality B has a higher average rainfall intensity — say 60 mm/h compared with 35 mm/h for locality A. It has been shown that the kinetic energy per mm of rain increases as the intensity increases so that for A the average kinetic energy may be 24 J/m² per mm of rain; and 28 J/m² per mm of rain at B.

The annual erosivity values are thus:

location A: 37.5 × 24 = 900 J/m²
location B: 600 × 28 = 14400 J/m².

The erosivity of rain falling on B is thus 16 times greater than that falling on A.

The value of the index is thus evident, but the actual erosion which occurs is also dependent upon

the characteristics of the soil, vegetation, and slope on which the raindrops fall.

The topographic factor is evident in the steepness and length of slopes. Nearly all of the experimental work on the slope effect has assumed that the slopes are under cultivation. In such conditions raindrop splash will move material farther down steep slopes than down gentle ones, there is likely to be more runoff, and runoff velocities will be faster. Because of this combination of factors the amount of erosion is not just proportional to the steepness of the slope, but rises rapidly with increasing angle (Fig. 5.23). Mathematically the relationship is:

$$E \propto S^a$$

where E is the erosion, S the slope in percent and a is an exponent. Values of a derived experimentally range from 1.35 (Musgrave, 1947) to 2 (Hudson and Jackson, 1959).

The length of slope has a similar effect upon soil

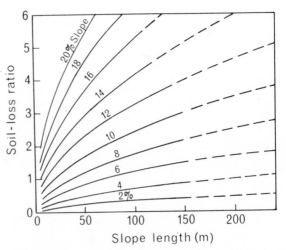

Fig. 5.23 The relationship between soil losses from bare soils and slope angle (in per cent) and slope length (after Wischmeier and Smith, 1965).

loss, because on a long slope there can be a greater depth and velocity of overland flow, and rills can develop more readily than on short slopes. Because there is a greater area of land on long than on short slope facets of the same width, it is necessary to distinguish between total soil loss and soil loss per unit area. The relationship between soil loss and slope length may be expressed as:

$$E \propto L^b$$

where E is the soil loss per unit area, L is the length of slope, and b is an exponent. In a series of experiments Zingg (1940) found that the values of b are around 0.6 but experiments elsewhere indicated that a rather higher value is more representative.

The *vegetation factor* largely offsets the effects on erosion of the other factors – climate, topography, and soil characteristics. The major effects of vegetation fall into at least seven main categories: they are (1) the interception of rainfall by the vegetation canopy; (2) the decreasing of velocity of runoff, and hence the cutting action of water and its capacity to entrain sediment; (3) root effects in increasing soil strength, granulation, and porosity;

(4) biological activities associated with vegetative growth and their influence on soil porosity; (5) the transpiration of water, leading to the subsequent drying out of the soil; (6) insulation of the soil against high and low temperatures which cause cracking or frost heaving and needle ice formation; and (7) compaction of underlying soil.

The interception of raindrops by the vegetation canopy affects soil erosion in two ways: (1) by preventing the drops from reaching the soil and allowing water to be evaporated directly from leaves and stems; and (2) by absorbing the impact of the raindrops and minimising the harmful effects on soil structure.

A close vegetation cover not only reduces runoff velocities by friction with plant stems, but it also prevents the runoff becoming channelled and so made able to cut into the soil. The slowing down of the runoff increases the time for infiltration.

A close forest cover often gives a virtually complete protection to soil. Not only do the canopy and understory prevent raindrops hitting the ground, but the accumulation of litter forms a complete protective blanket (Plates 5.3-5.7). Experimental

5.3 The complete sward of tussock grasses prevents splash or wash erosion.

5.4 The broken tussock grass cover permits frost action, splash, and wash erosion to occur between the tussocks.

5.6 Removal of cedar forest has permitted severe sheet wash erosion on these slopes, Atlas Mountains, Morocco.

work has shown, however, that if the understory and ground vegetation is destroyed by grazing animals then the energy of falling drops may actually be greater under forest than in the open. Rain falling with a low intensity may collect on the leaves of the canopy and then drip from the leaves in large drops. From a height of 10 m or so these drops will reach their terminal velocity so that even though some water is lost by evaporation and stemflow, the effect upon the soil may be considerable. This appears to be the situation in some tropical forests where leaf litter is rapidly decomposed and the soil left bare.

Within managed forests careful siting of roads and logging tracks to avoid channelling of runoff water, the preservation of a ground layer of plants,

the prevention of fire, and the avoidance of clear felling on steep slopes can provide for soil conservation and wood production. Both conservators and foresters would probably agree that the ideal protective combination is a regular stand of trees with a close ground cover of grass or herbs.

A well-managed high-yielding grassland is unlikely to be greatly affected by rain splash or sheet erosion, and poor grazing land is unlikely to be sufficiently productive to warrant expensive mechanical methods of erosion control. The methods which are used to decrease erosion on grasslands do not attempt to control soil movement directly, but

5.5 A cedar forest in the Atlas Mountains of Morocco. Much of the forest has been destroyed and the soil in the foreground is severely eroded. Grazing by goats prevents regeneration of the forest.

5.7 Leaf litter on the floor of a deciduous forest protects the soil from splash and wash erosion. The amount of protection varies with the seasons.

aim at improving the vegetation so reducing runoff and increasing infiltration.

The prevention of 'pugging' of the soil by animal trampling during wet periods, especially in gateways and other well-used places, and other farming practices which reduce soil compaction and exposure are obviously necessary on all soils. In some areas mechanical methods, such as pasture furrows and other soil depressions, are used to reduce runoff. Pasture furrows are shallow, follow the contours exactly, and are spaced only a few metres apart. The best methods, however, involve the maintenance of a dense grass sward.

General relationships are indicated in Table 5.3 which shows the effects of vegetation in inhibiting erosion in the northwest of USA. Ursic and Dendy (1965) found similar relationships in the midwest of USA. The erosion rate can clearly vary by factors

Table 5.3 The Relative Relationships between Erosion and Vegetation

Crop or Practice	Relative Erosion
Forest ground layer and litter	0.001–1
Pastures, humid region, and irrigated	0.001–1
Range or poor pasture	5–10
Grass/legume, hay	5
Lucerne	10
Orchards, vineyards with cover crops	20
Wheat, fallow, stubble not burned	60
Wheat, fallow stubble burned	75
Orchards, vineyards clean tilled	90
Row crops and fallow	100

of 5 to 100 000 depending upon the vegetation (Fig. 5.24).

The soil factor is expressed in the erodibility of the soil. Erodibility, unlike the determination of

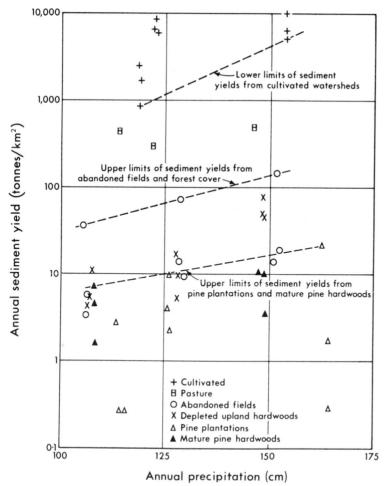

+ Cultivated
ᗺ Pasture
O Abandoned fields
X Depleted upland hardwoods
Δ Pine plantations
▲ Mature pine hardwoods

Fig. 5.24 Variation in sediment yields from individual watersheds in northern Mississippi under different types of land use and various amounts of precipitation (after Ursic and Dendy, 1965).

5.8 A laboratory rainfall simulator, and the result of a raindrop impact upon bare soil (inset).

erosivity of rainfall, is difficult to measure and no universal method of measurement has been developed. The main reason for this deficiency is that erodibility depends upon many factors, which fall into two groups: those which are the actual physical features of the soil; and those which are the result of human use of the soil (i.e. the management of the soil).

Many attempts have been made to relate the amount of erosion from a soil to its physical characteristics. Pioneer work in this field was done in North America in the 1930s. Bouyoucos (1935) suggested that erodibility is related to the sizes of the particles of the soil in the ratio

$$\frac{\text{per cent sand} + \text{per cent silt}}{\text{per cent clay}}$$

and Middleton (1930) used a 'dispersion ratio' based upon the changes in silt and clay content of the soil after dispersion in water. Other methods involve measures of the stability of soil aggregates and the rates of water transmission through the soil. Stability of soil aggregates, resulting from the cohesive properties of clay or organic matter, is the main inhibitor of erosion by raindrops.

The failure to find indices of erodibility has encouraged many research workers to subject soil samples to splash erosion in laboratory tests on trays of soil beneath simulated rainfall and to rank soil types in order of erodibility (Plate 5.8).

A second method is to lay out a series of plots in the field where natural rainfall can be measured, and the soil washed from the plots can be collected in troughs at the downslope end of each plot. The sediment yield can then be related to the rainfall and, by using a number of plots with various forms of land management and vegetation cover, it is possible to compare the efficacy of various forms of land use for inhibiting erosion. A variation of this method is to use plots in the field and to measure the natural features (i.e. properties of the soil, climate, vegetation etc.) and then to relate these properties statistically to the sediment yield in an attempt to determine which of the natural features influences the erosion (Plate 5.9).

A method recently suggested (Wischmeier and Mannering, 1969) is to establish a correlation between erodibility and a large number of variables which are soil physical properties and soil profile characteristics. From a statistical study of 55 US Corn Belt soils Wischmeier and Mannering derived an empirical equation to calculate the erodibility factor in the Universal Soil Loss Equation. It contains 24 variables representing soil properties which analysis has shown to be important. The equation predicted the erodibility of 11 bench-mark soils, whose erodibility had been previously established, by other means, very closely. It was found, however, that this equation was too complex for practical use, although valid for a broad range of medium-textured soils. A method using only five variables has been devised subsequently, and by use of a nomograph (Fig. 5.25) simply measured soil characteristics can be related to soil erodibility under rainfall impact and surface runoff.

In spite of the encouraging success of the Wischmeier and Mannering equation it is still true to say that none of the proposed measures conforms completely to the requirements of an erodibility index: that it be simple to measure, reliable in operation, capable of universal application, and capable of providing a quantitative measure of the erosion which will take place when the soil is subjected to rain of known erosive power.

5.9 An experimental plot with a trough to collect runoff and sediment. A rain-gauge stands beside the trough.

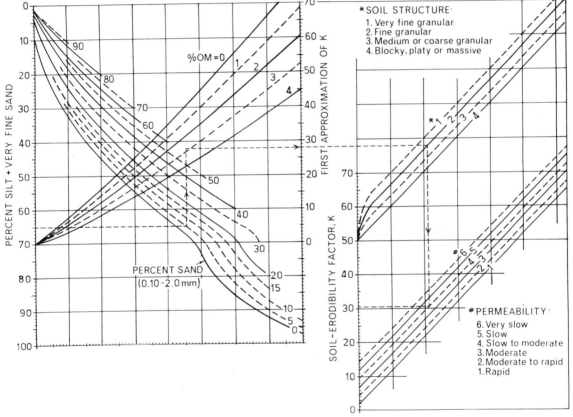

Fig. 5.25 A soil erodibility nomograph for use with cultivated soils. The procedure is: enter the nomograph on the vertical scale at the left with the appropriate percentage of silt plus very fine sand (0.1 to 2.0 mm); proceed horizontally to intersect the correct percentage sand curve, then vertically to the correct organic matter curve, and horizontally towards the right; (for many agricultural soils with a fine granular structure and moderate permeability the value of the erodibility factor *K* can be read directly from the first approximation-of-*K* scale on the right-hand edge of the first section of the nomograph and the procedure can terminate there); for soils with other than fine granular structure and moderate permeability continue the horizontal path to intersect the correct structure curve; proceed vertically downward to the correct permeability curve and move left to read the value of *K*. The dotted line illustrates the procedure for a soil with 65 per cent silt and very fine sand, 5 per cent sand, 2.8 per cent organic matter, 2 structure 4 permeability and hence erodibility of 0.31 (after Wischmeier *et al.*, 1971).

One conclusion of overriding importance, which has been reached as a result of field and laboratory studies in many countries, is that soil erosion is largely controlled by the type of management soil receives, and this conclusion is the basis of nearly all soil conservation practice. This point may be illustrated by a hypothetical example in which one soil might lose say 300 tonnes/ha/year when it is used for the cultivation of row crops which run up and down the slope, whereas an identical soil under the same climatic regime, under well-managed pasture, would lose only a few kilogrammes of soil per hectare per year. The erosion from two areas with the same soil but different management is commonly greater than the differences between two soils with the same management.

Management includes both the broad issues of land use and the details of crop management. The best management might be defined as the most intensive and productive use of which land is capable without causing any degradation.

Wash, rill, gully, and piping processes

Running water removes soil from slopes by a variety of processes — sheet wash, rilling, gullying, and piping.

Sheet wash

Overland flow or sheet flood give rise to the process of sheet wash. It occurs, where the rock or soil surface is smooth, as a continuous film of water, or on slightly rougher terrain as a series of tiny rivulets connecting one water-filled hollow with another. On a grassed slope wash is transformed into numerous threads of water passing around the stems. In forests where there is a thick litter surface flow may be virtually concealed among the decaying leaves and twigs.

Sheet flow is usually very shallow towards the crest of a hill but the flow depth increases downslope. Experimental work shows that falling raindrops retard the flow and disturb it so that while parts of a sheet flow may be laminar and have little erosive power, other parts are turbulent and more erosive (Palmer, 1965; Walker *et al.*, 1977). Maximum soil erosion in overland flows occurs when flow depth equals raindrop diameter, consequently it is greatest when flow depth is 3–6 mm. Soil particles are disturbed and transported by rainsplash even in the very shallow flows which occur at very low slope angles (Moss *et al.*, 1979).

The effectiveness of surface wash in transporting particles and soil aggregates is demonstrated by the accumulation of soil upslope of hedges at the bottom of cultivated fields. Wash is also very effective in removing soil loosened by needle ice or disturbed by animals (Plate 5.10). In cultivated soils sheet wash results in a thickening of the soil downslope and an increase of fertility at the foot, but a decrease at the head of the slope (Plate 5.11, Fig. 5.26).

5.10 The interior of a forest showing damage to the floor by browsing animals. All tree and shrub regeneration has been prevented and wash erosion is now active. (J. H. Johns: Crown copyright, by permission of the New Zealand Forest Service.)

5.11 Sheet wash erosion.

correlation between & rills?

It has been shown by Schumm (1962) that in semi-arid areas the velocity of overland flow is insensitive to slope angle. It appears that on steep hillslopes surface roughness reduces flow velocities and that on the footslopes, or pediments, velocities are just maintained by the lower slope angles. The upper and lower slope units are thus in equilibrium with the flows crossing them.

In spite of the relatively low energy of surface wash it can be very effective in transporting soil. In semi-arid areas of New Mexico, for example, surface erosion on unrilled slopes accounted for 98 per cent of sediment production from all sources (Emmett, 1978).

Rilling

Sheet wash rapidly becomes concentrated, as the water is diverted round objects, into very small channels or rills. The water in a rill has sufficient depth for considerable turbulence to develop in it

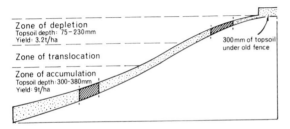

Fig. 5.26 Topsoil depth and yields of crops down a slope in Otago, New Zealand (data from New Zealand Soil Bureau).

and rill flows can therefore entrain larger particles than sheet wash. The head of a rill system may not extend all the way to a watershed divide, which remains unrilled because the depth of overland flow is not great enough to develop an erosive force equal to the cohesive strength holding the soil in place.

Parallel rills on a fresh surface become integrated into a drainage net by the breaking down of divides between rills with diversion of the water into the deeper rill, and the overtopping of rills and diversion of the water towards the lowest elevation (Plates 5.12 and 5.13). These two processes Horton (1945) called micropiracy and cross-grading. Their effect is to cause wider spacing of rills downslope. The greater runoff passing over lower slopes causes the channels to have more water and eventually a master rill becomes the head of an ephemeral stream, which may so cut down its bed that it captures the drainage of the area immediately around it and eliminates the signs of the original rills. This whole process has been known to occur on a sloping cultivated field during a single thunderstorm. It can frequently be seen in a sand-pit.

Rills can also form when subsurface water becomes concentrated into areas of deeper soil, called percolines, above the zone of normal channel flow. If the rainfall intensity is great enough to produce shear stresses that exceed the threshold of erosion of the soil and its plant cover, the cover will be ruptured and a channel formed.

Remedial measures are based on the knowledge

5.12 Rill erosion showing the development of a network.

5.14 Terraces for conservation purposes at an altitude of 3000 m on the equator in Rwanda, central Africa.

that runoff increases as slope length and slope steepness increase. The aim of good soil conservation measures is then to reduce the effective length of slopes by dividing them into sections using grass strips, hedges, walls, shallow drains, terraces, or furrows along the line of the contour. This inhibits both sheet wash and rilling as long as the barrier is wide enough and designed so that runoff cannot become channelled (Plates 5.14 and 5.15).

Slope steepness is reduced also by terracing which divides the slope into short, gently sloping sections separated by a terrace wall. As with all soil conservation measures the breakdown of a terrace can cause channelling of runoff and be the cause of very serious gully erosion.

5.13 Rill erosion expanding downslope into gully erosion. The man in the gully gives scale.

5.15 Conservation strips on the contour. The strips are well-grassed furrows below short-graded slopes in pasture, Transkei, southern Africa.

Rainsplash erosion is best inhibited by the preservation of a complete vegetation cover and where this is not possible by ensuring that soil is not bare in seasons when rainfall intensities are greatest. Mulches of straw, cut weeds, or plastic sheeting may be used when intensive cultivation can justify the use of such methods.

Gullying

A master rill may so deepen and widen its channel that it is classed as a gully — arbitrarily defined as a recently extended drainage channel that transmits ephemeral flow, has steep sides, a steeply sloping or vertical head scarp, a width greater than 0.3 m, and a depth greater than about 0.6 m (Brice, 1966, p. 290). Gullies may also form at any break of slope or break in the vegetation cover when the underlying material is mechanically weak or unconsolidated. Gullies are therefore most common in such materials as deep loess, volcanic ejecta, alluvium, colluvium, gravels, partly consolidated sands, and debris from mass movements.

Because they are very rapidly developed erosional forms they are usually not regarded as features of normal erosion, but the result of changes in the environment, such as faulting, burning of vegetation, overgrazing, climatic change affecting vegetation, extreme storms, or any other cause of a break in vegetation which will bare the soil.

Gully erosion nearly always starts for one of two reasons: either there is an increase in the amount of flood runoff, or the flood runoff remains the same but the capacity of water courses to carry the floodwaters is reduced. The most common causes of increases in runoff or deterioration in channel stability are changes in vegetation cover — especially removal of trees, increases in the proportion of arable land in catchments, excessive burning of vegetation, or overgrazing — or a climatic change with accompanying variations in rainfall periodicity and intensity.

The capacity of a stream channel depends on the cross-sectional area, the slope, gradient, and roughness. Changes in these factors can easily disturb an equilibrium between the channel geometry and the processes of erosion and deposition which have moulded it.

The relationship between the velocity of water in a channel and the geometry of that channel are expressed in Manning's equation:

$$V = \frac{1}{n} R^{\frac{2}{3}} S^{\frac{1}{2}}$$

in which:

V is the average velocity of flow in metres per second;

R is the hydraulic radius in metres;

S is the average gradient of the channel in metres per metre;

n is a coefficient, known as Manning's n or Manning's roughness coefficient. (Typical values of n are given in Table 5.4).

(Note: the hydraulic radius is given by $R = A/P$ in which A is the area of a transverse section of a stream, and P is the wetted perimeter or length of the boundary along which the water and channel bed are in contact in that transverse section.)

Table 5.4 Values of Roughness Coefficient n for Different Channel Conditions

Description of channel	Range of values		
	Minimum	Normal	Maximum
Concrete, trowel finished	0.011	0.013	0.015
Concrete, shuttering	0.012	0.014	0.017
Brickwork	0.012	0.015	0.018
Excavated channels:			
earth, clean	0.016	0.022	0.030
gravel	0.022	0.025	0.030
rock cut, smooth	0.025	0.035	0.040
rock cut, jagged	0.035	0.040	0.060
Natural channels:			
clean, regular section	0.025	0.030	0.040
some stones and weeds	0.030	0.035	0.045
some rocks and/or brushwood	0.050	0.070	0.080
very rocky or with standing timber	0.075	0.100	0.150

An increase in the resistance to flow by the growth of vegetation in or at the edge of a channel will cause an increase in the value of n and hence be accompanied by a decrease in the velocity of flow. This reduces the capacity of the channel to accommodate flood flows and increases the chances of flood waters spilling over channel banks and starting new erosion patterns. By contrast a reduction in vegetation along a waterway may cause a decrease in resistance to flow, and therefore a decrease in the value of n, and increased velocity with the possibility of the development of channel scouring.

Any local effect can upset the equilibrium which already exists. Such minor starting points are cattle tracks, pot-holes, or the diversion of drainage by a new road so that instead of spreading over a whole valley floor, flood waters are routed into confined areas with consequent smaller wetted perimeters and less resistance to flow.

Once gullying starts the gullied channel has a more angular and deep shape than the original bed (i.e. R increases). The gullied channel is rough and irregular so the value of n increases. For the velocity to remain constant the gradient must, therefore, decrease and this is what usually happens. The gradient of the floor of the gully is flatter than that of the original stream bed $- S$ decreases $-$ (see Fig. 5.27). The head of the gully then works back upstream and the height of the gully progressively increases. As the head gets higher, or the velocity increases, gullying is likely to become more rapid; once it has started gully erosion thus becomes increasingly difficult to control. It is always easier and cheaper to prevent than to stop once it has started.

There are two main types of gully, although compound types are also common: (1) continuous gullies and (2) discontinuous gullies. There are also

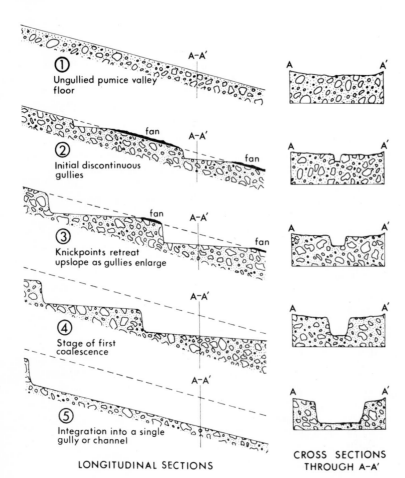

① Ungullied pumice valley floor

② Initial discontinuous gullies

③ Knickpoints retreat upslope as gullies enlarge

④ Stage of first coalescence

⑤ Integration into a single gully or channel

LONGITUDINAL SECTIONS

CROSS SECTIONS THROUGH A-A'

Fig. 5.27 Stages of development of discontinuous gullies (modified from Leopold *et al.*, 1964).

three main processes operating to form gullies, although these usually occur in combination: (a) surface flow, (b) mass movement, and (c) piping.

A continuous gully which forms by enlargement of a rill may have no headscarp because it forms in non-cohesive materials without a resistant capping. This type of gully is particularly common where erosion of shattered rock and talus gravels has resulted from a deterioration in the vegetation cover. Such gullies usually increase in width and depth downslope as the runoff increases and tributary gullies contribute their flow to the master gully, but in their lowest reaches where the mountain slope decreases, the gullies have zones of deposition which extend, as coalescing fans, beyond the gully mouth. The long profile of a typical gully therefore shows a slope near its head steeper than the mountain slope, but less steep near the base. Gullies may be cut by storm runoff and snow meltwater and therefore have ephemeral streams. Wind and frost action cause crumbling of the valley walls during dry periods.

Gullies without headscarps also develop within the scars and deposits of large mass movement features. Such features have a disrupted vegetation cover, and disturbed materials are then readily moved by surface flow or by minor mass movements

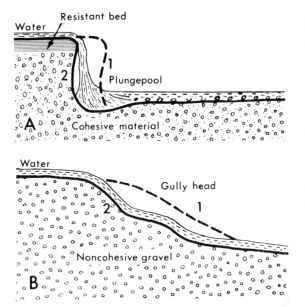

Fig. 5.28 Types of gully head. In A the materials are cohesive and have a resistant cap of root-bound soil supporting a head scarp. In B the non-cohesive material does not support a head scarp.

occurring at the head and sides of the gully (Fig. 5.28).

Gullies with headscarps will maintain that headscarp if (a) the gully originates at a scarp or steep face and the gully is cutting into cohesive materials and (b) if non-cohesive materials are capped by resistant materials such as root-bound soil. This is because the material making up the bed at the knickpoint has a resistance to shear stress greater than the stress provided by the flow. It is also essential for the flow to transport the eroded material away from the base of the headscarp so that undercutting processes can continue. Retreat results mainly from sapping as seepage water loosens material on the face of the headcut, or water under pressure flows out of the face after rain which has raised the water table above the foot of the face. A plunge pool resulting from surface flow may contribute both to erosion at the base of the headscarp and to downstream removal of the sediments.

Where the slope of the gully floor does not exceed more than a few degrees the ephemeral stream usually has a wide channel because it can easily remove sediments from the base of the gully sides. Such gullies therefore have a rectangular cross profile, with the side walls being kept steep by undercutting.

In many dry valleys discontinuous gully erosion may begin at several points where vegetation cover is broken. At each point a small headscarp forms and as this retreats headwards a trench is left downvalley with a debris fan spreading out from its toe to form an alluvial fan.

It can be seen from Fig. 5.27 that the newly developed gully floor has a lower angle of slope than the original valley floor, but as the gully deepens and retreats headwards the slope increases again so that when two gullies coalesce their floor is usually of nearly the same slope as the original valley floor (Plates 5.16 and 5.17). In the original valley rainfall is distributed over the whole valley floor where it infiltrates into the soil, but as the gully develops, the storage capacity of the valley floor sediments is reduced and the subsurface water moves towards the gully. Peak discharges in the gully therefore far exceed the previous discharges over the unchannelled valley floor. The floor of a discontinuous gully usually has a plunge pool at the foot of the headscarp, and downvalley from that a veneer of sediment, left by the falling flood of the last storm, over the original valley floor materials. The depth of the gully trenching is usually limited by the existence

5.16 Discontinuous gullies in valley floor infills, central North Island, New Zealand.

of a resistant bed in the valley materials.

Attempts have frequently been made to control gully erosion by building dams of concrete, stone, and wood (Heede, 1977) (Plates 5.18 and 5.19). But most dams are liable to decay, to be by-passed or undermined, and in many cases they do not modify the basic cause of gully erosion. By far the most effective control of gullies is vegetation. Vegetation protects the soil against further scour and it reduces the velocity of flow. It may be necessary for mechanical methods such are wire-netting, permeable rock, brushwood, or timber dams to be used to hold back silt and control drainage before vegetation can be established. Planting a catchment with densely rooting trees and developing a close ground cover of grasses, herbs, and shrubs are the usual conservation methods used (Heede, 1976). Such methods are only effective if the climate permits a close ground cover, and if grazing animals can be controlled. Because the discharge of a gully is controlled by catchment area, large gullies can sometimes be controlled by diverting water out of the gully along a controlled channel.

Gully erosion is particularly severe in semi-arid countries where there is pressure on soil and vegetation resources. Where wood is the only available fuel for heating and cooking, and where people cannot afford adequate fences for animal control it is very difficult to maintain a strong vegetation cover and to apply conservation measures, hence soil erosion is often found around settlements.

Piping

Subsurface tunnels in soil have been variously described by such terms as pot-hole erosion, suffosion, subcutaneous erosion, tunnel-gullying, or piping (Parker and Jenne, 1967; Crouch, 1976).

Natural pipes vary greatly in length and diameter from a few centimetres to lengths of hundreds of metres and diametres of up to two metres. They may be of little significance on some slopes but also they may be important routes for subsurface water movement and they may collapse to form gullies.

Among the factors which dispose a soil to piping are: a seasonal or highly variable rainfall; a soil subject to cracking in dry periods; a reduction in vegetation cover; a relatively impermeable layer in the soil profile; the existence of a hydraulic gradient

5.17 A fan formed at the lower end of a discontinuous gully.

5.18 Check dams of netting built to hold debris in a gully in the hope of reducing erosion rates.

5.19 This gully has cut down to bedrock, Transkei, southern Africa. Attempts are being made to reduce sediment transport by the erection of a porous dam of rock-filled wire baskets.

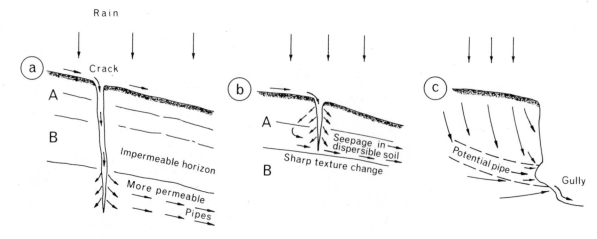

Fig. 5.29 Conditions favouring the formation of pipes: (a) cracking and a permeable horizon below an impermeable horizon; (b) an horizon of dispersible clay; (c) a gully head.

in the soil; and a dispersible soil layer.

The most commonly reported situation in which pipes develop is one in which a surface soil cracks as a result of desiccation. In a rainstorm water then infiltrates rapidly down the cracks and supersaturates a relatively permeable horizon in the subsoil. Lateral seepage may be fast enough to move soil particles and develop a channel, or, if the soil has dispersible clays, these may lose aggregation. Movement of water through subsurface cracks and voids is slow until water breaks through the soil surface farther down the slope, and rapid flow can then work headwards within the soil and form a gully (Figs. 5.29 and 5.30).

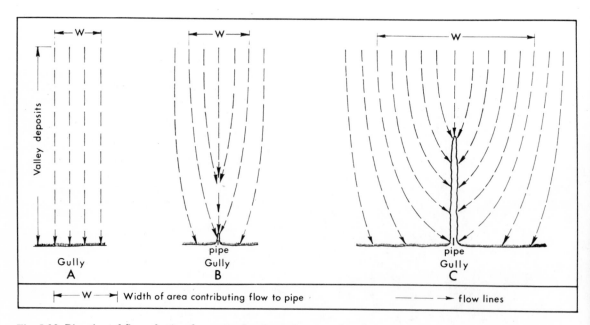

Fig. 5.30 Direction of flow of subsurface water showing an increase of catchment area (W) as a pipe erodes headwards. A, gully wall with no pipe; B, incipient; C, after a pipe has extended. These are plan views (modified from Terzaghi and Peck, 1948).

Dispersion in soils is the deflocculation of the clay fraction when water is added. The dispersion may be a result of weakening of chemical bonds between particles by ion exchange or by leaching.

Piping is particularly common in the walls of gullies and in the heads of landslides where the confined flow in soils is suddenly accelerated and seepage pressures permit particles to be washed out of the soil. Pipes can extend headwards very rapidly once they start to grow. Rates of 45 m/hour have been reported in pumice soils in New Zealand.

Reclamation procedures include the destruction of existing pipes, the establishment of vegetation, infilling of cracks, and may include chemical treatment, such as lime applications on soils with a high sodium content, to reduce soil dispersion.

Solution

Solution is a much more important process on slopes than is commonly recognised. In many catchments more than half of the material removed in erosional processes is carried in solution, and it is the transport of solutes from one part of the soil profile to another which permits continuation of weathering processes, the synthesis of secondary minerals, and the formation of pans of iron oxides or carbonates within the profile. With the exception of measurements of chemical erosion of limestones few studies have been made of the effect of solution

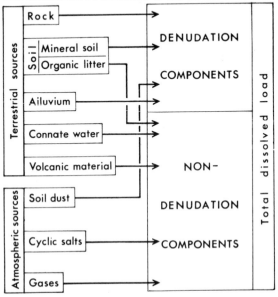

Fig. 5.31 Natural sources of material in the total dissolved load of rivers (after Janda, 1971).

on slope forms: most studies have been concerned with the contribution of solution to total denudation and with the composition of drainage waters.

The geochemical budget

Solutes derived from a slope or drainage basin may come from a number of sources (Fig. 5.31): only solutes derived from rock and soil weathering are truly part of the denudational component, but they may be exceeded in quantity by the non-denudational component from rainfall, windblown dust and salt, or deep-seated migration of ground water. If a catchment is virtually watertight so that there is no input or output in ground water, no inputs from fertiliser or sewage, and no outputs from cropping or water abstraction, then additions which are not denudational can usually be accounted for as inputs of dust or in precipitation and measured in collecting devices. Where vegetation is stable so that the incorporation of nutrients into plant tissue should, over a year, be equalled by the return from biological decay, then:

$$\text{solutes derived from weathering} = \text{output as stream solutes} - \text{input of solutes in precipitation}$$

The precipitation component can be very important, for in areas close to an ocean, or desert salt flats, catchments may obtain most of their sodium and chloride from the atmosphere. Net gains of potassium are also common, and nitrogen may be greatly increased in abundance by disturbance of vegetation, such as by the ploughing up of pasture or burning of forest; some nitrogen may also be added from the ammonium (NH_4^+) generated in the electrical activity of thunderstorms (Douglas, 1968).

In studies of soil water or river loads the total solutes are measured by evaporation of the water, by the electrical conductivity (which is a measure of the ionic activity), or by chemical analyses made of individual constituents. The results are usually expressed as concentrations in parts per million or milligrammes per litre (1 ppm = 1 mg/l). The distinction between solutes and suspended loads is, however, made on the basis of filtration, with all material passing a 2 μm filter being classed as solute. A proportion of solutes is adsorbed onto colloids and may be included in the suspended load. Consequently, and particularly in areas with a high output of colloids with high base exchange capacities, such as montmorillonite or humus, a considerable proportion of the cations may be classed with solid

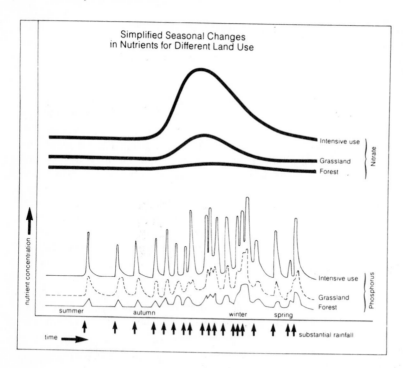

Fig. 5.32 Discharges of nitrate and phosphorus, during a year, in relation to rainfall. The curves are simplified (after Haughey, 1979).

loads and the significance of solutes thus under-estimated.

Variations in solute discharge

Concentrations and total outputs of solutes vary through the year, especially in areas with seasonal climates. During the growing season decay of soil organic matter and the production of CO_2 by plant roots increases the bicarbonate content of soil water in contact with mineral matter. Chemical weathering is thus more vigorous, and silica and bicarbonate ions show a strong peak in discharge, in the productive season. Plant nutrients such as phosphorus, nitrate, calcium, and potassium show a marked peak in the period of leaf-fall. The most soluble compounds are readily transported and show least fluctuation in output with runoff, but the more insoluble compounds are usually associated with solids in suspension and show marked association of concentrations with water discharge: Fig. 5.32 illustrates this point although both nitrate and phosphorus used in the example are probably not produced by rock weathering.

Discharges of the soil solutes may be very closely related, in time, to discharges of water because each storm may produce water which expels older soil water in the profile (where it has gained ions) and replaces it by freshly infiltrating water. Each storm is thus accompanied by the discharge of a mass of enriched water.

Solution and rocks

The ions, occurring in drainage waters, which are the result of chemical weathering of primary or secondary minerals may be only part of the product of weathering. The less soluble products of weathering remain within the regolith, and in arid climates drainage or soil waters may evaporate and leave pans of iron oxides, calcium carbonate, or other products. Interpretation of solute production from different rock types or the same rock type in different climates has, therefore, to be approached with care and variations in inputs in precipitation recognised. Some general trends do, however, appear to be detectable.

In an important study of sandstones (all consisting of about 92 per cent silica) Mainguet (1972) compared solutes from those rocks in four contrasting environments: the temperate forested area of the Vosges, eastern France; the tropical desert and semi-desert of Chad; the savanna woodlands of northern Central African Republic; and the tropical forests farther south. The mobility of the ions has strong seasonal variations, but silica is generally more soluble in the humid tropics and less soluble in arid and temperate areas. Concentrations of

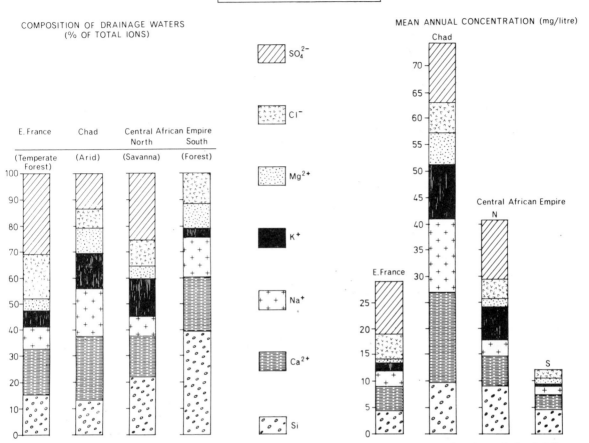

Fig. 5.33 Composition of drainage waters and concentration of ions in them from four areas of sandstone bedrock (after Mainguet, 1972).

solutes are greatest in semi-arid areas where water has prolonged contact with rock minerals between rainfall events (Fig. 5.33). Total solute discharges are, however, normally greatest in humid climates where solutions are not saturated, and the flow of water through the soil is high. Total solutional denudation, therefore, tends to increase primarily with rainfall and discharge (Dunne, 1978a), and to a lesser extent with temperature – the latter in response to greater rates of chemical reactions (Fig. 5.34).

The conversion of solute output data to estimates of rates of ground surface lowering can be misleading, for the removal of solutes may be

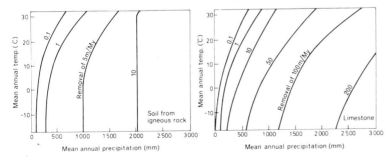

Fig. 5.34 Estimated chemical removal from an igneous rock soil and a pure limestone in metres per million years. Calculated for cations only. As this is a theoretical study it should be regarded as a working hypothesis rather than an established fact (after Carson and Kirkby, 1972).

compensated for by a decrease in soil bulk density. Furthermore not all data are representative of long periods nor corrected for precipitation inputs, changes in land use, or variations within a catchment, and some solutes may come from depth where they have no effect upon slope profiles. The data from published reports (Tables 5.5 and 5.6) suggest that there is a general increase in solute discharges from Precambrian igneous and metamorphic rocks such as granite, quartzite, and schist, to ancient and Mesozoic or Tertiary sedimentary rocks, with limestones being by far the most soluble common rocks.

Table 5.5 Estimates of Ground Lowering by Solution Alone

Rock type	Ground lowering mm/1000 year
Precambrian igneous and metamorphic	0.5–7.0
Precambrian micaceous schist	2.0–3.0
Ancient sandstones	1.5–22.0
Mesozoic and Tertiary sandstones	16–34
Glacial till	14–50
Chalk	22
Carboniferous limestone	22–100

Source: derived from Waylen, 1979.

Table 5.6 Estimates of Ground Lowering in Central Europe where Annual Rainfall is 800 mm/year

Rock type	Ground lowering mm/1000 year
Crystalline rocks	0–2
Sandstone	2–14
Shales and slates	6–18
Limestones and dolomite	24–42
Gypsum and anhydrite	*c.* 400
Sand and gravel	0–36

Source: data from Hohberger and Einsele, 1979.

In their study of data from 80 catchments in Central Europe Hohberger and Einsele (1979) emphasise that ground water carried the greatest part of the chemical load supplied to rivers, so that areas with a high relative yield of surface runoff have low chemical denudation rates. In Central Europe rivers from the Alps transport nearly similar quantities of solid and dissolved material, but rivers of hills and lowlands carry more solutes than solids.

Very few comments can be made on the effect of solute production on slope forms. It might be expected that solution would be at a maximum in areas of rapid interflow and high soil moisture contents, and at a minimum in areas of low permeability and moisture retention. The contributing areas of an expanded saturated zone would thus lose mass by solution more rapidly than spurs and upper slopes, while footslope areas in semi-arid zones of Hortonian overland flow would lose least, and may gain from deposition of salts. Much more field measurement will have to be done before such conclusions are verified or refuted.

Further reading

Among the most useful general texts on slope erosion and processes are those of Carson and Kirkby (1972), Young (1972), and Hudson (1971). Soil conservation is treated by Hudson (1971), in FAO Soils Bulletin 33 (1977), and in the older texts of Bennett (1939) and Jacks and Whyte (1939). Pereira (1973) provides a useful review of land use and management in relation to water. The volume edited by Rapp *et al.* (1973) provides a number of case studies of erosion which illustrate the impact of land use upon processes. Current research in the USA is well reviewed in the volume *Soil Erosion: Prediction and Control* published by the Soil Conservation Society of America (1977), and the use of soil loss equations in conservation planning is discussed at length by Wischmeier and Smith (1978).

6

Mass wasting of soils

Mass wasting is the downslope movement of soil or rock material under the influence of gravity without the direct aid of other media such as water, air, or ice. Water and ice, however, are frequently involved in mass wasting by reducing the strength of slope materials and by contributing to plastic and fluid behaviour of soils.

A great variety of materials and processes is involved in mass wasting with a consequent variety of types of movement. Distinguishing between these types requires considerations of at least the following criteria: velocity and mechanism of movement; material; mode of deformation; geometry of the moving mass; and water content.

Types of mass wasting

With so many criteria available for distinguishing mass wasting it is perhaps not surprising that there are many classifications in use and conflicts in applications of terms. The earliest widely used classification is that of Sharpe (1938) and most workers since then owe some debt to him for his pioneer effort. More recent classifications are those of Varnes (1958, 1975), Hutchinson (1968), and Nemčok et al. (1972). Sharpe's scheme is shown in Table 6.1, that of Varnes (1958) in Table 6.2, and that of Hutchinson in Table 6.3. The classification of Varnes (1958) is of landslides only and excludes creep and frozen ground phenomena, while Hutchinson's is the most complete scheme. Because Varnes's terminology for landslides is the most commonly used it will be followed here, but it has to be realised that this, like all other classifications, can be difficult to apply. Recognising the nature of a process from the debris it leaves behind is not always easy and the group of processes – debris slide, debris avalanche, debris flow (Fig. 6.1), collectively called translational slides – are not

always distinguishable from each other. Slides frequently break up into avalanches and may become flows towards the base of a slope if sufficient water is available. A second source of confusion in classifications is the varied uses attached to the words debris, soil, earth, and mud. The 1975 classification of Varnes tries to overcome this problem by using 'debris' to include only coarse material, and 'earth' to denote sand, silt, and clay (Table 6.4). Even this distinction is not always possible because many failed regoliths include a range of materials from boulders to clay.

Soil creep is the slow downslope movement of superficial soil or rock debris which is usually imperceptible except to observations of long duration. In environments with seasonal variations of soil moisture and soil temperature, soil creep is predominantly an episodic process dependent upon heaving and settling movements in the soil produced by solution, freeze–thaw, warming and cooling, and wetting and drying cycles. Creep may also be induced by the activity of soil microfauna, plant roots, and plastic deformation of clay soils.

The mechanics of creep have been investigated experimentally and theoretically (Terzaghi, 1953; Goldstein and Ter-Stepanian, 1957; Saito and Uezawa, 1961; Culling, 1963; Haefli, 1965; Bjerrum, 1967; Carson and Kirkby, 1972). Movement is quasi-viscous, occurring under shear stresses sufficient to produce permanent deformation but too small to result in discrete failure. Movement of the soil mass is mainly by deformation at grain boundaries and within clay mineral structures. Both interstitial and absorbed water appear to contribute to creep movement by opening the structure within and between mineral grains and so reducing friction within the soil mass. In many soils therefore rates of soil creep are greater during the wet season.

Measurements indicate that soil creep movements

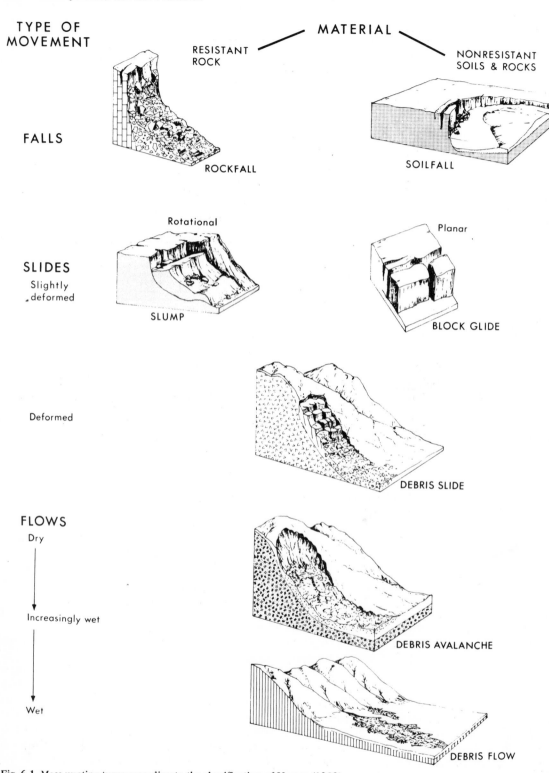

Fig. 6.1 Mass wasting types according to the classification of Varnes (1958).

Table 6.1 Classification of Sharpe (1938)

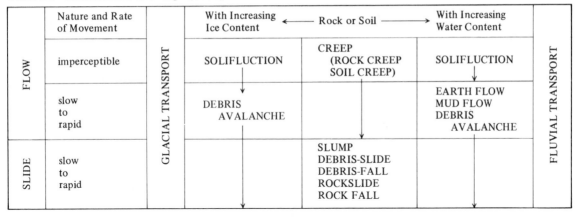

	Nature and Rate of Movement	GLACIAL TRANSPORT	With Increasing Ice Content ← Rock or Soil →		With Increasing Water Content	FLUVIAL TRANSPORT
FLOW	imperceptible		SOLIFLUCTION	CREEP (ROCK CREEP SOIL CREEP)	SOLIFLUCTION	
	slow to rapid		DEBRIS AVALANCHE		EARTH FLOW MUD FLOW DEBRIS AVALANCHE	
SLIDE	slow to rapid			SLUMP DEBRIS-SLIDE DEBRIS-FALL ROCKSLIDE ROCK FALL		

Table 6.2 Classification of Varnes (1958)

Type of Movement		Type of Material			
		Bedrock		Soils	
FALLS		ROCKFALL		SOILFALL	
SLIDES	few units	rotational SLUMP	planar BLOCK SLUMP	planar BLOCK GLIDE	rotational BLOCK SLUMP
	many units		ROCKSLIDE	DEBRIS SLIDE	FAILURE BY LATERAL SPREADING
FLOWS		All Unconsolidated			
		rock fragments	sand or silt	mixed	mostly plastic
	dry	ROCK FRAGMENT FLOW	SAND RUN	LOESS FLOW	
				RAPID EARTHFLOW DEBRIS AVALANCHE SLOW EARTHFLOW	
	wet			SAND OR SILT FLOW DEBRIS FLOW MUDFLOW	
COMPLEX		Combinations of Materials or Type of Movement			

diminish progressively with depth and are most marked in the upper metre or so of most soils, although they can occur at depths of up to 10 m (Kojan, 1967). The restrictions in depth are understandable as temperature and moisture changes in the soil diminish rapidly with depth. Directions and rates of movement are also variable as soil properties change both vertically and laterally. In a homogeneous soil, heaving processes followed by settling should push soil particles upward, and then be followed by a downslope movement parallel to the slope surface and a settling back into the slope.

Table 6.3 Classification of Hutchinson (1969)

CREEP	(1) Shallow, predominantly seasonal creep;
	    (a) Soil creep
	    (b) Talus creep
	(2) Deep-seated continuous creep; mass creep
	(3) Progressive creep

FROZEN GROUND PHENOMENA	(4) Freeze-thaw movements
	    (a) Solifluction
	    (b) Cambering and valley bulging
	    (c) Stone streams
	    (d) Rock glaciers

LANDSLIDES	(5) Translational slides
	    (a) Rock slides; block glides
	    (b) Slab, or flake slides
	    (c) Detritus, or debris slides
	    (d) Mudflows
	        (i) Climatic mudflows
	        (ii) Volcanic mudflows
	    (e) Bog flows; bog bursts
	    (f) Flow failures
	        (i) Loess flows
	        (ii) Flow slides
	
	(6) Rotational slips
	    (a) Single rotational slips
	    (b) Multiple rotational slips
	        (i) In stiff, fissured clays
	        (ii) In soft, extra-sensitive clays; clay flows
	    (c) Successive, or stepped rotational slips
	
	(7) Falls
	    (a) Stone and boulder falls
	    (b) Rock and soil falls
	
	(8) Sub-aqueous slides
	    (a) Flow slides
	    (b) Under-consolidated clay slides

Observations, however, suggest that most movements are far more irregular than this with local upslope, downslope, and vertical movements being rather erratic, although the trend of movement is always downslope (Fig. 6.2).

Observations, over a twelve-year period, in northern England (Young, 1978) have shown that the linear downslope component of movement within the top 20 cm of mineral soil was 0.25 mm/year, but this was exceeded by a component of 0.31 mm/year inwards towards the rock and perpendicular to the ground surface. The inward movement is interpreted as being caused by loss of weathered material in solution with rearrangement and settling of the remaining particles. The greatest loss at the surface, with progressively smaller losses with depth in the profile, is consistent with the suggestion that chemically undersaturated rainwater enters the soil from the surface, and takes ionic materials into solution until it reaches chemical equilibrium at depth. As solution represents ground loss, and also causes rearrangement of soil particles, and hence soil creep, it is geomorphically more important than creep. Solution is thus probably the most important geomorphic process in many low-energy vegetated environments.

Many types of field evidence have been held to demonstrate the existence of soil creep including (Fig. 6.3) outcrop curvature, tree curvature, tilting of structures, soil accumulations upslope of retaining structures, turf rolls, and cracks in the soil (Plate 6.1). Most of these phenomena can, however,

Table 6.4 Classification of Varnes (1975)

Type of Movement			Type of Material					
			Bedrock	Soils				
				coarse		fine		
FALLS			ROCKFALL	DEBRIS FALL		EARTH FALL		
TOPPLES			ROCK TOPPLE	,, TOPPLE		,, TOPPLE		
SLIDES	rotational	few units	,, SLUMP	,, SLUMP		,, SLUMP		
	translational		,, BLOCK GLIDE	,, BLOCK GLIDE		,, BLOCK GLIDE		
		many units	,, SLIDE	,, SLIDE		,, SLIDE		
LATERAL SPREAD			,, SPREAD	,, SPREAD		,, SPREAD		
FLOWS			,, FLOW (deep creep)	,, FLOW (soil creep)		,, FLOW		
COMPLEX			Combination of 2 or more types					

also be produced by other processes such as wind, slope wash, or tilting under the weight of the object. The only reliable method of determining the rate of creep is by direct measurement. This is usually done by placing pins, or acrylic rods, in the walls of trenches which are then refilled; by inserting columns of beads, blocks, or tubes into the soil; by attaching cones to piano wire which is then led to the surface; by inserting sensitive tilt bars into the surface soil; or by using strain gauges. All methods have errors associated with them as the soil is disturbed during insertion of instruments, but there are now sufficient measurements available to indicate that common creep rates downslope are between 0.1 and 15 mm/year in vegetation-covered soil, but on exposed talus slopes, or in cold climates where freeze–thaw processes are common, higher rates up to 0.5 m/year have been recorded. (See Young, 1972; Carson and Kirkby, 1972; Swanson and Swanston, 1977, for reviews.)

Deep-seated or continuous creep occurs below the depth affected by seasonal climatic changes and may be a rock rather than a regolith phenomenon. It is so slow as to be imperceptible and there is usually little direct evidence of its existence. It is usually assumed to be of significance in the formation of deep-seated failure planes. Where slope materials are subjected to such high shear stresses that they are close to failure, creep movements at depth can become relatively rapid and continuous — up to or even exceeding 20 cm/year. In over-consolidated clays creep may cause realignment of clay minerals so that their long axes are parallel to the failure plane, and this predisposes the slope to

failure by landsliding. Progressive creep at depth can also cause fracture of overlying more rigid strata and so permit creep to be followed by rapid failure (Hutchinson, 1968).

Terracettes are, perhaps, the most prominent surface features attributed to soil creep. They are interlacing networks of low but long tracks (see Vincent and Clarke, 1976, for a review). The networks may form continuous staircases up some hillslopes, or they may occur intermittently. It has been estimated that on a 24° slope, on soils from loess, they may occupy as much as 11 per cent of the ground surface and on a 37° inclination on steep mudstone-based hillslopes 40 per cent.

It is evident that many short discontinuous

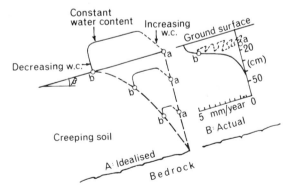

Fig. 6.2 A, an idealised displacement profile of soil particles moving from *a* to *b* in a soil with seasonal moisture changes; w.c. indicates water content (after Fleming and Johnson, 1975); B, a commonly observed type of movement showing the depth of observed creep (note the different vertical and horizontal scales).

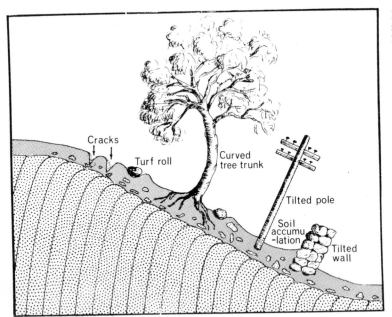

Fig. 6.3 Types of field evidence commonly, and not always accurately, held to be indicative of soil creep.

6.1 Small trees, presumably tilted during a period of slope movement, have subsequently grown vertically.

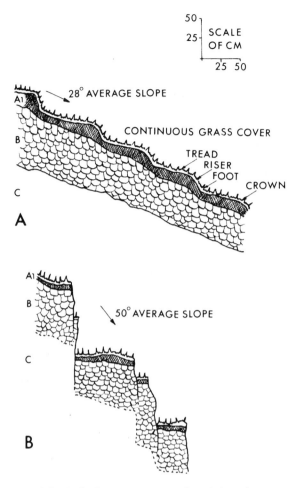

50 — SCALE
25 — OF CM

25 50

28° AVERAGE SLOPE

A1

B

CONTINUOUS GRASS COVER

TREAD
RISER
FOOT
CROWN

C

A

A1

B

50° AVERAGE SLOPE

C

B

Fig. 6.4 Idealised terracettes on gentle and steep slopes.

terracettes form beneath forest by the washing of soil and accumulation of leaf litter behind tree roots. When forest is cleared and the land sown in pasture grasses an incipient terracette pattern is therefore already in place. It is equally clear, however, that animal treading integrates several different kinds of slope irregularities into continuous networks which roughly follow the contours. In cold climates frost action and flow in thawing soils may produce terracettes.

Animal treading is almost certainly primarily responsible for continuous terracette systems which may have been built upon natural forms developed beneath forest and by irregular soil creep. Terracettes become repositories for animal dung and urine which encourage plant growth, particularly along the edge of a terracette, and this factor may also assist in the preservation of a sharp edge to the feature.

In morphology many terracettes appear to belong

to two main classes: on slopes of less than about 30° the grass cover is not broken; on slopes steeper than about 30° soil cracking increases; and on slopes of 50° or more the cracks occur at the back of each tread and a minor slumping process may sometimes appear to be occurring (Fig. 6.4).

Falls in soil or soft rocks usually involve only small quantities of material because steep slopes in weak materials are necessarily very short. These falls are usually the result of undercutting of the toe or face of a slope by a river or by wave action. They are facilitated also by weathering and the opening of fissures near a cliff top as a result of freeze–thaw, wetting and drying, earthquake shocks, or tension.

Slumps have curved failure planes and involve rotational movement of the soil mass. They have received much attention in the engineering literature because they are common in the walls of cuttings in soft rocks such as shales, mudstones, and over-consolidated clays. They occur also under entirely natural conditions, especially where the toe of a slope has been undercut by river or wave action. Slumps may be single rotational failures with shearing occurring rather rapidly on a well-defined curved surface. Backward rotation of the sliding mass often causes fracturing of further blocks at the crown of the slump and sometimes is accompanied by ponding of water in the depressed, or backwards-tilted, area of the crown. Such water may promote continuing movement. In many materials, especially with high water content, the toe of a slump may be so remoulded that it becomes a flow (Fig. 6.5; Plates 6.2 and 6.3).

Some slumps, and especially those in clays under-lain by hard impervious strata, and overlain by porous caprocks which form water reservoirs, develop multiple rotational slides with two or more blocks tilting backwards and moving on the same slide plane at the sole of the slumps. In stiff fissured clays forming relatively high relief the face of a slope may be formed of numerous shallow rotational slides. They develop from the base of a slope where the lowest slump removes lateral support from the slope immediately above it, which then fails. This type of retrogressive failure may gradually extend up the slope. Deep-seated multiple retrogressive failures appear to be most common in sea cliffs experiencing continuing removal of the toe, and shallow multiple failures are most common in the soil forming in a stiff clay as weathering reduces the shear strength of the superficial materials (e.g. Brunsden and Jones, 1976).

6.2 The crown of a rotational landslide.

Failures by lateral spreading are special classes of slumps which are virtually confined to clay-rich sediment deposited in the shallow seas and lakes around the edges of ice sheets. Nearly all known examples come from southern Norway, the St Lawrence lowlands of eastern Canada, and the Alaskan coast. The failures usually begin with a single rotational slide in a bank undercut by a stream. The slide movements remould the clay along the slide plane very rapidly, and it turns to a dense liquid which supports moving blocks of overlying clay or sands to leave a chaotic topography of small horsts and grabens (Fig. 6.6; Plate 6.4). The movements are retrogressive as each block removes the lateral support of the soil upslope of it. The soil close to the river is sometimes more weathered than that farther inland and the plan of the failure then becomes bottle-necked in shape with the narrow neck in the river bank.

Most failures by lateral spreading occur in

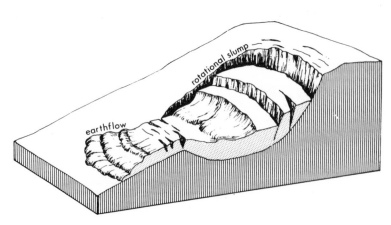

Fig. 6.5 Features of a slump.

6.3 A large slump in mudstones showing the hummocky lower area with disturbed drainage pattern and ponding. The debris has dammed the river and continuing erosion of the toe is causing repeated movements.

sensitive soils. Sensitivity is defined as the ratio of peak strength of a soil when undisturbed to its strength after being remoulded, so sensitive soils undergo a very large decrease in strength once they are disturbed. Once slight movement occurs failure can then be rapid as formerly stable soil becomes increasingly liquefied.

Sensitive soil behaviour is usually explained as a

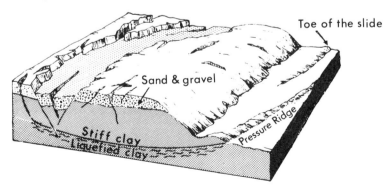

Fig. 6.6 A lateral spreading failure (after Hansen, 1965).

6.4 A large spreading failure in sensitive soils; the Nicolet landslide of November 1955, Quebec, Canada (From a paper by Jacques Béland, courtesy of Roger Bédard, 'Nicolet Landslide, November 1955', The Proceedings of the Geological Association of Canada, Vol. 8, Part 1 (Nov. 1956). By permission of M. Béland.)

fabric change on remoulding (see Fig. 4.41). In some Norwegian soils sensitivity develops as marine 'quick' clays are leached of their salt content, but in many Canadian sediments this is not usually the cause of sensitivity (Kerr 1963; Crawford and Eden, 1969; Torrance, 1975; Gadd, 1975; R. J. Mitchell, 1976; Mitchell and Klugman, 1979).

Many of the St. Lawrence basin marine glacial deposits are lightly over-consolidated and have exceptionally high undrained shear strengths. The clays contain high proportions of unweathered primary minerals, of less than 2 μm size, as quartz and feldspar glacial flour, with few expanding lattice clay minerals (Cabrera and Smalley, 1973; Smalley and Taylor, 1972; Smalley, 1976). In their undisturbed state the soils are naturally well cemented, but once remoulded become fissured and reduced to gravel-sized aggregates.

Block glides are slides in which material remains largely undeformed as it moves over a planar slide plane. These are uncommon features in soils, except at a very small scale.

Translational slides are by far the most common form of landslide occurring in soils. They are always shallow features and have essentially straight slide planes with some curvature towards the crown, and hence some rotational movement can occur (Plates 6.5 and 6.6). The distinction between debris slides, debris avalanches, and debris flows is based upon the degree of deformation of the soil material and the water content of the sliding mass. As both deformation and water content frequently increase downslope what may be a debris slide at the crown of the landslide — with relatively large undeformed blocks of soil sliding downhill — may become an avalanche of small blocks and wet debris in midslope and a thoroughly liquefied flow at the base of the slope, especially if the flow mass descends into a river.

Unlike falls, slumps, and glides which may occur

6.5 A slump formed towards the base of the slope where a stream undercut the toe and water tables are high. It has progressively extended up the slope. Another failure is developing to the right of the photograph where cracking and slight dropping of the head and bulging of the toe of a failure can be seen. Note the terracettes on the remoulded soil.

as a result of deep percolation of water, and hence at a considerable time after a rainfall, translational slides nearly always occur during heavy rain. Rainstorms with sufficient intensity or duration are required to raise the water table to near the soil surface or fill pre-existing tension cracks. In low-intensity rainfalls the removal of water from the soil by interflow can keep pace with infiltration, and in short-duration falls the field capacity of the soil may not be exceeded. Only when the capacity of the soil to drain is exceeded for long enough for water pressures to rise substantially can the soil lose sufficient strength to fail.

Flows occur when coarse debris, fine-grained soil, or clay are liquefied (Plate 6.7). The terms 'debris flow', 'earthflow', and 'mudflow' are used to distinguish between these three classes of material. The tendency for flows to develop may be encouraged by a number of factors: remoulding of soils during landsliding; the presence of clays with high liquid limits in areas where rainfalls are high; the

presence of soils with low liquid limits in areas of low rainfalls — in such soils little water is required to make the soil behave as a liquid; the presence of soils with open fabrics resulting from flocculation

6.6 Numerous translational landslides formed on these slopes during an abnormally wet winter, Wairarapa, New Zealand. The soils are in loess overlying mudstones. Note the debris which has flowed into the valley floors.

6.7 A small debris flow. Note the levees.

during deposition; or the thawing of soil ice. It has been suggested by Hutchinson and Bhandari (1971) that some flows may be promoted by the collapse of soil from surrounding cliffs or steep slopes on to the upper part of a concave moving mass, thus raising pore water pressures and promoting flow which, in turn, causes loading on the debris farther down the flow and rapid movement. They called this mechanism 'undrained loading'. A feature of many mudflows is that they can occur on slopes of very low angle. Mudflows may be relatively slow-moving (1-20 m/year) or very fast, as in the catastrophic flows described in Chapter 7.

The term solifluction literally means soil flow and is usually used to refer to the flow failures of high latitudes and high elevations, although some authors prefer to use the term gelifluction for the slow flows occurring in thawing soils.

Dry flows range in size from large catastrophic failures of thick deposits of loess (wind blown silt-sized deposits) in China to small flows of dry sand and silt in quarries and sandpits or down the faces of dunes.

Field study of landslides

Most geomorphic studies of landslides attempt to assess the quantitative significance of landslides in total denudation and in modifying the ground surface, or attempt to determine the causes and mechanics of the process. Studies usually begin with the preparation of a map of the area relating the sites of landslides to lithology, slope angle, pre-existing slope disturbances, hydrological conditions,

and locations on the slope. A study is also made of the climatic or seismic conditions before and at the time of the slide. Where an attempt is to be made to understand the mechanisms of the failure the geometry of the slide has to be determined and the appropriate method of shear strength testing adopted. The following notes are intended as a guide to appropriate procedures and the use of a common terminology.

One of the difficulties of attempting to discover the significance of erosional events is that many different forms of reporting are used and the information given for each case study is varied. In Table 6.5 four landslide-producing storms and their effects are compared. If this method were used more commonly a better understanding could be obtained of mass wasting processes and their importance.

Morphometry

The shape, or morphometry, of landslides has its own terminology which is outlined in Fig. 6.7 in which a translational slide is displayed. If the terms were being applied to a slump the head of the displaced mass would be much closer to the crown of the slide and the length of the exposed slide plane would be small. Also shown in Fig. 6.7 are the measurements which are made to determine the area, depth, and volume of the scar and the displaced mass. This information is needed for quantitative studies of the stability of hillslopes against landsliding.

Attempts have been made to use the ratio between the depth and length of landslides as a means of interpreting the processes which gave rise to the scar. Crozier's (1973) compilation of data for landslides in New Zealand showed that the ratios could be used to distinguish types of process but, by contrast, Blong (1973) concluded that simple morphometric indices are not of value for distinguishing the processes which have given rise to a landslide, as similar ratios may be associated with widely differing behaviour among failures. Because the process of erosion affects both the shape of the scar and the fabric and strength of the deposited material it is advisable to study all available evidence before categorising a landslide.

Shear strength testing

In all strength tests an attempt is made to simulate the magnitude of loads, the loading rates, and drainage conditions which occur in the slope failures

Table 6.5 Landslide Records

Area	Mgeta, Tanzania	Tarfala, Lappland	Longyear Valley, Spitsbergen	Mangawhara Valley, New Zealand
1. Latitude, Longitude	7° S; 37° E	68° N; 19° E	78° N; 16° E	37° S; 175° E
2. Altitude (m.a.s.l.)	1100	1130	37	150–350
3. Mean annual temperature (°C)	24.3	−4.3	−6.2	14
4. Annual precipitation (mm)	1058	950	203	1800
5. Rainfall records since	1951	1947	1915	1956
6. Relief range (m)	1100–2000	1200–2000	50–500	50–500
7. Bedrock	Gneiss/granulite	Amphibolite schists	Sandstone, schists	Sandstone/siltstone
8. Regolith type, depth (m)	Sandy/silty regolith	Bouldery debris	Bouldery debris on permafrost	Clay to sandy, with volcanic ash 1.5–2.0
9. Vegetation	Cultivation, grass fallows	Mountain tundra	Mountain tundra	Pasture grasses
10. Rainstorm date	23.II.1970	6.VII.1972	10–11.VII.1972	28.II.1966
11. Rainstorm total (mm)	101	45	31	150–230
12. Rainstorm duration (hours)	2	2	10	24
13. Rainstorm type	Convective	Convective	Cyclonic	Cyclonic with local convection
14. Rainstorm area[1] (km²)	50	11	>30	250
15. Rainstorm return period (years)	10	>27	>53	10 or >10
16. Slope erosion type[2]	Debris slides	Gullies	Debris slides	Debris avalanches/ debris slides
17. Slope gradient (degrees)	33–44	12–30	30–34	32
18. Depth of scars (m)	0.8–3	0.5–4	0.6–1	0.9
19. Width of scars (m)	5–30	1–13	5–20	12
20. Denudation rate/ catchment area[3] (mm/ km²)	14 mm/20 km²	5 mm/11 km²	1 mm/4.5 km²	80 mm/0.5 km² to 10 mm/20 km²
21. Mass transport form	Mudflow	Debris flow	Debris flow	Debris flow
22. Mass deposition form	Total fluvial removal	Colluvial fan	Colluvial fan	Colluvium reworked in stream bed
23. Mass max. depth (m)	–	3	1.5	1.5
24. Mass gradient (degrees)	0	10–15	8–15	5–25

Notes: (1) Area in which mass movement occurred
(2) Classification of Varnes (1958)
(3) Vertical downwearing of the landscape
    Denudation rate is calculated from volume of erosion scars
Source: data from Rapp, 1975 and Selby, 1976.

being studied. Three types of test are commonly carried out in laboratories.

(1) In *unconsolidated undrained tests* no drainage of pore water is allowed at any stage and the sample is sheared at the moisture content it had in the natural state. For fully saturated soils there should be no volume change during testing. The sample is always sheared immediately after the application of a normal load in a shear box test, or immediately after the application of cell pressure in a triaxial test. A test may be carried out quickly – in 15 minutes or less – but as pore water pressures cannot redistribute themselves through the sample in this time pore water pressures can neither equalise

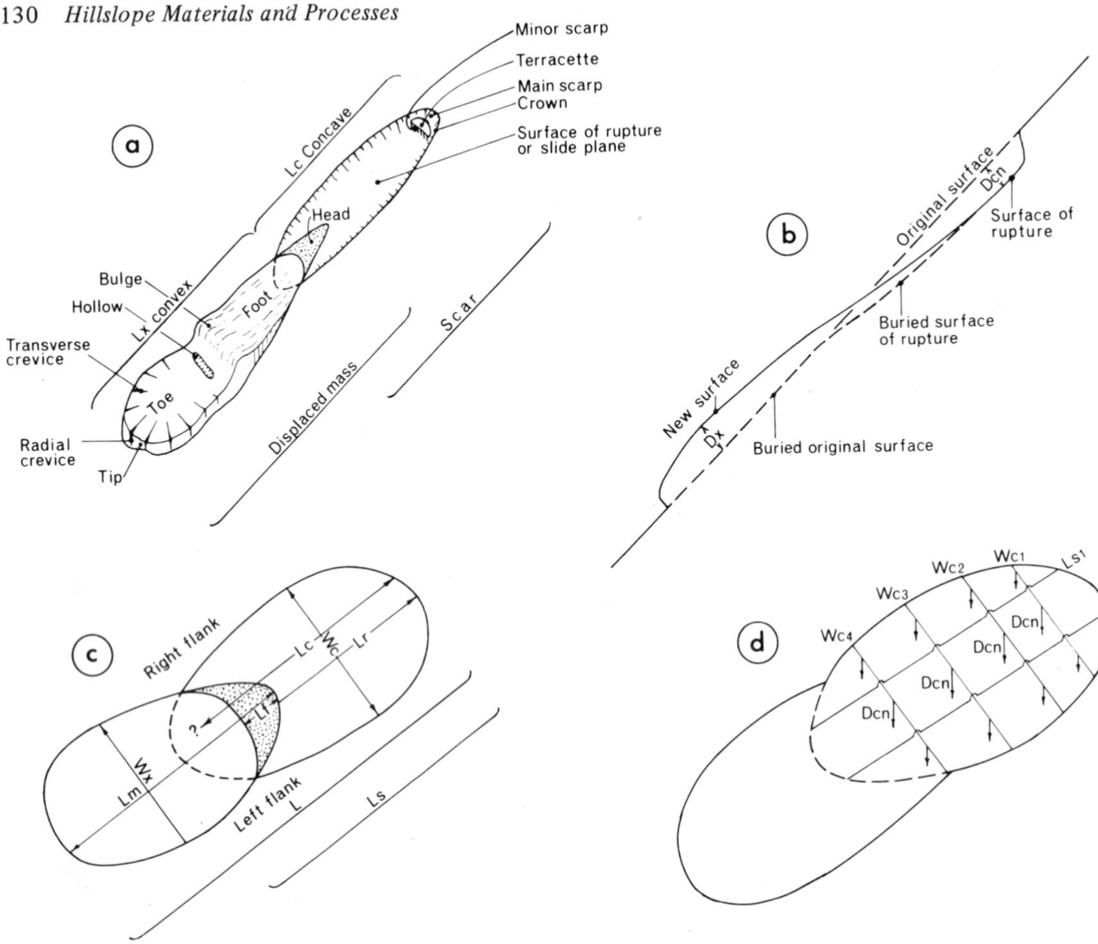

Fig. 6.7 The terminology of a landslide and the indices of landslide size and volume (adapted in part from Crozier, 1973).

nor be measured, and the results of a test can only be expressed in terms of total stress parameters (c_u, ϕ_u).

(2) In *consolidated undrained tests* the sample is allowed to consolidate under an effective stress corresponding to the effective stress *in situ*. Thus initial drainage is permitted during the consolidation phase which is completed when the reduction of volume, or expulsion of water, is complete. The sample is sheared under conditions of no drainage and pore water pressures are therefore induced and measured. The rate of shear must be slow enough to allow pore water pressures to distribute themselves evenly through the sample. For many samples a compression rate of 0.05 mm/minute is satisfactory and tests are completed in two to three hours. The results of the test are expressed in terms of effective stress (c', ϕ'), or total stress (c_{cu}, ϕ_{cu}).

(3) A *drained test* is performed so that the sample is allowed to drain at all times, consequently there is a continuous change of moisture content and continual volume changes during the test. Another condition of the test is that loading is applied so gradually that there is no build-up of pore water pressures in the sample, but this requires a duration of testing of three to fourteen days for a sample of clay, although it may be rapid for sands. Results of the test are expressed in terms of the parameters c_d, ϕ_d: drained tests should give the same values for the strength parameters as undrained tests in which pore pressures are measured.

In choosing the type of test which best simulates field conditions consideration has to be given to all relevant criteria. The situations and suggested parameters given in Table 6.6 should be regarded as a guide only. For 'first-time' landslides in which soil has not been remoulded peak shear strength values (ϕ_p') are usually most appropriate, but

Table 6.6 Shear Strength Test Chosen for Selected Field Problems

Problem	Parameter	Test
Shallow translational landslide formed in prolonged wet season	c_p' ϕ_p'	Consolidated undrained laboratory test, or field shear box test in saturated conditions
Shallow translational landslide formed in dry season rainstorm	c_p' ϕ_p'	Consolidated undrained laboratory test, or field shear box test at natural moisture content
Long-term stability of clay slope subject to gradual toe removal, progressive failure, or first time failure in fissured clays	c_r' ϕ_r'	Drained test, but for fissured over-consolidated clays $c_r' = 0$ and ϕ_r' parameters are appropriate
Long-term creep of clay slope or slow mud flow	c_r' ϕ_r'	Ring shear test
Failure of slope in sand	c_d ϕ_d	Drained test

where prolonged creep or flow has occurred, or soils have been involved in an earlier period of instability, residual values (ϕ_r') are appropriate. For shallow landslides, or those in very variable soils, pedological structures and root materials may have a major effect upon shear strength. Field shear box tests may then be useful and attempts should be made to simulate moisture conditions and rates of strain which occurred during the landslide. For discussion of test procedures and landslides see Bishop and Bjerrum (1960), Bishop (1971), and Skempton (1970).

Stability analyses

Hillslopes upon which landslides are active, or slopes which are being deeply undercut at the base, are oversteepened with respect to an angle for long-term stability. In the investigation of slope instabilities it is desirable to know what causes the instability and what effect various remedial measures, such as draining a slope or planting trees on it, may have. Quantitative studies of this kind have been part of engineering practice for 50 years but they have only recently been introduced to geomorphology, and then mainly used for studies of the common translational slides.

Factor of safety

The stability of a slope is usually expressed in terms of a *factor of safety*, F, where

$$F = \frac{\text{sum of resisting forces}}{\text{sum of driving forces}}.$$

Where the forces promoting stability are exactly equal to the forces promoting instability $F = 1$; where $F < 1$ the slope is in a condition for failure;

where $F > 1$ the slope is likely to be stable. There is no such thing as absolute stability, only an increasing probability of stability as the value of F becomes larger. Most natural hillslopes upon which landsliding can occur have F values between about 1 and 1.3, but such estimates depend upon an accurate knowledge of all the forces involved and for practical purposes design engineers always adopt very conservative estimates of stability (Table 6.7).

Table 6.7 Values of Minimum Overall Safety Factors

Failure type	Item	F
Shearing	Earthworks	1.3 to 1.5
	Earth retaining structure	1.5 to 2.0
	Foundation structures	2 to 3
Seepage	Uplift, heave, slides	1.5 to 2.5
	Piping	3 to 5

Source: after Meyerhof, 1969.

It can be seen that the greatest uncertainties are usually associated with soil water, especially with its local variability of pressure and seepage.

Hillslopes which have too low an angle for landsliding to occur have F values much greater than 1.0 and are theoretically being modified entirely without the operation of landslides. Such slopes, however, may occur only at very low angles — less than 8° or so in many clays — and even on very gentle slopes mass wasting may occur in periglacial environments where soil ice and soil saturation in an active zone above permafrost can be effective.

Stability analyses of shallow translational slides

Translational slides are usually analysed by the infinite slope method which is a two-dimensional analysis of a slice on the sides of which the forces are taken as being equal and opposite in direction and magnitude. It is assumed that the mobile slice is uniform in thickness and rests on a slope of constant angle and infinite extent. This dispenses with the need to consider side and end effects, and is justified as translational slides are long in relation to their depth and width and are often uniform in cross-section. This mode of analysis was employed by Skempton and De Lory (1957).

The forces acting at a point on a shear plane of a potential shallow slide are illustrated in Fig. 6.8. The gravitational stress acts vertically, the normal stress is normal to the shear plane and is partly opposed by the upthrust or buoyancy effect of pore water pressure; the shear stress acts down the shear plane and is resisted by the shear strength of the soil.

Where a rectangular block of soil rests on a slope (1 in Fig. 6.9a), the resisting force holding the block in place is given by W (which is the mass $\times$ the gravitational force) multiplied by the cosine of the angle of slope. The driving force is $W\sin\beta$.

For an infinite slope analysis the soil block is within the regolith and the value of W has to be determined indirectly. The easiest method is to make measurements of the vertical thickness of the block. This is also realistic as cracks in the soil are approximately vertical. The block of soil therefore has the sectional form of a parallelogram ABCD (2 in Fig. 6.9a). For computation this is converted to the equivalent rectangle so that the parallelogram ABCD is represented by the rectangle AEFD. The angles EAB, and FDC in 2 are equal to β and in 3 (Fig. 6.9a) $\cos\beta = \dfrac{p}{z}$ thus $p = z\cos\beta$.

Assuming a unit width of the soil block so that we are dealing only with a two-dimensional problem:

$W = l\gamma p$ (substituting $p = z\cos\beta$) $= l\gamma z\cos\beta$,

where γ is the unit weight of the soil at natural moisture content.

The normal stress on the shear plane is:

$\sigma = W\cos\beta$, hence $\sigma = l\gamma z\cos\beta\cos\beta$.

The shearing stress on the shear plane is:

$\tau = W\sin\beta$, hence $\tau = l\gamma z\cos\beta\sin\beta$.

(Note: (a) stress must be expressed in force per unit area, so a two-dimensional analysis is only realistic when it includes the assumption of a unit width to the soil block; (b) because the length of the block (l) is not relevant in an infinite slope analysis it can now be dropped from the equations.)

Effective shear strength at any point in the soil is given by the Coulomb equation as:

$$s = c' + (\sigma - u)\tan\phi',$$

where s is the effective shear strength at any point in the soil;

c' is the effective cohesion, as reduced by loss of surface tension;

σ is the normal stress imposed by the weight of solids and water above the point in the soil;

u is the pore water pressure derived from the unit weight of water (γ_w) and the piezometric head ($\gamma_w mz$) (see Fig. 6.9b, 4);

ϕ' is the angle of friction with respect to effective stresses.

Substituting for σ the Coulomb equation can be rewritten as:

$$s = c' + (\gamma z\cos^2\beta - u)\tan\phi'.$$

Because $F = \dfrac{\text{sum of resisting forces}}{\text{sum of driving forces}} = \dfrac{s}{\tau}$;

then $\quad F = \dfrac{c' + (\gamma z\cos^2\beta - u)\tan\phi'}{\gamma z\sin\beta\cos\beta}$.

It is convenient to express the vertical height of the water table above the slide plane as a fraction of the soil thickness above the plane and this is denoted by m. Then if the water table is at the ground surface $m = 1.0$, and if it is just below the slide plane $m = 0$. Pore water pressure on the slide plane, assuming seepage parallel to the slope, is then given by:

$$u = \gamma_w mz\cos^2\beta$$

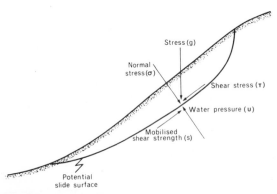

Fig. 6.8 Forces acting at a point on a potential failure plane.

thus

$$F = \frac{c' + (\gamma - m\gamma_w)z\cos^2\beta\tan\phi'}{\gamma z \sin\beta\cos\beta}.$$

For example if:

$\phi' = 12°, c' = 11.9 \text{ kN/m}^2, \gamma = 17\text{kN/m}^3,$
$\beta = 15°, z = 6 \text{ metres}, m = 0.8, \gamma_w =$
$9.81 \text{ kN/m}^3,$

then:

$$F = \frac{11.9 \times (17 - 0.8 \times 9.81)6 \times 0.92 \times 0.2}{17 \times 6 \times 0.25 \times 0.96} = 0.9.$$

Thus the slope is prone to failure when the water table is about a metre below the ground surface. If the water table can be lowered, by drainage, to just below the slide plane (when $m = 0$) then $F = 1.3$ and the slope will be stable in the longer term.

Where there is not a continuous water table with flow parallel to the soil surface an alternative form of analysis must be used.

Under most conditions the free soil water level cannot rise above the soil surface and the pore pressure (u) is given by:

$$u = \gamma_w h,$$

where h is the piezometric height or the height to which water will rise in a stand pipe, inserted in the soil to the depth of the failure plane. Where seepage is not uniform, and directed out of the slope, it is convenient to use the ratio r_u between pore pressure and the weight of a vertical column of soil:

$$r_u = u/\gamma z, \text{ thus } u = r_u . \gamma z.$$

The equation for determining the factor of safety then becomes (Haefli, 1948):

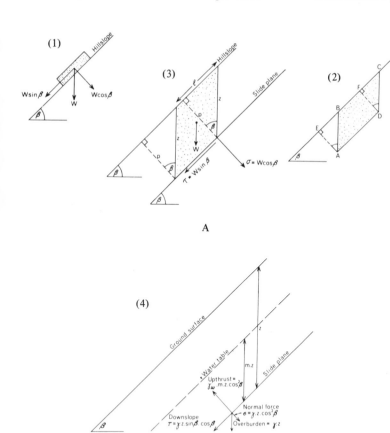

Fig. 6.9 Stresses acting on a slope, which are identified in an infinite slope analysis of a translational landslide. In (1) a block sits on a slope; in (2) the block is part of the soil profile and the area of a block with a parallelogram section is shown to be equal to that with a rectangular section; (3) the stresses acting on a slide plane; in (4) these stresses are analysed in a form suitable for inclusion in an infinite slope analysis.

134   *Hillslope Materials and Processes*

$$F = \frac{c' + (\gamma z \cos^2\beta - r_u \cdot \gamma z)\tan\phi'}{\gamma z \sin\beta \cos\beta},$$

substituting $\gamma_w h/\gamma z$ for r_u this becomes:

$$F = \frac{c' + \left(\gamma z \cos^2\beta - \dfrac{\gamma_w h}{\gamma z}\cdot\gamma z\right)\tan\phi'}{\gamma z \sin\beta \cos\beta}.$$

By dividing through by γz this may be rearranged to give:

$$F = \frac{\dfrac{c'}{\gamma z}\cdot + \left(\cos^2\beta - \dfrac{\gamma_w h}{\gamma z}\right)\tan\phi'}{\sin\beta\cos\beta}.$$

Under most conditions the highest pore pressures will exist when ground water level is at the surface. As γ is usually about $2\gamma_w$ the corresponding value of r_u is approximately 0.5. The extreme upper limit is the 'geostatic' value due to weight of soil and the pore pressure ratio $r_u = 1.0$, which implies a piezometric level rising above the slope to a height approximately equal to a depth z — a state clearly impossible under ordinary conditions, but possible if frozen soil is thawed so rapidly that the entire weight of overburden is transferred to the pore water, without any of the water being able to escape. This condition occurs in periglacial environments, and fossil landslides in southern England have been explained in this way (Skempton and Weeks, 1976; Chandler *et al.*, 1976).

Very high pore water pressures may also be produced where water fills a deep tension crack during a dry season rainstorm and moves laterally at depth below more impervious soil. The piezometric head downslope of the tension crack may then tend towards the geostatic value. Such a condition is relatively common in parts of California and New Zealand where severe summer storms have often been responsible for landsliding (Selby, 1967a, b, 1976; Campbell, 1975).

In one case study Rogers and Selby (1980) demonstrated that the piezometric surface rose 0.2 m above the ground surface causing high uplift pressures at the failure plane where a silty clay, with relatively high permeability, underlies a less permeable clay (Fig. 6.10). Solutions of the stability equation are given in Fig. 6.11. They illustrate how the factor of safety against landsliding varies with changes in the value of the parameters. F is very sensitive to changes in values of c' and h, moderately

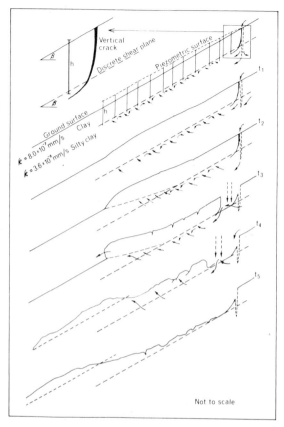

Fig. 6.10 Diagrammatic representation of an eye-witness account of a shallow landslide in Matahuru Valley, New Zealand. The slide began with a bulging of the toe and the formation of a tension crack at the crown. Over a period of 15 minutes the moving mass slid over the failure plane and burst through the sod at the toe as a saturated flow of debris. The upper part of the displaced mass then slid rapidly, but discrete soil blocks stabilised in the failure zone once the saturated material drained. k is the coefficient of permeability and h the piezometric head (after Rogers and Selby, 1980).

sensitive to values of z and β and rather insensitive to values of ϕ' and γ. The value of h may be reduced by better soil drainage but c' may be modified only by increasing apparent cohesion through a denser plant root network.

The type of analysis given above can be of great value to engineers, foresters, and planners who wish to determine what effect altering land use, draining a slope, or planting trees may have upon stability.

Spring flows from highly fractured and closely jointed rocks may also cause local very high pore water pressures at the base of the soil and trigger landslides during or after storms. Shallow translational landslides in much of the Appalachian region, USA,

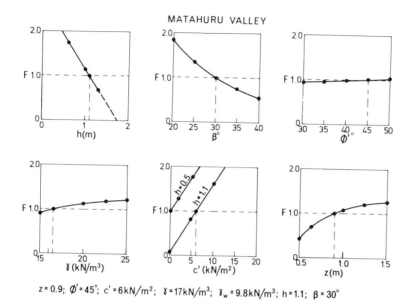

MATAHURU VALLEY

Fig. 6.11 Variations in the factor of safety against shallow landsliding as one parameter only varies. Parameter values are as given except where one parameter is varying.

$z = 0.9$; $\phi' = 45°$; $c' = 6 kN/m^2$; $\gamma = 17 kN/m^3$; $\gamma_w = 9.8 kN/m^3$; $h = 1.1$; $\beta = 30°$

in 1972 are attributed to such a process (Everett, 1979). Very high local water pressures may cause a burst or 'blow out' of the soil which may become the initiating event leading to a larger landslide.

Submarine translational landslides are very common in the weakly consolidated sediments of submarine deltas and they may be studied in the same way as terrestrial landslides. Such a study of slides up to 10 km long and 33 m deep (z) has been made by Prior and Suhayda (1979). They have shown that failure of sediment slopes, under gravitational stresses alone, can occur at angles as low as 0.5° where very high internal pore water pressures exist. Such pore pressures have been measured directly using piezometers. Using the assumption that $F = 1$, and hence that the slope is in a condition for failure, the pore pressure (u) needed to initiate failure can be calculated from the rearranged stability equation:

$$u = \frac{c' - F(\gamma z \sin\beta \cos\beta)}{\tan\phi'} + \gamma z \cos^2\beta.$$

In the Mississippi delta front measured and calculated values have been shown to correspond closely. The pore water pressure is so high that is constitutes a strong artesian pressure. Calculated values in some places approach, or even slightly exceed, geostatic pressures and indicate a condition of almost zero effective stress.

Stability analyses of rotational landslides

Deep rotational landslides are confined to clays and clay-rich soil, and do not occur in sands, because the strength of a soil due to cohesion only is not controlled by overburden pressure. Values of frictional strength for both clay and sand increase in proportion to the normal stress acting on a potential failure plane within a soil, thus for a frictional soil strength increases with depth. In frictional materials the rate of strength increase with depth exceeds the rate at which shear stresses increase and deep failures cannot occur. For clays shear stresses may increase more than strength for each increment in depth, hence deep-seated failures are possible, especially for clays in which $\phi_u = 0$.

Rotational failures may be treated as a series of vertical slices for each of which a modified infinite slope analysis is carried out and the values for each slice are then summed. A toe of the slope which resists failure, is treated as providing a negative driving force and because the length of the base of each slice may vary the value of l_A etc. (Fig. 6.12) is included in the formula thus:

$$F = \frac{\sum\limits_{O}^{A} [c'l + (W\cos\alpha - ul)\tan\phi']}{\sum\limits_{O}^{A} [W\sin\alpha]}.$$

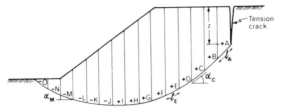

Fig. 6.12 Division of a rotational failure into vertical slices. Those resisting failure are treated as having negative values.

This method of analysis by slices was first proposed by Fellenius (1936) and is named after him; alternatively it is called the Swedish Method of Slices or Conventional Method. A worked example using the method is provided in Fig. 6.13. The original profile is reconstructed and the shape and slope of the slide plane are plotted on graph paper from field measurements. The centre of rotation of the slide can then be established. The failed mass is divided into slices (five or six are sufficient for a simple failure) and these are drawn on the graph paper so that the centre of gravity of each slice and its area can be obtained from the drawing. The length of the base of the slice is taken from the drawing and the angle of inclination of the base of the slice is found from an extended radius of the circle and a vertical through the centre of gravity of the slice. All slices are assumed to have a thickness of 1 m and thus a weight given by the product of the area (A in m^2), the unit weight of soil, and the thickness. Samples are collected from the unfailed soil in the sides of the landslide scar for strength testing and determination of unit weight of the soil.

The Fellenius Method is widely used because of its simplicity and because the calculations are easy. Unfortunately the value of F derived this way is often 10 to 15 per cent below the value derived by more rigorous methods and may be even more in error in certain cases (Whitman and Bailey, 1967). Most of the error occurs in the treatment of pore water pressures; some error occurs because of the method's assumption that all side forces on each slice act in a direction parallel to the failure plane and that normal forces are assumed to act at right angles to the failure plane (Fig. 6.14). The Fellenius Method treats each slice as though it were nearly rectangular, but with increasing curvature of the failure plane this becomes an untenable assumption.

For a slice with a curved base and an upper surface which is not parallel to the failure plane corrections have to be made. As a result an alterna-tive method was proposed by Bishop (1955) and this was simplified by Janbu *et al.* (1956). The Simplified Bishop Method of Slices assumes that forces acting on each slice are in a horizontal and vertical direction. This assumption is not entirely valid but the method has been shown to provide values of F which are in the range of values derived by more rigorous methods and are seldom more than 2 per cent in error. The correction factor is:

$$\cos\alpha\left(1 + \frac{\tan\alpha\tan\phi'}{F}\right)$$

and the formula for the Simplified Bishop Method is then:

$$F = \frac{\sum\limits_{i=1}^{i=n}\left[c'b+(W-ub)\tan\phi'\right]\left[1\bigg/\left[\cos\alpha\left(1+\frac{\tan\alpha\tan\phi'}{F}\right)\right]\right]}{\sum\limits_{i=1}^{i=n}\left[W\sin\alpha\right]}$$

where b is the horizontal width of the slice.

It will be seen from the equation that F appears on both sides, and a trial-and-error procedure is required with values being assigned to F and the equation solved repeatedly until it balances. This disadvantage is overcome by using a computer. A number of other more rigorous methods of slices are in use. They differ in their handling of interslice forces, but the established methods give closely corresponding values of F (Fredlund and Krahn, 1977).

Tension cracks frequently develop on slopes as a result of desiccation of the soil. For analysing a natural failure it is reasonable to assume that a crack developed at the head scarp of the failure and extended down to the failure plane. Where this cannot be determined the maximum possible depth of the crack (h_c) may be estimated from:

for cohesive soils ($\phi_u = 0$) h_c (metres) $= \dfrac{2c_u}{\gamma}$

in a soil with both frictional and cohesive properties

$$h_c = \frac{2c'}{\gamma}\left(\tan 45° + \frac{\phi'}{2}\right).$$

Tension cracks have the effect of decreasing the length of a failure plane over which cohesive resistance can be mobilised; where cracks are filled with water a hydrostatic pressure (V) may be exerted

BY THE SWEDISH METHOD OF SLICES

From field measurement angle of slope: $\beta = 34°$
From shear strength testing: $c' = 45$ kN/m^2,
  $\phi' = 27°$
Soil unit weight at natural moisture content:
  $\gamma = 20$ kN/m^3
Unit weight of water: $\gamma_w = 9.81$ kN/m^3
From scale drawing, and assuming a water table at
  the ground surface:

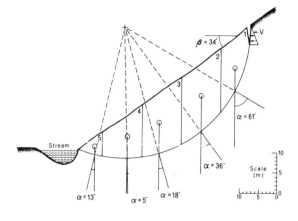

Slice	Area (m^2)	h(m)	$\alpha°$	l(m)
1	82	13.6	61	17.5
2	154	16.0	36	13.1
3	151	15.0	18	10.9
4	97	11.2	5	10.0
5	18	4.3	13	6.1

Slice	c'	l	$c'l$	W	$\cos\alpha$	$W\cos\alpha$	u	ul	$\tan\phi'$	$c'l + (W\cos\alpha - ul)\tan\phi'$	$\sin\alpha$	$W\sin\alpha$
1	45	17.5	787	1640	0.48	787	92	1610	0.51	367	0.87	1427
2	45	13.1	589	3080	0.81	2495	108	1415	0.51	1140	0.59	1817
3	45	10.9	490	3020	0.95	2869	101	1101	0.51	1392	0.31	936
4	45	10.0	450	1940	0.99	1921	76	760	0.51	1042	0.09	175
5	45	6.1	274	360	0.97	349	29	177	0.51	362	0.22	− 79
										$\Sigma 4303$		$\Sigma 4276$

$$F = \frac{c'l + (W\cos\alpha - ul)\tan\phi'}{W\sin\alpha} = \frac{4303}{4276} = 1.00$$

(By the Simplified Bishop Method $F = 0.99$)

Consequently the slope is in equilibrium with the water table at the surface.

BUT if the tension crack is filled with water to a depth of 4.5 m
  $V = \frac{1}{2}\gamma_w h^2 = 99$ kN/m^2

Then: $F = \dfrac{4303}{4276 + 99} = 0.98$, hence the slope is liable to failure.

[Notes: $W =$ Area $\times$ 1 m $\times \gamma$ (assuming a 1 m thick slice).
  $u = \gamma_w h \cos^2\beta$ (because seepage is downslope, parallel to the ground surface (see fig. 6-9(4)).

A negative value is assigned to $W\sin\alpha$ for slice 5 because it resists sliding.
The water table was assumed to be at the ground surface from the field observations that tension cracks were filled and seepage was occurring from the base of the slope.
If the water table is not at the surface a flow net should be used to determine pore water pressure for each slice (see Lambe and Whitman 1979, p. 359).]

Fig. 6.13 A worked example of a post-mortem stability analysis of a rotational failure.

related to the depth of water (h) in the crack then:

$$V = \tfrac{1}{2}\gamma_w h^2$$

V is an addition to the driving forces acting against the slice downslope of the tension crack.

Alternative methods of analysis are often used by engineers for analysing the stability of slopes against rotational failure. Few of these methods, however, are appropriate to natural landslides as they were devised for study of the short-term stability of embankments and other constructions. Some methods assume total stresses are active; others a homogeneity of slope forms and materials which seldom apply in natural slopes.

Charts for investigating the stability of slopes may be useful where it is required to study the effect of variation in one variable, such are pore-water pressure or surface slope angle, upon other variables and upon the factor of safety. Charts for investigating the stability of homogeneous soil slopes, for soils with cohesion and frictional strength, have been published by Taylor (1937), Bishop and Morgenstern (1960), Spencer (1967), Janbu (1954), Hoek and Bray (1977), and Cousins (1978). All charts have limitations which restrict their use. Taylor's charts are based on total stresses and thus do not take into account pore water pressures; Bishop and Morgenstern's charts are based on effective stresses but are for a limited range of slope angles only (11–27°), and no information is given on the location of the critical slide planes; Spencer's charts extend the range of slope angles up to 34° but are for toe circles only, as are Janbu's charts, although the latter, and the charts of Cousins, are of wider applicability than most other charts. The charts of Hoek and Bray are for a number of conditions of water tables and are easy to use.

Culmann wedge analysis

Where streams or waves undercut a soil slope rapidly the type of failure which results may be wedge-shaped, with a failure plane passing through the toe of the slope. The slope will fail when a certain critical height H_c is exceeded (Fig. 6.15). Wedge-shaped failures are most common in materials such as loess, glacial till, volcanic ash, and alluvial silts. Culmann (1866) has shown that

$$H_c = \frac{4c}{\gamma} \cdot \frac{\sin\beta\cos\phi}{[1-\cos(\beta-\phi)]} \cdot$$

As the slope gets steeper the critical height for

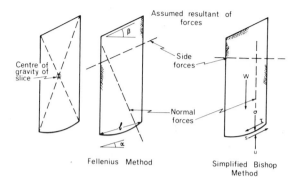

Fig. 6.14 Location of the centre of gravity of a slice, and the resultants of side and normal forces assumed in the Fellenius and Simplified Bishop methods.

stability will decrease, or as a valley is deepened the critical slope angle for stability decreases. At higher angles of surface slope (β) the potential failure plane is inclined at a higher angle (α) so that a smaller depth of slope is necessary to produce a landslide.

It is common for a tension crack to develop behind the slope face and if this extends down to a potential failure plane the critical height is reduced by that depth (z). The theoretical maximum possible depth of a crack is about half of the critical height of the slope, but because the actual depth is related to its position behind the cliff the usual procedure is to solve the equation for H_c and then to determine the critical height of the slope with a crack as $H_c - z$.

Analyses of Culmann wedge failure in loess have been carried out by Lohnes and Handy (1968) and in glacial tills by McGreal (1979). Their analyses show that the heights of stable natural slopes do not exceed the values predicted in a Culmann analysis.

Limitations of stability analyses

Stability analyses are two-dimensional studies of three-dimensional problems. They are also concerned with phenomena which may have extremely variable properties which are sometimes difficult to measure. Sampling soils, deciding whether peak or residual strengths are most appropriately used in an analysis, and defining the shear plane are examples of these difficulties. A very common problem is that of determining the nature of pore water fluctuations, as it is commonly observed that slides occur when pore water pressures rise rapidly, and that failure is less frequent under stable high water tables. Consequently the calculated factor of

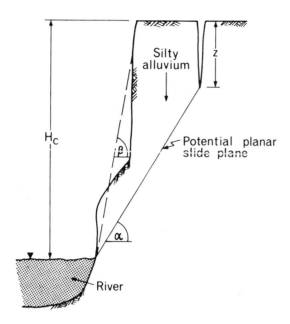

Fig. 6.15 Culmann analysis of a wedge
failure in stiff soil.

BY THE CULMANN METHOD

From shear strength testing: $c = 8.0 \text{ kN/m}^2$, $\phi = 25°$
Soil unit weight at natural moisture content: $\gamma = 22 \text{ kN/m}^3$.
From ground survey: $\beta = 80°$, $z = 0.8$ m.

Then $H_c = \dfrac{32}{22} \cdot \dfrac{0.98 \times 0.91}{(1-0.57)} - 0.8$

$H_c = 2.2$ m

Where $\beta = 90°$ and $\phi = 0$, because $\sin 90° = 1$ and $\cos 0° = 1$
then $H_c = \dfrac{4c}{\gamma}$

By substituting in the Culmann equation it can also be shown that
the critical height for stability will increase as the slope angle
(β) decreases.

safety may be within only ±10 per cent of the
true value.

   Where such uncertainties exist better estimates
of slope stability may be obtained from mapping
the susceptibility of slopes to landsliding than from
an approach based upon a study of the forces
involved in instability. Landslide susceptibility maps
seek to identify potential failure areas by mapping
old landslide features and factors likely to cause
failure in the future — such as slope undercutting,
drainage ponding, weak rocks, and recently de-

forested slopes. The most probable sites for new
landslides are areas where sliding has occurred
already, or areas which are similar to those where
landslides have occurred. Old landslide debris is in a
remoulded state, and hence of lower strength than
the original soil, and is weakened further by weathering. It is also likely to be a site of impeded or disturbed drainage. Modification of such a site by
drains cutting across a toe and removing support can
therefore lead to renewed movement, as can periods
with unusually high water tables.

Factors contributing to landsliding in soils

Many factors influence the development of landslides, and a particular slide can seldom be attributed to a single definite cause although it may be possible to identify a dominant or a triggering effect.

Possible causes and contributing factors are listed in Table 6.8 but, as many of these have been discussed already, mention here is confined to the effects of vegetation, vibration and ground acceleration, soil ice, weathering, and the role of water.

Vegetation

Vegetation change is a very important influence upon slope stability in areas where forests are being removed from hillslopes as part of a programme of agricultural expansion or as part of a regular cycle of cropping a forest before replanting the slopes with trees. Deforestation of slopes has frequently been followed by severe shallow landsliding in areas such as New Zealand, Alaska, British Columbia, the Himalayan foothills, and Japan (Plate 6.8).

At least four effects of trees upon slope stability can be identified (Gray, 1970; O'Loughlin, 1974; Brown and Sheu, 1975).

(1) Wind throwing and root wedging occurs as trees are overthrown by strong winds and under heavy snowfalls. This effect is probably most noticeable on very steep slopes such as the walls of formerly glaciated valleys where soils are very shallow and root penetration into rock is limited. The wind effect on most forests is very small and

Table 6.8 Factors contributing to Mass Movement in Soils

A. Factors contributing to high shear stress

Types	Major mechanisms
1. Removal of lateral support	(i) Stream, water, or glacial erosion (ii) Subaerial weathering, wetting, drying, and frost action (iii) Slope steepness increased by mass movement (iv) Manmade quarries and pits, or removal of toe slopes
2. Overloading by	(i) Weight of rain, snow, talus (ii) Fills, wastepiles, structures
3. Transitory stresses	(i) Earthquakes – ground motions and tilt (ii) Man-made vibrations
4. Removal of underlying support	(i) Undercutting by running water (ii) Subaerial weathering, wetting, drying, and frost action (iii) Subterranean erosion (eluviation of fines or solution of salts) (iv) Mining activities
5. Lateral pressure	(i) Water in interstices (ii) Freezing of water (iii) Swelling by hydration of clay

B. Factors contributing to low shear strength

Types	Major mechanisms
1. Composition and texture	(i) Weak materials such as volcanic tuff and sedimentary clays (ii) Loosely packed materials (iii) Smooth grain shape (iv) Uniform grain sizes
2. Physico-chemical reactions	(i) Cation (base) exchange (ii) Hydration of clay (iii) Drying of clays
3. Effects of porewater	(i) Buoyancy effects (ii) Reduction of capillary tension (iii) Viscous drag of moving water on soil grains
4. Changes in structure	(i) Spontaneous liquefaction
5. Vegetation	(i) Removal of trees (a) reducing normal loads; (b) removing apparent cohesion of tree roots

6.8 Cutover and burned forest five years after clearance.

adds less than 1 kPa to the shearing stresses even when a 90 km/h wind is blowing down the slope (Hsi and Nath, 1970).

(2) Trees have the effect of increasing the surcharge, and hence the shearing stresses, on a slope by $T\sin\beta$ while the normal force is increased by $T\cos\beta$ (where T is the weight of the trees). The effect of the surcharge is thus expressed by:

$$\frac{T\cos\beta\tan\phi'}{T\sin\beta}$$

where, for example, $\phi' = 36°$ and tree weight is 2 kPa (Fig. 6.16a), the trees increase stability on slopes of less than 34°, but where slopes are greater than this angle they may be detrimental to stability if the increase in shearing stress produced by the weight force is not offset by root strength. Large, closely spaced trees on low-angle slopes can increase the normal stress by 5 kPa while increasing the shearing stress by about half that (Bishop and Stevens, 1964).

(3) Mechanical reinforcement of the soil is provided by root networks (Plate 6.9). Some roots grow downwards through the potential failure zone into the underlying soil or rock but in shallow soils most trees have shallow root networks which interlock (Plate 6.8) and provide an apparent cohesion to the soil which usually falls in the range of 1.0 to 12.0 kPa (Table 6.9). Lateral reinforcing effects are particularly important around the perimeter of a potential failure.

(4) Trees modify soil moisture by lowering the water table as a result of transpiration and inter-

6.9 The shallow but laterally spreading root system of a fallen tree.

ception and they thus delay or prevent soil saturation. They also cover the soil with leaf litter and may prevent the soil drying out and cracking in dry seasons. By contrast trees provide conditions for rapid infiltration and interflow along root channels and cracks: the latter mechanism has been identified as promoting landsliding on steep slopes with shallow soils in Brazil (de Ploey and Cruz, 1979).

In studies of landsliding on forested slopes soil shear strength is measured with a large field shear box in which pedological structures and tree roots are retained in the sample. The effect of roots is incorporated as apparent cohesion so the infinite slope equation then becomes:

$$F = \frac{C'_{s+t} + [(\gamma z + T) - (mz\gamma_w)]\cos^2\beta\tan\phi'}{(\gamma z + T)\sin\beta\cos\beta}$$

Table 6.9 Strength added to Soil by Plant Roots

Plant	Soil	Increase in apparent cohesion (kPa)
Conifers (pine, fir)	glacial till	0.9–4.4
Alder	silt loam	2.0–12.0
Birch	silt loam	1.5–9.0
Podocarps	silty gravel	6.0–12.0
Poplar	silt loam	2.0–9.0
Alfalfa (lucerne)	silty clay loam	4.9–9.8
Barley	silty clay loam	1.0–2.5
Clover	silty clay loam	0.1–2.0

Note: in any soil the strengthening effect of roots varies greatly with depth and laterally.
Source: data from unpublished sources and O'Loughlin, 1974 and Waldron, 1977.

where C'_{s+t} is the total cohesion derived from soil effective cohesion plus the apparent cohesion due to tree roots (kPa),

 T is the tree weight (kPa).

Upon deforestation many of the beneficial effects of trees upon slope stability may persist for a few years because the tree roots decay gradually (Fig. 6.16b), with about half the root strength being lost in two to five years. The decay of roots usually leaves many root channels and the soil is then more permeable and the water table can respond more rapidly to storms. Any actions which impede soil drainage, such as diverting runoff from forestry roads into gullies or depositing spoil into depressions, and thereby raising the water table in depressions, then become more likely to cause instability.

The overall effect of trees on hillslopes with shallow soils is to increase soil shear strength by 60 per cent or more. The effects are summarised in Figs. 6.16 and 6.17. These effects are in accordance with the common experience that deforestation is followed by shallow landsliding in many upland areas, but only after an interval of some two to ten years. It is also in accordance with the observation that landsliding is far less common on forested slopes than on adjacent slopes under cultivation or grass (Selby, 1967a, b; Swanston, 1970, Pain, 1971).

Earthquakes

Vibration which triggers landslides most commonly occurs as a result of ground motions produced in an earthquake, although in rare sensitive soils it is suspected that vibrations from traffic may produce shearing stresses which lead to failure. Ground motions have their most dramatic effect upon unconsolidated sediments, which are largely of sand composition, with a high water content.

Sands liquefy after a certain number of cycles of shaking depending upon the shear stress induced by each shaking cycle and the void space of the sand. In the example shown in Fig. 6.18 liquefaction is likely to occur in about 30 cycles under a cyclic shear stress of about 20 kPa. This scale of shear stress variations was exceeded in the Alaskan earthquake of 1964. At Turnagain Heights, near Anchorage, a large complex spreading failure developed in which large houses were rafted 150 to 200 m towards the sea. A graben developed in the slide, thus reducing lateral support for the material inland of it and blocks were progressively incorporated into the crown of the slide. One of the features which promoted the landsliding was the long duration of shaking (4.5 to 7.5 minutes compared with a more usual 1.0 to 1.5 minutes) as well as its intensity.

Under the influence of repeated low-magnitude vibrations, which are not strong enough to induce liquefaction, sands may become more resistant to subsequent earthquakes without significant changes in their density. The stress history is thus significant for liquefaction behaviour of sands (Seed *et al.*, 1977).

Earthquake effects are, of course, far greater in the tectonically active mountains of the world. Large rock landslides will be mentioned in Chapter 7. Smaller translational slides are also commonly triggered by earthquakes on steep slopes with a soil and vegetation cover. Because such landslides can occur in very shallow coarse soils at low-water contents they can develop on very irregular rupture zones.

Examples of earthquake-triggered landslides have been reported from rapidly rising and forested ranges of Papua-New Guinea (Simonett, 1967; Pain and

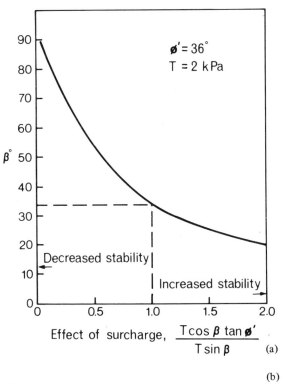

(a)

(b)

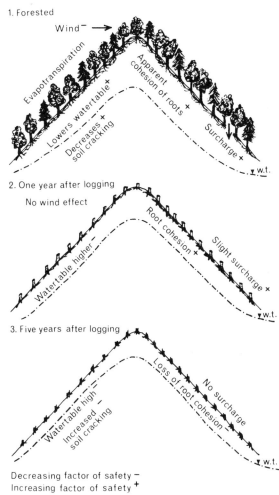

Fig. 6.16 (a) The effect of tree weight on slope stability. On slopes steeper than 34°, for the given soil and tree cover, tree weight decreases stability; (b) The effect of trees upon soil shear strength, and its decline after the forest is cut down. Note that the effect of surcharge may be detrimental on steep slopes.

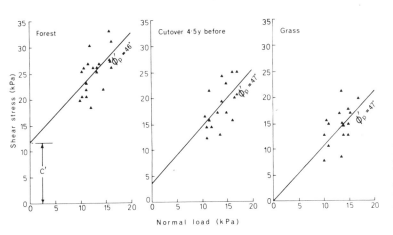

Fig. 6.17 The decline of shear strength of a soil after the forest is cut is caused by a loss of apparent cohesion. The scatter in the data points is caused by the high variability of structure and texture in steepland soils (supplied by D. Parker).

Bowler, 1973; Pain, 1975). The area affected by each earthquake is commonly a few hundred km² and the average downwearing of the landscape 70 to 400 mm. It seems probable that such earthquakes occur about every 200 years so the landscape is being lowered on average 0.35 to 2.0 mm/year by this process alone.

Some areas of frequent earthquakes such as Turkey and southern Italy have landscapes which have evolved largely as a result of landsliding. In Calabria, Italy, a combination of factors have contributed to massive and widespread terrain failures (Cotecchia and Melidoro, 1974). Seismic activity there is among the highest in the world: the rocks are argillaceous marine formations of late Cenozoic age which have been uplifted more than 1000 m during the last two million years. As a result of uplift stream incision is very active with each wave of erosion working up the valleys from the coast. The slopes are thus being steadily undercut. Furthermore the winter rainfall is often intense and prolonged. These factors acting in combination produce severe remoulding of soils and of rocks along fault planes, so that landsliding is frequent but at its peak during winters after each major earthquake.

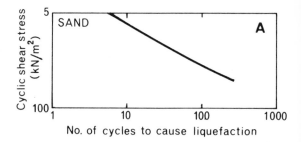

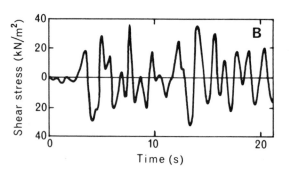

Fig. 6.18 A, Stresses inducing liquefaction of sand under cyclic loading. Liquefaction is likely to occur after 30 cycles when the cyclic shear stress is 20 kN/m²; B, Shear stress variations within a regolith at a depth of 25 m during the Alaskan earthquake, 1964 (after Seed and Wilson, 1969).

Ice

Soil ice has a most important effect upon flows and slides of soils occurring in severe cold climates of the periglacial zone. During the onset of winter, soil drainage becomes blocked by freezing at the surface and as a result the soil water is confined. When spring thaws are rapid the entire weight of the overburden may be transferred to the pore water without any of the water being able to escape. Pore water pressures may then be artesian (i.e. the piezometric surface is above the soil surface) and the soil will flow at very low angles of slope (Morgenstern and Nixon, 1971). The solifluction debris of cold climates is frequently found at angles below 5° which would not normally be attained under landsliding processes. It is believed that many landslides in Western Europe are relict from periglacial conditions and that they account for the very low angle failures in the Lias Clay of central England (Chandler, 1971), and in the Weald Clay where solifluction occurred on slopes as low as 1.5° (Skempton and Weeks, 1976). Under present climates the lowest slope angles at which slides occur in these areas is about 8°.

Weathering

Weathering is obviously of great importance in the long-term conversion of strong rock to weaker soils. It is also of importance in the progressive weakening of marine clays, mudstones, shales, and related argillaceous rocks. Lightly over-consolidated clays usually contract during shearing, but heavily over-consolidated clays dilate. Dilation involves swelling of the clay mass and the extension of small fissures throughout the weathered soil. Fissuring has important implications for long-term changes in slope stability. In both the East Midlands and London area of England the results of weathering of marine clays have been studied for some years.

The Upper Lias Clay of the East Midlands is of Jurassic age and has been heavily over-consolidated in the past by an overburden of at least 1000 m of sediment. Like all over-consolidated clay its particles are closely packed, the clay has low water content and low permeability. As a result of weathering the upper surface of the clay is so fissured that it has the characteristics of a breccia to a depth of 10 m or more. The clay mass consists of clay lumps in a matrix of remoulded clay, and the water content

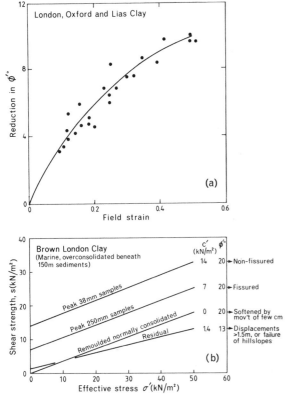

Fig. 6.19 (a) Loss of strength in sedimentary clays related to field strain (after James, 1971); (b) Coulomb plots for direct shear box strength tests on weathered London Clay showing the effect of sample size and degree of remoulding: small samples lack fissures (data from Skempton, 1977).

the failure plane and a reduction in the apparent cohesion. Large field strains (defined as the ratio of the amount of slide movement occurring to the length of the slide plane) may also cause a reduction in ϕ', but most of the loss of strength at the fully softened condition is from loss of apparent cohesion (James, 1971; see Fig. 6.19). Residual shear strengths are reached most commonly in movements along bedding planes and in shallow failures on natural slopes where final failure has been preceded by large displacements along the shear plane.

Similar weathering effects of argillaceous rocks are common on the Great Plains of central USA and Canada. The rocks were over-consolidated beneath about 600 m of sediment, which has been removed since Eocene times, and subsequently suffered rebound, swelling, and fissuring. In early postglacial times meltwater cut deep channels into the weathering rocks and many landslides occurred along the flanks of the channels. More recently water tables have dropped and the slopes have stabilised in many areas, but where water levels are still high translational slides are common and often associated with bentonite-rich beds (Mollard, 1977).

Progressive weakening through weathering implies that recently eroded cliffs, undercut valley sides, and sites of older landslides will retain their steep angles for a period of time and will then fail by landsliding. Skempton (1948) found, for example, that slopes in London Clay, 6 m high and with an inclination of 25°, failed 10-20 years after the removal of excavated soil, and where the slope was 18° after 50 years. There are consequently progressive changes occurring in the angle at which slopes are stable against landsliding, but the adjustments occur only after a critical decrease in strength associated with fissuring and increases in water content. The fully softened condition is reached when there is no further increase in water content.

Water

Water is by far the most important contributor to slope failure. Its operation may be through hydration of clay, undercutting of slopes, weight of rain, as an agent in weathering, as soil ice, in spontaneous liquefaction, but more than these by buoyancy effects, reduction of capillary tension, decrease in aggregation, and by viscous drag on soil grains.

Decreases in shear strength as the water content of a soil increases towards saturation can reach 30 per cent or more (Fig. 6.20a) although 20 per cent

has risen to about the plastic limit. At failure the weathered material is well below peak strength and close to that of 'fully softened' clay with the same strength values as remoulded, normally consolidated soil (Chandler, 1972).

The London Clay is of Eocene age and was mostly deposited in a moderately deep marine environment with an overburden of 150 m of sediment. Many deep landslides occur in this material because of numerous excavations for railways, roads, and canals in the London area. All slides occur in the brown weathered clay and none penetrate the underlying unweathered blue clay to any appreciable depth (Skempton, 1977). Unweathered and unfissured London Clay exhibits peak strength at failure, but progressive slight movements during formation of fissures cause work softening along parts of potential slide planes so that there is a non-uniform mobilisation of shear strength along is a more usual figure for rocks and soils. Most of this change is in the cohesion as capillary tension is reduced but friction angles also may be changed (Fig. 6.20b).

Buoyancy effects from pore water pressures have been discussed already (see Fig. 4.14).

Aggregation of soil particles commonly increases the internal friction of soil. In the rather extreme case quoted by Yee and Harr (1977) soils under forest in western Oregon (USA) had friction angles of 40°. During severe rainstorms the soils were saturated and lost aggregation so that there was a reduction of about 10° in the friction angle. Such a reduction may help to produce instability in wet conditions. This mechanism does, however, apply only to the humic horizons and would promote very shallow translational landsliding only.

One important effect of lateral changes in water pressure results from friction created by molecular attraction between the water and the solid particles

between which it passes. This drag of moving water exerts a force on particles known as seepage pressure (Terzaghi and Peck, 1948). Where the pressure gradient is steep, seepage pressures may become great enough to trigger landslides.

The effectiveness of seepage pressure depends on the particle size of the material on which it acts. In coarse gravel and other fast-draining materials, resistance to flow is slight and large seepage pressures seldom develop. Pressures caused by seepage in clays are also usually minor because of the impermeability of the clay, but materials in the size range of silt and fine sand are most affected by seepage pressures. Rapid large drawdowns in reservoirs, or along river banks or lake shores, can create steep pressure gradients in the ground water and seepage pressures large enough to trigger bank collapses. As the water level falls in a stream or reservoir the friction between the alluvial particles of the bank and the escaping water creates an internal stress which acts towards the free face of the alluvium, at the same time pore pressure in the alluvium is no longer balanced by water pressure in the adjacent water body, and the unbalanced pressure promotes the bank collapse (Plate 6.10).

Ultimate stability against landsliding

Chemical weathering and the production of cohesive soils tend to be at a maximum towards the base of slopes. In cohesive soils the stability of a slope against landsliding is controlled by the height of the slope, the soil strength, and the angle of the slope. Mechanical weathering is of increasing importance at high elevations and many soils on upland slopes are essentially non-cohesive and the maximum angle at which the hillslope is stable against landsliding is equal to the angle of internal friction of the soil.

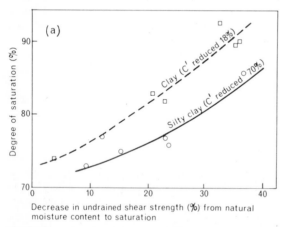

Fig. 6.20 (a) Decrease in effective cohesion as the moisture content of a soil increases (data supplied by N. W. Rogers);

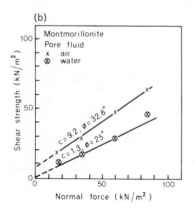

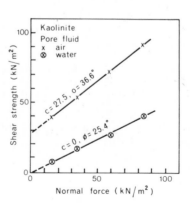

(b) Coulomb plots for consolidated drained shear box strength tests of statically compacted montmorillonite and kaolinite, showing the effect of water on shear strength (data from Sridharan and Venkatappa Rao, 1979).

The reasoning for this conclusion is as follows:

the forces acting parallel the slope $= W\sin\beta$
the forces acting perpendicular to the slope $=$
$$W\cos\beta,$$

where W is the weight of soil above a slide plane and β is the angle of slope, Fig. 6.9.
For conditions of equilibrium stability the forces promoting sliding

$$= \frac{\text{restraining forces}}{F}$$

thus $W\sin\beta = \dfrac{W\cos\beta\tan\phi'}{F} \left[\text{as } \dfrac{\sin\beta}{\cos\beta} = \tan\beta\right]$

then $\quad F = \dfrac{\tan\phi'}{\tan\beta}$

when $\quad F = 1, \tan\beta = \tan\phi'$

then $\quad \beta = \phi'$.

This is true whether the slope is dry or submerged, and it may be of any height, provided that $c' = 0$ and there is no lateral seepage of water.

With soils on slopes, however, with a water table at the ground surface and seepage occurring parallel to the slope, at equilibrium ($m = 1$, $F = 1$) when shearing forces = resisting forces, from the infinite slope analysis:

$$\gamma z\sin\beta\cos\beta = (\gamma - \gamma_w)z\cos^2\beta\tan\phi'$$

which simplifies to:

$$\frac{\sin\beta}{\cos\beta} = \left(\frac{\gamma - \gamma_w}{\gamma}\right)\tan\phi'$$

as $\quad \dfrac{\sin\beta}{\cos\beta} = \tan\beta$

then, $\tan\beta = \left(\dfrac{\gamma - \gamma_w}{\gamma}\right)\tan\phi'$.

As the unit weight (γ) of most granular soils is close to 20 kN/m^3 and the unit weight of water (γ_w) is 9.8 kN/m^3:

$$\tan\beta = \left(\frac{20-9.8}{20}\right)\tan\phi'$$

thus $\tan\beta = \frac{1}{2}\tan\phi'$ (approximately).
Therefore for saturated cohesionless soils with seepage parallel to the slope and the water table at the surface the maximum angle for slope stability approximates to $\frac{1}{2}\phi'$.

6.10 Seepage pressures have caused small failures in the banks of this lake.

Three characteristic maximum hillslope angles for stability have been recognised (Carson, 1976):

(1) a frictional threshold slope angle when the soil is a dry rock rubble and $\beta = \phi$;

(2) a semifrictional threshold slope angle for cohesionless soils when the water table can rise to the surface and seepage downslope is parallel to it, then $\beta \approx \frac{1}{2}\phi'$;

(3) an artesian condition in which the piezometric surface is above the soil surface. This condition has yet to be analysed in enough situations for a general statement to be established, but β will be less than $\frac{1}{2}\phi'$.

It is implicit in this discussion that pore pressures, at their maximum, will be uniform downslope along a potential failure plane. In situations of shallow soils over bedrock of lower permeability this is probably common. In many areas, however, pore pressures will increase downslope from a ridge and this requires that the threshold angle must decrease downslope. This provides one hypothesis for the development of concave basal slopes.

The concept of threshold slopes applies primarily to straight slope segments between upper convexities and lower concavities and it relates only to those slopes which are subject to landsliding. In areas of rapid uplift and deep incision by streams, slope angles may be steeper than the threshold angles until landslides reduce the slope angle to the threshold angle. In high tectonically active mountains landslides are common. Once the rate of uplift is low enough for landsliding processes to adjust slopes to a threshold angle, this angle may be maintained for a while, but will eventually be reduced below the threshold value by creep, wash, and other processes. Threshold angles are thus temporary features of the landscape.

Across any hillslope the characteristic slope angles will vary because of variations in the strength of the soils and the pore water conditions. Faces of spurs may be at frictional angles and hollows containing fine-grained colluvium may have high pore water pressures and thus be close to semi-frictional angles. This may be the reason why many shallow landslides occur inside old landslide scars: for scars are collecting places for water and may be the heads of the drainage network.

Threshold slope angles are extremely varied. Examples in the literature include angles of 6–14° for soils from weathered clays and shales in the semifrictional condition with the possibility of high water tables; 19–28° for semi-frictional sandy soils in upland England and Wyoming, and 33–55° for frictional soils in Colorado and California (e.g. Skempton, 1948; Chandler, 1970; Prior, 1977; Carson, 1976).

Landslides will be able to adjust slopes towards threshold conditions only when trigger mechanisms set off movements — as during storms, earthquakes, or wet seasons. Longer-term changes such as deforestation cause a regrading of the slopes of entire uplands. In many parts of the temperate latitudes slopes may have evolved to angles appropriate to the climates of glacial or periglacial environments during glacial times. In areas like northern New Zealand, where the uplands retained a full forest cover throughout the last glacial, slopes have been subjected to disturbance of equilibria in the period of human settlement and severe mass wasting is a characteristic of most uplands. Such processes are, of course, also promoted by the tectonic instability of the area.

In discussions of threshold slopes it is often assumed that soils are non-cohesive. This is only an approximation as most soils have some cohesion, even if it is only the apparent cohesion provided by plant roots. Measurements of friction angles are also approximate as the collection of large undisturbed soil samples from angular materials is difficult, and few very large shear boxes are in use. With these reservations the concept of threshold slopes does suggest why hillslope angles are related to the strength of rock and soils forming them, and why sets of characteristic slope angles occur in many landscapes.

Preventive and remedial measures

Slope erosion is a natural process which will continue unless man interferes in an extreme manner, such as by covering a slope with concrete. In some situations, as where wave or stream action is undercutting a cliff with buildings on its crest, it may be economic to undertake preventive engineering work: but such solutions are extremely expensive. It is nearly always easier and cheaper to avoid hazardous sites, or to undertake conservation measures before erosion has started, than to heal erosion scars once they have formed.

Induced erosion is usually the result of either vegetation clearance and change, to permit the spread of farming, or the construction of buildings or roads.

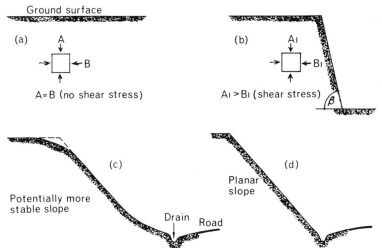

Fig. 6.21 (a) On flat ground stresses within the soil may be equal or unequal but no failure can occur because of the lateral confinement of surrounding soil; (b) On sloping ground, and especially on a steep slope, the stress A_1 is greater than the stress B_1 and shearing failure may result. This could be reduced by: (c) decreasing the angle of slope; (d) by removing soil from above A_1, or by loading the toe of the slope.

Landslides are far less common under forest than under other forms of land use so the most common remedy is to replant severely eroded slopes with trees (Plates 6.11a, b). Where large scale, deep-seated, landslides occur the debris may be gullied and the shear planes may well be below the depth at which tree roots can be effective. In such cases reafforestation may slow erosion rates by modifying infiltration into the soil but it will not control erosion. Deep-seated failures can then only be controlled by diverting slope drainage. This may involve boring horizontal pipe drains into a hillside to reduce pore water pressures, or the diversion of surface water off a slope. Infiltration into a slope may be reduced by compacting soil with machinery, grassing bare areas, and building surface drains. Where buildings and roads or railways are involved it may be economic to load the toe of a failure to increase resisting forces, or to reduce the slope angle by removing soil and thus decrease gravitational stresses.

In some clay soils it is possible to increase soil shear strength by changing the dominant cations adsorbed on the clay particles. Sodium montmorillonite, for example, may have values of $\phi_r = 4°$, compared with calcium montmorillonite with $\phi_r = 10°$. The cation exchange may be accomplished by drilling into the soil to at least the depth of the potential failure plane, and preferably deeper, and then filling the drill holes with a grout of hydrated lime, $Ca(OH)_2$. The calcium will exchange with the sodium on the clay and the sodium will be leached away in the drainage water.

The development of urban areas in hilly terrain frequently involves the cutting of roads or platforms for buildings. Commonly this results in the undercutting of the toe of a slope above the road, and the increase of load on the slope below it if spoil is tipped down the slope (Fig. 6.22). Debris is also frequently deposited in valley floors thus raising the local water tables (Fig. 6.23) and reducing slope stability. The sealing of surfaces, water from roofs,

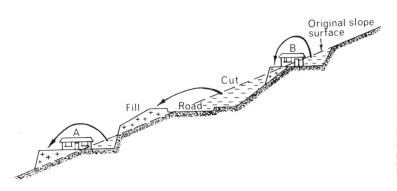

Fig. 6.22 Cutting and filling sites for first a road and then houses on a slope. House site A was formed by cutting into old, poorly designed fill, house site B was formed by cutting into a slope already steepened by cutting for the road (after Taylor *et al.*, 1977).

6.11 In (a) can be seen a deeply gullied complex slide-flow which has just been planted with pine trees. In (b) a similar feature in mudstones is no longer active since the upper slopes have been forested. Note the deep infilling of the valley floor with the debris, Mangatu Forest, New Zealand (Crown copyright. By permission of the New Zealand Forest Service).

and irrigation of lawns can all cause local increases of runoff.

All such modifications can produce potential hazards, especially if slopes are at, or steeper than, a threshold angle for a particular process. Many of these hazards can be reduced by careful geomorphic investigations of potentially unstable areas. This usually involves the mapping of unstable areas, the

mapping of water seepage lines and drainage channels, and consideration of the effect of long-term weathering upon exposed soils and rocks.

Good engineering design will avoid increasing gravitational stresses and decreasing resisting forces on slopes, as well as providing for adequate drainage of surface and soil water. It is a matter of engineering judgement as to what factor of safety is accept-

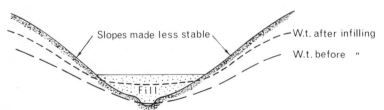

Fig. 6.23 Deposition of fill in a depression causes a rise in the water table at the base of surrounding slopes.

6.11 (b)

able in hazardous areas, and of political and economic judgement as to what preventive measures can be afforded by the community. A review of corrective methods is given by Hutchinson (1977) and by Schuster and Krizek (1978).

7

Processes on rock hillslopes

Bare rock slopes exist for a number of reasons. They may be too high, as a result of uplift or deep incision, for debris to accumulate and bury them; there may be active processes at their bases removing debris so that it cannot accumulate; or they may be too steep or the climate too severe, as a result of cold or aridity, for chemical weathering and vegetation to maintain a regolith. In many environments bare rock faces exist where slope angles are steeper than about 45°, for this is approximately the maximum angle maintained by rock debris, but in the humid tropics weathering and vegetation establishment may be so rapid, on rocks such as mudstones and basalts, that a regolith may form on slopes as steep as 80°. Examples of such slopes occur on Tahiti and in Papua-New Guinea where bare rock may be exposed for only a few years after a landslide.

The form of rock slopes is determined both by the properties of their rocks and by the processes acting to modify them. An arbitrary distinction may be made between those rocks which have such high internal strength that they fail, almost invariably, along joints and fractures and those rocks of lower intact strength or intense fracturing which behave more like soils. The first class of rocks may be spoken of as 'hard' rocks and the second class as 'soft' rocks. In general bare rock slopes are formed on hard rocks, but this is not an invariable condition. Soft rocks such as mudstones and shales are often raised to high elevations by tectonic processes and steep slopes can be retained on them by regular undercutting of the slope. The rate of denudation on such slopes, however, is very rapid compared with that on hard rocks and they acquire a soil and vegetation cover far more rapidly than do hard rocks.

Factors in rock resistance and failure

Hard intact rocks have strengths controlled by their internal cohesive and frictional properties, but few rocks forming hillslopes are intact. Their strength is largely controlled by the size, spacing, and continuity of partings within the rock mass, and their stability by the angle at which partings dip with respect to the hillslope angle (see Chapter 4).

Partings may form for a variety of reasons. In sedimentary rocks, joints form along bedding planes or at detailed structures within the bedding, such as lenses of coarser or finer material, or as desiccation cracks. Cooling joints are common in igneous rocks, and in regional metamorphics, such as schists and slates, partings may open along the planes of oriented mineral grains. In all rocks, joints open as a result of release of tectonic stresses or of overburden pressures.

Natural rock is thus a flawed discontinuous material and its strength is more related to the spacing of the discontinuities than to any other single factor. In most rocks the maximum dimensions of joint blocks are 2-3 metres, although granite, in particular, can have joint blocks up to 1.5 km wide (Twidale, 1976a) and dimensions of tens of metres are not rare. Exposures in deep quarries indicate that joint spacing tends to be greatest at depth and least near the ground surface, suggesting that the opening of joints as overburden stresses are released is a common process.

Joint opening is believed to develop along the microscopic flaws which are present in all brittle solids, and around the tips of which are concentrated rather high local stresses. These flaws, known as Griffith cracks, serve as zones of weakness which are extended by tensile failure and elongation of

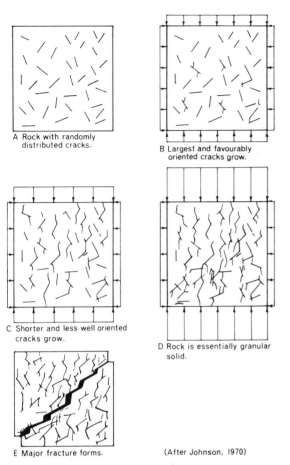

A Rock with randomly distributed cracks.

B Largest and favourably oriented cracks grow.

C Shorter and less well oriented cracks grow.

D Rock is essentially granular solid.

E Major fracture forms.

(After Johnson, 1970)

Fig. 7.1 Stages in the extension of Griffith cracks. The length of arrows is proportional to the stress (after Johnson, 1970).

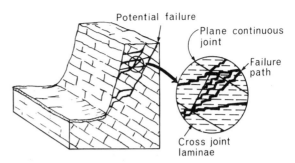

Fig. 7.2 The extension of fractures in a jointed mudstone both with and across the lamination.

their tips until contiguous cracks join up to form a composite failure surface with a stepped form (Fig. 7.1). The cracks in rock samples subjected to compressive stresses are propagated in the direction controlled by the largest stress. In rock outcrops they have a tendency to propagate parallel to the exposed rock face so that slabs of rock are commonly produced (see Johnson, 1970).

In a well-jointed rock cracks may be propagated across intact joint blocks either by following lineations in the rock, such as bedding features or mineral alignments, or by fractures forming to relieve shear and tensile stresses (Fig. 7.2).

Weakening of hard rocks on hillslopes may lead to rock failures occurring by falling, toppling, or sliding of discrete joint blocks and rock fragments, or by the development of landslides. No matter what the scale of the failures the causes can usually

be attributed to geological, climatic, weathering, or human factors.

Geological factors

Nearly all of the world's largest landslides occur in the tectonically and seismically active belts of rising mountain chains formed at the boundaries of the major crustal plates — particularly around the rim of the Pacific Ocean and along the Alpine–Himalayan chains (Kingdom-Ward, 1955). High mountainous terrains, with plateaux and ranges towering above steeply incised river and glacial valleys, have multiple layered, jointed, and fractured rock masses exposed on steep valley walls, ground water pressure fluctuations, severe physical weathering, and steeply dipping discontinuities.

Tectonic activity not only increases the relief but growing folds cause increases in hillslope angles, and faulting may produce growing scarps (Fig. 7.3). Many of the largest and most devastating landslides have followed earthquakes which, by adding a horizontal force to ground accelerations, cause fracturing of joint blocks and loss of strength along joints. In Alaska, for example, 24 earthquakes of Richter magnitude 7.0 or greater occurred between 1898 and 1975, causing many thousands of landslides and large rock avalanches (Voight and Pariseau, 1978). The distance at which earthquakes can trigger landslides depends upon many factors. These include the stability of the slopes, the orientation of the earthquake in relation to the slide mass, earthquake magnitude, focal depth, seismic attenuation, and after-shock distribution. Large earthquakes can be effective at a considerable distance. That which triggered the Mount Huascarán rock avalanche of 1970 in Peru was located 125 km from the mountain and had a focal depth of about 54 km.

The total effect of earthquakes may well be underestimated for we commonly note only those

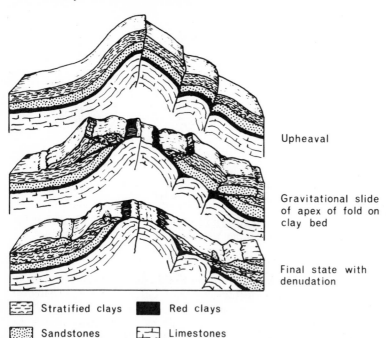

Upheaval

Gravitational slide
of apex of fold on
clay bed

Final state with
denudation

Fig. 7.3 Development of a large
gravitational landslide as a result of
an increase in folding and uplift
(modified from Zaruba and Mencl,
1969).

| | Stratified clays | | Red clays |
| | Sandstones | | Limestones |

which are the final 'trigger' events which produce
failure. The long-term cumulative effect of many
low-magnitude earthquakes may be as important as
the rarer high-magnitude event.

   Swelling of rock and the opening of joints can be
a major cause of failures on rock slopes or surfaces.
Swelling may be caused by three main processes:
(1) the release of locked-in stresses; (2) reduction
of the internal forces which make a rock competent
(Lindner, 1976); and (3) weathering.

   (1) Overburden and tectonic stresses cause compression of the mineral fabric of rock so that intergranular stresses and strains gradually increase over
a very long period and rock masses therefore acquire
a considerable amount of 'locked-in' strain energy.
A substantial proportion of this energy may be lost
in the permanent deformation of mineral grains and
can no longer be recovered, but much of the strain
energy will be released when confining pressures are
reduced by erosion of adjacent rock, or wasting of
overlying glacial ice. The rocks most obviously
affected are those, like laminated shales or closely
bedded sedimentary rocks, which have inherent
planes of parting which can open, but even massive
rocks, with no pre-existing planes of parting, can
develop large-scale joints through stress release, and
the greater the *in situ* stresses the more severe will
be the effect of stress change caused by unloading
(Lajtai and Alison, 1979).

   It was once assumed that the dominant stress
would be that caused by overburden and that the
major unloading joints would be approximately
horizontal, but measurements of stress directions
show that over large areas the ratio of horizontal
to vertical stress is greater than 1.0 and can be as

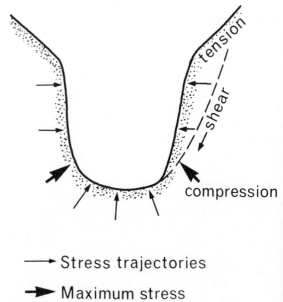

⟶ Stress trajectories

➤ Maximum stress

Fig. 7.4 The distribution and types of stress in rocks of a
recently deglaciated valley.

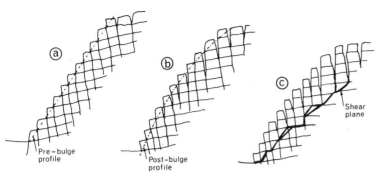

large as 4.0 (Lo and Lee, 1975). In areas of very high horizontal stresses removal of overburden can cause buckling and upward arching of rock, and valley deepening can cause large-scale opening of joints parallel to valley side-slopes.

Valley deepening can occur at a much faster rate than general land surface lowering on plains and plateaux. Consequently, rocks which have been deeply buried can have confining pressures reduced rather rapidly. In a valley the rock at the base of a slope is under the greatest stress and is in compression (Fig. 7.4) (Sturgul and Scheidegger, 1967). Granite may expand its volume by as much as 1.5 per cent upon stress release without fracturing if the release occurs slowly, and the base of a slope may bulge as a consequence (Zaruba and Mencl, 1976). More commonly, valley walls may bulge as a result of joints opening under tension and this may be followed by the linking of tension joints to form a shear plane along which a landslide may form (Fig. 7.5).

Stress release joints often open parallel to the hillslope or ground surface and so may be straight, concave, or convex, although convex joints are probably most common (Plate 7.1). Consequently there is a tendency for the formation of dome-shaped hills or outcrops. These extensive joints can form in any massive rock and are particularly common in granite and sandstone. They normally terminate at intersections with weaker or laminated strata (Fig. 7.6). In addition to preparing rock slabs for spalling these joints permit water to penetrate the rock and so promote weathering processes.

Stress release may occur in some mountain valleys which were formerly glaciated. Retreat of glaciers has been known to permit the sudden fracturing and up-arching of rock on valley floors (e.g. Gage, 1966) and the occurrence of slab-like joint blocks on valley walls is commonly thought to be evidence of stress release joint formation after the removal of laterally supporting ice (Kieslinger, 1960). Persuasive evidence of stress release is seen in some walls of formerly glaciated valleys where opening of joints exactly follows the change of hillslope from a cliff to an extending erosional slope of lesser angle below the cliff (Plate 7.2). Such joint patterns are particularly well developed in granites of formerly glaciated valleys in Antarctica. It has been pointed out by Whalley (1974), however, that tectonic stresses which have not been released by recrystallisation of rock far exceed those which result from an overburden and so it is improbable that all slab jointing or rock failures are attributable to relief of overburden and lateral confinement.

(2) Internal forces holding rock intact may be substantially reduced by the addition of water to rocks which are not chemically altered in this process. The effect is most important in mudstones, shales, weakly cemented sandstones, and micaschists, but even strong igneous rocks may decrease in strength, from the dry to the saturated state, by up

7.1 Convex joints in well-bedded sandstone, Drakensberg, South Africa.

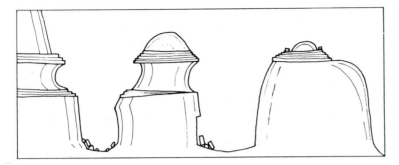

Fig. 7.6 Concave and convex joints developed in massive sandstones (modified from Bradley, 1963).

to 30 per cent (Broch, 1979).

The addition of water reduces the attractive forces holding rock together and increases the internal repulsive forces. It involves cation exchange, hydration, the production of negative electrical force fields, the attraction of mineral surfaces for water, and capillary tensions. Natural cementation is strong enough to resist repulsive forces in many rocks, but those which contain high proportions of clay minerals may be readily disrupted. Atterberg Limit tests on samples of swelling rock of this kind nearly always indicate high liquid and plastic limits and, commonly, the presence of montmorillonite.

(3) A third cause of rock swelling is weathering in which chemical alteration by hydration, oxidation, or carbonation creates by-products which may occupy a larger volume than the original rock. The conversion of sulphides to sulphates by the addition of water is a common cause of swelling, and the slow processes by which olivine-bearing rocks are converted to serpentine are also accompanied by the development of high pressures in the rock mass.

Unfavourable angles of dip of bedding and joints are, perhaps, the most common cause of rock slope weakness. The angle at which a cliff will stand is controlled by the angle of dip of the joints compared with the friction angle between the blocks. Because of the effect of the overburden pressures the friction angle will decrease as the cliff height increases. This is probably because the increase in normal stress with depth is associated with an increase in the number of fractures across rock grains, which reduces the resistance to sliding due to interlocking between grains on opposite sides of a joint, and increases fracture intensities (Roš and Eichinger, 1928). It follows then that stable slope angles decrease as the height of a slope in weathered and jointed rock increases. This relationship is well

7.2 Granites in Antarctica. Major joints trending parallel to the slope surface even where there is a sharp change of slope angle.

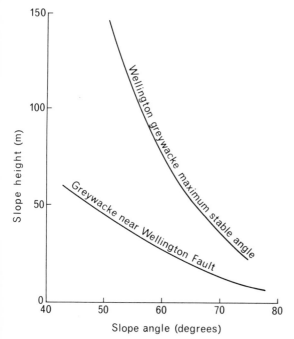

Fig. 7.7 The maximum stable angles of slope in relation to slope height and nearness to the Wellington Fault for greywacke, New Zealand (after Grant-Taylor, 1964).

illustrated for New Zealand greywacke (indurated siliceous sandstones with beds of argillite) in Wellington where the highly sheared and shattered rock within 1 km of the Wellington Fault will support a stable slope lower in angle and height than the less shattered rock farther from the fault (Fig. 7.7).

Stratified sedimentary rocks have bedding planes, which are surfaces of minimum resistance, separating rock layers of various thickness. Bedding planes may become preferred surfaces of shearing because: internal erosion occurs along a sand–clay junction; tectonically induced bedding plane creep or sliding produces a residual strength condition; stress relief by erosion permits shear deformation; shear strength of the rock below a bedding plane may be greater than that of the rock above a plane; or thin seams of weak material, such as clay, may occur at a bedding plane.

An example of the effect of a clay bed upon mass stability occurs in Natal (Sugden *et al.*, 1977) where well-bedded sandstones and shales with a total thickness of 2000 m dip into the sea, at about 15°, from the flanks of a monocline. Along the bedding planes are occasional layers of clay 1–20 mm thick. Very large masses of rock move slowly

above the lenses of clay and produce serious instability problems, especially during wet periods and in areas where the failure planes outcrop. The sedimentary rocks have strength parameters of $\phi = 21°$ and $c = 1.5$ kN/m^2, but the clay has $\phi_r' = 9.5°$ to $12.5°$ and is much weaker, but controls the stability of the whole mass. In this case the clay is at residual strength values because it was probably sheared during folding of the monocline, but a residual strength value has to be assumed in all such cases.

Strata are not continuously intact but are broken by cross-joints which may be regular in their angle of intersection with the bedding planes or nearly random. The stability of a slope in a rock with regularly dipping bedding depends upon the orientation of the bedding planes with respect to the hillslope, because the rock mass will have an effective strength controlled entirely by friction along the bedding planes if the joints are continuous and lacking cohesion.

If the bedding planes are horizontal no landslide can occur and the critical hillslope for long-term stability is theoretically vertical, although in reality weathering processes will reduce this. Some theoretical solutions with respect to slope angle (β), friction angle (ϕ), the dip of the strata (α), and the relative spacing of the bedding joints (L) and cross-joints (W) are given in Fig. 7.8 (modified by Young, 1972, from Terzaghi, 1962).

These theoretical solutions are, however, greatly oversimplified because they ignore other important geological phenomena. The dip of strata and cross-joints together with the width of the joints (assuming that joints are not sealed by impermeable infill) largely determine the rate at which water and weathering processes can penetrate the joints. Suitably inclined permeable beds permit water to enter a slope and develop cleft water pressures which promote instability and lower stable hillslope angles (Fig. 7.9). Furthermore joint friction angles are not constant but change as joint roughness is modified by shearing of the asperities.

Joint surface roughness is composed of three components: (1) the basic friction angle of the rock material (ϕ); (2) the average angle i of the large asperities along the joint with a plane through the base of the asperities (Fig. 7.10); and (3) the average angle of the small, or second-order, asperities. The second-order asperities are of minor significance because they contribute to strength only at very low normal stresses which are induced by overburdens

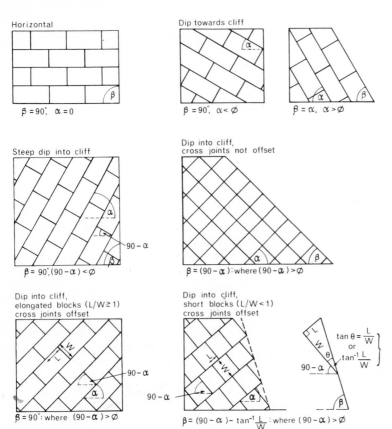

Fig. 7.8 The theory of critical angles in hard, bedded, and jointed rocks (modified by Young, 1972, after Terzaghi, 1962).

a few metres thick. For a joint surface with no cohesion the shear strength is thus represented by:

$$s = \sigma\tan(\phi+i) = \sigma\tan\phi_j$$

It follows that the rougher the joint the greater is the value of i and the greater is the frictional resistance to shearing along the discontinuity. A common value of i in exposures on rock slopes is 3°–5°. At very high normal stresses even the large asperities will be crushed or sheared through and values of i will decline so that the residual angle of friction of the joint (ϕ_{jr}) decreases to the basic

friction angle of the rock (Barton, 1973).

Investigations of over 300 slopes in the Rocky Mountains, by Patton (1966), showed that slopes are generally stable where the dip of the discontinuities ($\alpha°$) is less than the residual angle of sliding resistance (ϕ_{jr}) and that slopes are seldom unbuttressed where $\alpha > 45$–50° (that is, α is much greater than ϕ_{jr}). Potentially unstable slopes are those for which values of i are high (i.e. ϕ_j is much greater than ϕ) as shearing through asperities can lead to failure.

On many rock slopes the toe is removed by

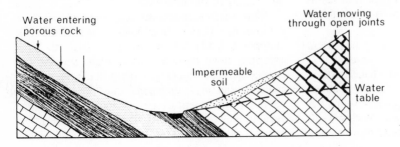

Fig. 7.9. The effect of bedding, jointing, and soil cover on the percolation of water into rock masses.

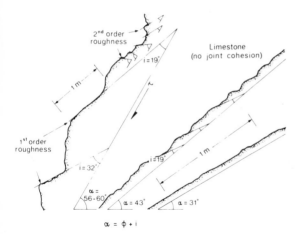

Fig. 7.10 The measurement of *i* angles along joints.

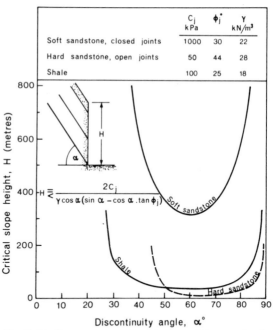

Fig. 7.11 Curves representing solutions of the equation for stable cliff heights for rock slopes with bedding dipping out of the slope.

erosion. The stability of the slope is then largely controlled by the depth of the cut. The height of a stable cliff is controlled by the cohesive and frictional strength along the bedding planes, the dip of those bedding planes out of the cliff and the unit weight of the rock (Fig. 7.11). The force, per unit area of the bedding plane, which tends to produce sliding along the discontinuity through the base of the cliff is:

$$\gamma H \cos\alpha \sin\alpha.$$

The force resisting this disturbing force is:

$$c_j + \gamma H \cos^2\alpha \tan\phi_j.$$

The cliff will be stable if resisting forces are equal to, or greater than, the disturbing forces:

$$H \leqslant \frac{2c_j}{\gamma\cos\alpha(\sin\alpha-\cos\alpha\tan\phi_j)}.$$

Thus the potential height of a stable cliff becomes very large as α approaches 90° (unless strata fail by buckling) or α is less than the value of ϕ_j. (It should be noted that this simple analysis ignores cleft water pressure.) Any increase of H beyond a critical height would be followed by a rock slide with its foot at the base of the cliff (Terzaghi, 1962). The importance of cohesion across the bedding joints is indicated in Fig. 7.11, where it is shown that a weak sandstone with strong cementation across the bedding planes is capable of supporting vertical slopes over 300 m high even when the bedding joints dip at 60° out of the slope, but a hard sandstone and a shale with the same dip but with little cohesion across the joints, can only support vertical

slopes 10-20 m high: as the dip decreases so the stable vertical cliff height increases. Nearly vertical bedding is also theoretically stable, although this theoretical situation is usually modified by bulging or buckling of the strata.

Displacements along a shear plane can cause very large decreases in shear strength. The shear strength of intact rock is commonly 10 to 200 times that of soil, yet the residual strength after large displacements is commonly the same for a smooth joint surface in rock as for soil. The loss of strength on displacement is thus many times greater for rock than for soil, and consequently cumulative minor displacements along discontinuities can lead to unexpected catastrophic failures. Fig. 7.12 is a cross-section of a rock slope with an irregular joint plane with unfavourable dip and an uncemented fault plane. The shear strength-displacement figure for the joint indicates that initially the strength along the joint, with small displacements only, far exceeds the shear stresses so that the slope is stable. Only when considerable displacement has occurred will shear strength fall to a level where failure can occur. If, however, the valley floor is lowered by erosion, failure can occur along the fault without prior small displacements. These examples assume no cleft water pressures.

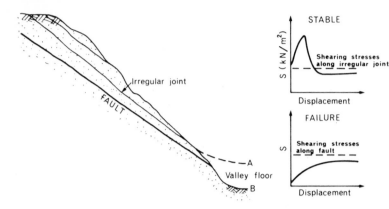

Fig. 7.12 The significance of pre-existing displacements, along an irregular joint, for joint strength (modified from Patton and Deere, 1971).

Displacements may have unexpected effects in some mudstones. Many clay-rich rocks have coatings of iron oxides, silicic acid gel, or manganese around joint blocks (Claridge, 1960; Nankano, 1967; Weaver, 1978). These coatings prevent water readily penetrating the blocks but when mudstones and shales undergo creep, or other slow or limited displacements, the blocks may fracture, be ground against each other and slickensided, and the coatings broken. Water may then penetrate the rock mass. In some mudstones the water is adsorbed, and cannot be removed by drainage, and causes rock swelling. The most severe displacements may occur near the ground surface where alternate wetting and desiccation causes deep cracking, slaking of the rock, and major losses of strength. Many of the severe landslides in Tertiary mudstones in Japan and northeastern New Zealand have been attributed to such a mechanism.

The mode of slope failure is greatly influenced by both inclination and intensity of jointing (Fig. 7.13). Failures by toe bulging may occur where cross joints permit thrusting upwards and outwards to release stresses, even though the dominant joint set is at a potentially stable angle (Fig. 7.13a). The shape of failure planes is controlled not only by the dip of joints but also the strength across the strata compared with strength along the dominant joints (Fig. 7.13b, c). Where joints are very closely spaced, but dipping very steeply into the slope, failure may be by rotational toppling of individual blocks, or, if jointing is sufficiently intense, the whole rock mass may behave as a soil and fail by deep rotational slumping (Fig. 7.13d, e).

Geological structures favourable and unfavourable to large-scale slope failure are difficult to classify because of their extreme diversity and the presence of many local conditions which reduce the validity of broad generalisations.

Certain structures which commonly inhibit the development of large gravitational failures and therefore ensure the predominance of superficial creep, fall, wash, solutional and fluvial channel processes, include the following: (1) rock complexes of lower strength overlying those of greater strength such as occur in the normal stratigraphic sequence with older, more deeply buried and indurated, rocks underlying younger less indurated rocks; (2) very thick rock complexes that have approximately the same strength characteristics throughout, such as large intrusive bodies like granites, or thick strata of limestones, dolomite, sandstones, and greywacke; and (3) rock complexes only slightly affected by tectonic folding, faulting, or tilting.

Three general geological-tectonic conditions favourable for large-scale slope failures have been identified by Nemčok (1977).

(1) Rock complexes with high strength overlying weaker rock units are subject to undermining or failure along deep slide planes. The stronger rocks may be rigid and of constant volume, have high shear strength, be resistant to weathering, be impermeable, and capable of maintaining steep slope faces. However, where they are underlain by soft, plastic, expandible, low strength, impermeable, easily weathered, crushed, or highly fissured beds the upper rocks may fail along very thin weak rock units, even when these have low angles of slope. Interbedded clays, silts, tuffs, fault pug, hydrothermally altered rock, or intruded bentonite layers provide examples of such weak units. If the overlying rocks are permeable or sufficiently porous to maintain a reservoir of water above the weak unit, failure along the toe of a slope can occur as earth-

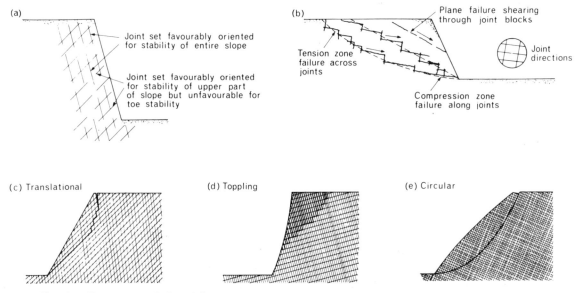

Fig. 7.13 Joint effects upon rock slope failures.

flows rafting blocks of coherent strong rock. A very large example of such a failure has occurred on Samar Island in the Philippines (Wolfe, 1977). Here a limestone block with an area of 18 × 25 km, and a vertical relief of 300 m or more, is underlain by argillised tuff which has been saturated by water percolating through the limestone. The whole block has glided about 5 km on a slope of only 0.6° during the last 10 000 years.

(2) Alternate strata, having higher and lower mass strength characteristics, are particularly prone to failure because of deformation, weathering, and water pressures along the weaker beds. Flysch complexes and their metamorphic equivalents characteristically have such variability.

(3) Rock units severely deformed by tectonism not only undergo progressive steepening of their major joint systems, but also seismic disturbance, the opening of joints, intensification of joints and formation of shatter zones, and the deepening of valleys where landmasses are rising.

Climatic factors

Climate influences slopes in hard rock mainly through the water pressure within the slope. In many high alpine areas cliff faces are frozen for much of the winter; in the European Alps the wave of winter cold may penetrate 10 to 20 m into the rock (Whalley, 1974). This freezing of the face raises water tables inside the cliff as well as causing freeze-thaw action on the face. In the succeeding spring and early summer the water table will be lowered causing high seepage pressures acting towards the cliff face and thus reducing the strength within joints (Fig. 7.14). Heavy rainstorms can raise water tables in well-jointed rocks. Seepage towards a cliff face may also be caused by rapid deglaciation of a valley during which water tables in valley slopes will be lowered. The vertical height over which tension cracks may open is increased by these processes and this encourages shear planes to form.

Seasonal concentrations in the frequency of rockfalls, with maxima in the spring snow-melt period and the autumn (which is the wettest season) have long been recognised in the Norwegian mountains (Bjerrum and Jørstad, 1957). The very cold period of the seventeenth and eighteenth centuries, known as the Little Ice Age, was also a period of marked increase in rockfalls in European Mountains (Grove, 1972).

Ground water conditions depend upon the spacing and openness of the joints. Where there are closely spaced, intersecting, open joints the water pressure within the rock mass can be treated in the same way as that used for soil slopes. The shearing resistance (s) along the joint is determined by the normal force and coefficient of friction between the joint walls:

$$s = \sigma \tan \phi_j = \gamma z \tan \phi_j$$

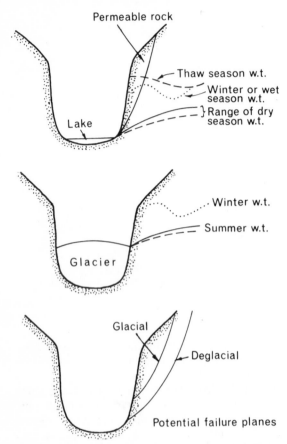

Fig. 7.14 The height of seasonal water tables in rocks of a valley without and with a glacier, and the location of shear planes in the rock mass. Cleft water pressures in the dry season are too low to affect stability, but in the wet season open joints may permit pressures to rise and failures to occur (modified from Terzaghi, 1962).

but cleft water pressures reduce this to:

$$s = (\gamma - \gamma_w) z \tan\phi_j$$

where γ and γ_w are the unit weights of rock and water respectively, and z is the thickness of the rock mass above the shear plane. As rock is commonly three or more times as dense as water, rock mass strength may be reduced by 30 per cent where clefts are filled with water to the ground surface.

Where the joints are widely spaced, and of differing width and continuity, very variable distributions of water pressure may result and consequently the shearing forces also vary from one joint to the next. This is illustrated in Fig. 7.15 where the water level is much lower in joint b than in a. As a result the magnitude of the stress Va due to the hydrostatic water pressure along the joint a is several times the

stress Vb. The figure also indicates the difficulty of obtaining reliable information on joint water pressures without an array of piezometers.

Ground water levels fluctuate much more in rock slopes than in soils because the void space in rocks is much less. Fig. 7.16 shows the effect of 250 mm of rain which entirely infiltrates into a soil and into a slope of nearly non-porous rock. Where soil porosities vary from 33 to 10 per cent the rainfall may cause a rise in the ground water table of 750 to 2500 mm respectively. In the rock slope the joints may be filled and the rise of the ground water level will depend upon their volume, it may be many metres and produce very high water pressures.

The effect of the openness of joints is illustrated in Fig. 7.17 where open joints drain readily and produce low water pressures near the cliff face, and hence a relatively high factor of safety against sliding, while a closed bedding joint permits high cleft water pressures and a much lower factor of safety. The high water pressure may induce movement of the outer block and so lower pressures that failure may occur only after a series of small movements. Alternatively small movements along the sub-horizontal bedding joint could reduce the cleft volume, so reducing seepage but increasing water pressure, thus leading to failure.

Weathering factor

Weathering is an important factor in all humid environments where the opening of joints and the

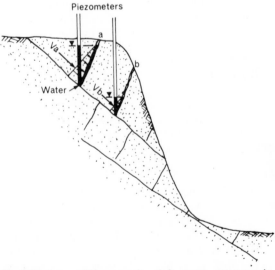

Fig. 7.15 Large differences in cleft water pressures in adjacent joints (modified from Patton and Deere, 1971).

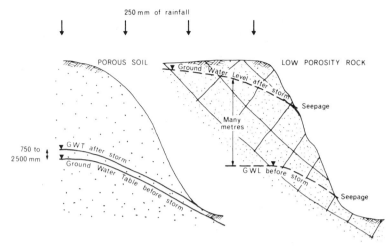

Fig. 7.16 Comparison of ground water rises in soil and rock slopes for the same storm.

formation of clay-rich joint infills can occur (see Chapters 2, 3, and 4).

Human factor

Human interference with the stability of hillslopes in hard rock is still a minor factor compared with geological factors, but locally it can be of great importance. It usually results from the undercutting of slopes or removal of their lateral support during quarrying or excavations for canals, roads, and railways. Its effect is similar to that of natural undercutting by waves or downcutting by rivers and glaciers, but it can also be enhanced by blasting which may open existing joints and cause further rock fracturing.

Types of rock slope failure

The rate at which joints and other fissures open upon a rock face, which is not mantled by debris, controls the rate of change of that face. Hence, it can be said to be weathering-limited. This contrasts with the rate of change of regolith-covered slopes which can only change at a rate controlled by the speed with which regolith materials are removed from the slope: such slopes are transport-limited.

There are many types of failure that occur on weathering-limited slopes. They range from very large features of tectonic-scale dimensions to movements of individual rock particles, with processes ranging from very fast falls, to slower slides and flows and very slow creep. In each group phenomena

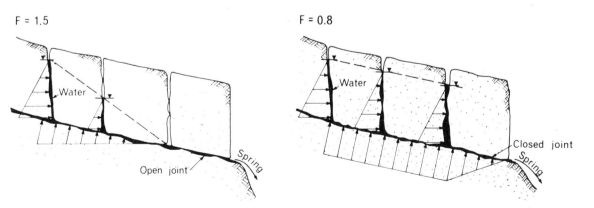

Fig. 7.17 Effects of joint openness on cleft water pressures and slope stability.

can be further subdivided according to processes operating and the morphology of the moving material. The classification adopted here is based upon Nemčok *et al.* (1972) (Fig. 7.18).

Falls

Hillslopes dominated by fall processes are characterised by the production of weathered rock which is removed almost instantaneously under the influence of gravity. The size and shape of the rock fragments varies from whole joint blocks to individual rock grains and this factor is the criterion for distinguishing between rockfalls, slab failures, and rock avalanches.

Rockfalls are relatively small landslides confined to the removal of individual and superficial blocks from a cliff face. Compared with rock avalanches they may be relatively frequent events with several falls occurring each year, or day, on one major face. In some arid areas, however, falls may be very rare and large joint blocks may slowly separate from a cliff face over a period of hundreds or even thousands of years before a series of wet seasons or other events leads to final failure (e.g. Schumm and Chorley, 1964) (Plate 7.3).

Most rockfalls are promoted by hydrofracturing, stress release, the wedging action of tree roots, and other weathering processes. The actual removal of debris may be promoted by sliding snow, heavy rains, or by earthquakes. A common cause of rockfalls is the undercutting of a face by streams or the more rapid weathering of an underlying weak rock such as shale or mudstone to leave an unsupported overhang below the harder rock. The unsupported mass will fail when shearing stresses cause failure along fractures which relieved the tensional stress.

Granular disintegration occurs in weak rocks in which grains are released by weathering. Sandstones with weak cements are particularly prone to this kind of weathering which takes place over entire exposed surfaces. Removal of debris may be by fall or wash processes.

Slab failures are the common form of weathering on steep valley walls in hard rock. The release of lateral confining pressures permits the opening of tension joints which cut across geological structures and closed cooling joints or bedding planes. Slab failures in their initial stages of development are most obvious when they occur at the tops of cliffs where tension joints may be seen extending parallel to the cliff face. Over a period of time tension cracks will extend vertically, lengthening the slab,

and sideways, widening it, until the tensile strength of the rock at the edges of the slab is exceeded and a fall occurs. Extension of the crack is frequently aided by water seepage, especially where this is under a high pressure head, by debris falling into the crack and acting as a wedge, and by ice. Slab failures on cliff faces frequently leave overhanging 'roofs' of rock above the scar from which they have fallen and this may become a site of further slab development (Plate 7.4).

Slab formation is a very slow process and even large vertical joints may exist for centuries before failure occurs. Many valley walls which have been deglaciated for 10 000 years, or more, still show no sign of slab failures developing. Once a failure has occurred a further slab may start to form behind the scar of the first failure. Sandstones, granites, and chalk are rocks which are most commonly reported as being subject to slab failure.

Toppling failures are common where bedding

7.3 Small joint blocks are separating from this cliff in Antarctica and the scar of a recent rockfall can be seen below the overhang.

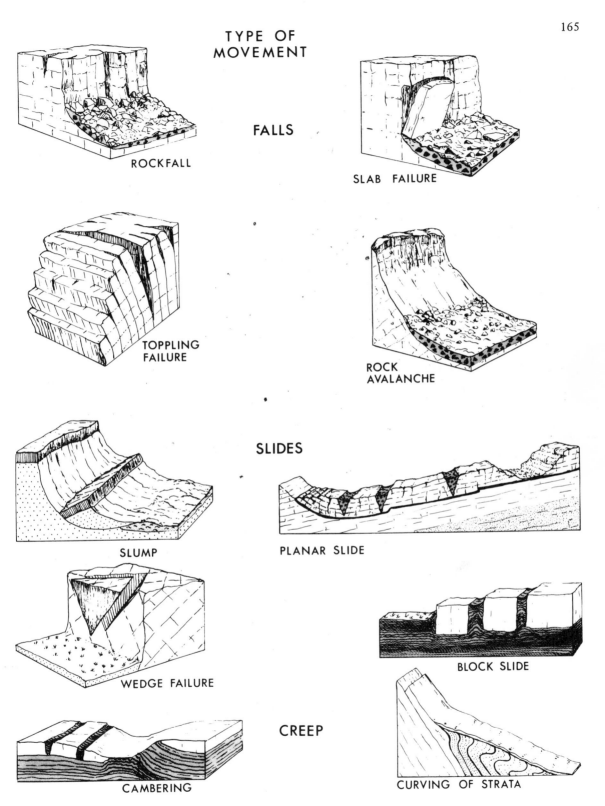

Fig. 7.18 A classification of landslides in rock.

7.4 A cliff face up to 300 m high on Mount Boreas, Antarctica. To the right of centre can be seen the scar of a huge rock-fall which has left an overhang below the high peak and massive debris on the slope below. The largest debris blocks are as big as a medium-size house.

planes and joints are inclined into a hill so that failure takes place along the cross-joints which dip towards the valley. Toppling is particularly common in slates and schists, but also occurs in thinly bedded sedimentary rocks, and in columnar-jointed igneous rocks such as basalt and dolerite. Toppling is a failure mode of slopes which involves overturning of columns. The distinction between toppling and sliding failures is based upon the ratio between the width of the joint block and its height (Fig. 7.19). Wide low blocks slide but tall narrow blocks overturn (topple).

(1) A block is stable where $\alpha < \phi$ and $b/h > \tan\alpha$.

(2) A block will slide but not topple where $\alpha > \phi$ and $b/h > \tan\alpha$.

(3) A block will not slide but will topple where $\alpha < \phi$ and $b/h < \tan\alpha$.

(4) A block will both slide and topple where

$\alpha > \phi$ and $b/h < \tan\alpha$ (Freitas and Watters, 1973; Goodman and Bray, 1976).

Rock avalanches result from the common condition in which well-jointed rock loses internal cohesion by joint propagation and becomes a mass of closely fitting masonry whose strength is entirely derived from friction between the blocks. Friction angles for discontinuous rock vary from about 43–45° for rock in a loose aggregate state, to about 65° (Silvestri, 1961) in a densley packed state with a maximum of about 70° (MacDonald, 1913).

A highly fractured rock will remain stable until weathering, or removal of lateral support, reduces resistance. Water seepage or pressure will reduce the effective angle of friction below the critical value and deepening of a valley will increase the shear stress: ultimately failure will occur. Rock avalanches may be massive catastrophic failures occurring at

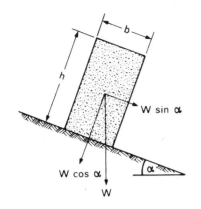

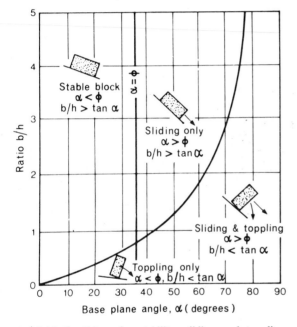

Fig. 7.19 Conditions for stability, sliding, and toppling failure of a block (modified from Hoek and Bray, 1977).

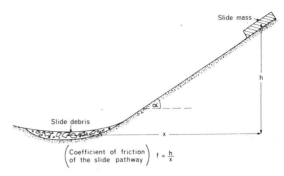

$$\left(\begin{array}{c}\text{Coefficient of friction}\\\text{of the slide pathway}\end{array}\right) \; f = \frac{h}{x}$$

Fig. 7.20 Calculation of the coefficient of friction of a landslide pathway.

long intervals or they may be minor more frequent events (Plate 7.5).

Most rock avalanches start as rockfalls or as sliding rock masses on steep slopes. They achieve high velocities and appear to flow as flexible debris sheets which cover a large area compared with the area of the source. This high efficiency of transport is the most distinctive feature of rock avalanches compared with rockfalls and rock slides. This efficiency is indicated by the ratio of maximum height dropped to the maximum distance travelled; the coefficient of basal friction of a landslide pathway is defined similarly (Fig. 7.20). Very mobile rock avalanches may have efficiencies of 0.11.

The volume of the moving mass commonly exceeds 10 Mm^3 and may be greater than 100 Mm^3. The velocity of the moving mass is commonly in excess of 90 km/h and may reach 350 km/h with individual boulders being propelled through the air at up to 1000 km/h. The material is mostly bedrock, although in some mountains snow and ice may be incorporated in it and water may be acquired from lakes and rivers as the mass travels. There is a general absence of gravity sorting of debris; normally the material of the avalanche is chaotic, and large blocks may either fit closely together or be separated in a matrix of finer material. The debris sheet is commonly thin, being a few tens of metres thick, but it usually has a well-defined distal rim, lateral ridges, and transverse surface patterns of hummocks and ridges. The surface pattern of the debris thus has many features in common with glacial moraines.

The size of features and energy involved in large avalanches may be gauged from a brief description of what is probably the world's largest known landslide – the Saidmarreh in southwestern Iran (Harrison and Falcon, 1937). The slide occurred on the northern flank of Kabir Kuh, an elongate anticlinal ridge of Asmari limestone dipping at about 20° and resting on thin bedded Eocene marl and limestone. A segment of the ridge 15 km long, about 5 km wide, and about 300 m thick, slid off the mountain into an adjacent valley (Plate 7.6). Part of the moving mass had sufficient momentum to rise 600 m above the valley floor, cross the nose of a neighbouring anticlinal ridge, and come to rest in the next valley 20 km distant from its source. The dammed Saidmarreh River formed a lake 40 km long and, on average, 5 km wide. The mass involved was about 20 km^3 and covered an area of 166 km^2 to a maximum depth of over 300 m and an average

7.5 A rock avalanche which briefly dammed the Buller River and destroyed the road, West Coast, New Zealand. (Photo: by permission of N.Z. Aerial Mapping Ltd., Hastings, New Zealand)

thickness of about 130 m. The edges of the deposits are sharp fronts at least 50 m high.

A well-known example of a devastating rock avalanche is that from the east face of Turtle Mountain which destroyed the southern end of the town of Frank, in the Crowsnest Pass of Alberta, in 1903 (McConnell and Brock, 1904). The Frank avalanche is variously estimated at involving about 30 million to 90 million tonnes of fissured limestone which moved as a fluidised mass for a distance of up to 4 km, and travelled 140 m up an opposing slope at speeds calculated as exceeding 160 km/h. Approximately 2.6 km² of the valley floor was covered with debris to an average depth of 20 m.

The crest of Turtle Mountain is anticlinal and the avalanche material came from a dipping limb of the anticline, so it probably took place along bedding surfaces. The surface of rupture close to the toe of the avalanche followed a minor thrust above the Turtle Mountain Fault. The base of the mountain is formed in compressible and impervious shales, and underground coal-mining at the toe of the slope may have added to instability (Fig. 7.21).

Day and night temperatures preceding the avalanche were respectively about +21 °C and −17 °C, indicating that regular freeze–thaw processes were occurring and failure may also have been triggered by earthquake activity during the preceding months.

A recent devastating rock avalanche is that which fell from the crest of Mount Huascarán (6654 m) on 31 May 1970, and killed 18 000 to 20 000 people when it obliterated the towns of Ranrahirca and Yungay (Fig. 7.22). Mount Huascarán is the highest mountain in the Peruvian Andes. The upper part of its peak, which failed, is composed of granodiorite and the avalanche debris included a mass of glacial ice and glacial till which it picked up at the base of

7.6 Scar and debris of the Saidmarreh landslide, Zagros Mountains. The slide plane follows a bedding plane and the hummocky debris covers the valley floor. Some of the debris which rose up the opposing slope can be seen on the right of the anticline in the foreground. (Photo: by permission of Aerofilms.)

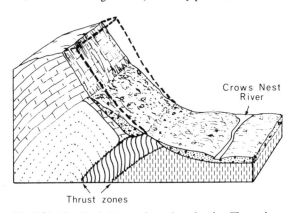

Fig. 7.21 The Turtle Mountain rock avalanche. The geology is from a study by Cruden and Krahn (1973). Shear tests on bedding planes in blocks from the slide debris gave for peak strength, $c = 220$ kN/m², $\phi = 32°$; for residual strength, $c_r = 124$ kN/m², $\phi_r = 16°$.

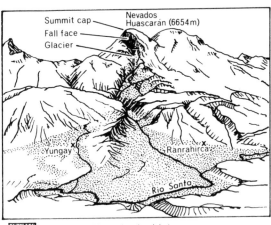

Fig. 7.22 Sketch of the Huascarán rock avalanche, 1970 (modified from Plafker and Ericksen, 1978).

the cliff (Browning, 1973).

The avalanche was triggered by an earthquake which occurred off the coast about 125 km from Mount Huascarán. The same earthquake triggered many thousands of small shallow landslides in the mountains.

The cliff face has an angle of 70-90°: the rock mass of 50 to 100 Mm^3 fell for approximately the first 600 m and then descended a further 2700 m along a valley with a slope of about 23° for 14.5 km. This part of the journey took three minutes and was accomplished at average velocities of around 280 km/h. The avalanche had a low coefficient of friction of 0.22, and although it left virtually no debris in the valley it must have moved as a flow of at least 100 m thickness. Air escaping from the debris carried dust and mud high up the valley walls and boulders with a mass of several tonnes were hurled as much as 4 km through the air. The valley down which the avalanche moved was incapable of containing all the materials and, at a narrow bend, the debris split with one part travelling up and over a 200 m high slope before plunging down on the town of Yungay.

From Yungay the avalanche continued down valley for a further 50 km at an average velocity of 25 km/h. This second part of the movement was presumably a wet flow mobilised by water picked up from the river valley and from melting snow and ice.

The material which covered the two towns where the avalanche lost its velocity is composed of un-sorted clay and boulders. Many boulders have a mass of 700 tonnes and a few exceed 14 000 tonnes. Some sorting occurred down valley in the area of flow with the largest boulders being dropped highest up the valley.

The Huascarán rock avalanche appears to have been an event with a greater height of fall, velocity, and probable volume than any avalanche known to have occurred in historic times. Nevertheless there is clear geological evidence that a considerably larger avalanche from the north peak of Huascarán devastated the same area some time before the arrival of the Spaniards (Ericksen *et al.*, 1970).

The extreme efficiency of transport in rock avalanches has been explained in a number of ways. The first fundamental paper on the subject was published by Heim (1882) who reported the Elm rock avalanche in Switzerland (see Hsü, 1978). Heim recognised that the debris flowed, that sliding was of little or no importance in the transport, that

the large distance of travel was a result of low internal friction, and that the reduction of internal friction is velocity dependent. Thus the principles of fluid mechanics can be applied to the study of rock avalanches and the velocity of the slide is proportional, among other things, to the square root of the thickness of the debris wave. Velocities will, therefore, be highest in confined valleys and greatly reduced as the debris spreads out.

The idea of fluidisation of debris was taken up by Kent (1966) who postulated that trapped air is the fluid medium, and by Shreve (1968) who suggested that the moving mass is carried on a cushion of air. Both of these ideas appeared to be supported by observations of a blast of displaced air which precedes the debris mass. A related hypothesis is that of Habib (1975) who suggested that the rate of shearing at the base of the moving debris is high enough to generate sufficient heat to vapourise water so that the debris flowed on a cushion of steam. Alternative, but certainly not universally applicable hypotheses, include the involvement of lubrication by underlying weak layers, or a layer of rolling debris (Johnson, 1965).

Perhaps the most important recent observation is that rock avalanches occur on the moon where neither fluidisation by gas nor by water can occur. The original theory of Heim has thus been restated and applied to fluidised rock masses, which Hsü called 'sturzstrom'. Heim noted that an individual block travels in a zig-zag bounding path through elastic impacts with its surroundings. A large aggregate of small blocks behaves quite differently because each block is confined to bouncing back and forth between its neighbours, and only the outer ones may fly away. Thus kinetic energy is exchanged between particles by elastic collisions, and the same energy keeps the particles separated during countless elastic contacts. The mass therefore behaves as a fluid with very low internal friction.

The presence of fine particles increases the frequency of collisions, and also the dispersion of large blocks which may then pass one another. Rock avalanches thus have high fluid viscosities when the mass is thick (perhaps around 4 MPa/s at a thickness of two or three metres) but as little as 100 kPa/s when the flow is less than this in thickness. Such high fluid viscosities prevent internal turbulent mixing and internal deformation, so the mass moves as a thin flexible sheet of material with plastic behaviour and with shearing at the base of the debris.

Slides

Slides in rock or soil are characterised by movement above a sharply defined shear plane. In rocks such as slate, schist, and many sedimentary formations the shear plane follows a structural plane within the rock – such as a plane of foliation or bedding – and it is often straight. The resulting failure is then a planar slide when movement is rapid and deformation of the rock mass occurs, or a rock glide when movement is slower and the hard rock mass is rafted on soft clay beds. Well-jointed blocks of hard rock may be tilted and deep fissures opened between them but the gliding blocks remain intact.

In clay-rich rock, such as mudstone, planes of failure are frequently curved and the resulting rotational movement is called a slump. Such failures are most frequently considered with soil failures but because they can occur beneath overlying hard rocks, and can cause tensional and shear failures in those rocks, slumps have to be recognised also as rock failures. Undercutting of basal slopes in clay by rivers, or the sea, frequently promotes slow but continuing slumping (Fig. 7.23).

Rock slides may be very large and catastrophic in mountain regions where the large available relief permits accelerations of rock debris to velocities as great as those of rock falls and rock avalanches. An example is the huge landslide near Flims, Switzerland, which probably occurred during the last glacial (Heim, 1933). This is probably the largest Pleistocene landslide in Europe. It involved marly limestones dipping towards the Rhine Valley at 7-12° and may have been triggered by the retreat of a valley glacier which removed lateral support from the base of the slope. The scarp at the head of the slide is over 1000 m high: the volume of rock involved in the slide was 12 km³ and it blocked the Rhine to form a temporary lake over 15 km long and 400 m deep. The slipped rocks covered an area of about 49 km² and travelled over 150 m up the opposite valley wall.

A study by Cruden (1976) has identified many very large rock slides in the Canadian Rocky mountains. Simple planar slides tend to take place in sedimentary rocks where the bedding dips down slope at about 30°. Limestones are frequently involved in such slides because groundwater can circulate in the caves and open joints produced by limestone solution, and cause cleft water uplift pressures on the failure planes.

A devastating planar slide into the lake impounded by the Vaiont Dam, Italy, occurred on 9 October 1963. The debris largely filled the lake and swept waves of water 100 m high over the dam. This surge of water carried away two high-level bridges in the valley below the dam and was still 70 m high at the confluence with the Piave Valley 1.6 km away. A compressive airblast preceded the flood down the valley and assisted in the destruction and death of 2600 people.

The Vaiont Valley is in the Italian Alps. It was occupied by a valley glacier until about 14 000 years ago, and since then a deep gorge 200-300 m deep has been cut below the glacial valley. The dam is in this gorge. The rocks of the valley are mostly limestones, with frequent clayey interbeds, and a series of clay-rich formations. They have a synclinal form dipping into the valley. The limestones are notably affected by solution. The joint pattern is related not only to the bedding, but to tectonically induced and stress release patterns which extend beneath the floor and walls of the glaciated valley to form a weak zone of highly fractured and layered rock 100-150 m thick (Kiersch, 1965). Furthermore the inner canyon below the glaciated valley was cut so quickly that a second set of stress release joints formed parallel to the canyon walls and they intersect the older set of joints at the top of the walls (Figs. 7.24 and 7.25).

The steep profile of the gorge is conducive to gravitational creep and sliding; the toe of the sedimentary formations is unsupported and these

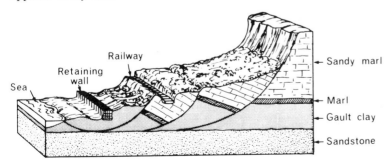

Fig. 7.23 Repeated slumping threatens the railway line near Folkstone, southern England. Marine abrasion has removed the toe of the slump and cylindrical failure planes have developed in the Gault clay which has been weakened by water seeping through the overlying glauconitic sandstones. Corrective measures include an extensive concrete apron to load the toe and protect it from further erosion.

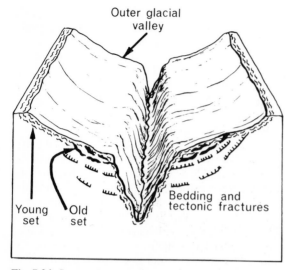

Fig. 7.24 Stress release joints in the postglacial gorge and the formerly glaciated valley of Vaiont (after Kiersch, 1965).

formations are faulted and contain weak clay beds subject to water pressures through the limestone. The geological conditions were thus highly conducive to failure, and the raising of the water table, by the dammed lake, to a critical level was followed by accelerating creep of the slide mass from 1 cm/week in April to 10 cm/day or more by early October. The creep gave warning of the slide and attempts were made to lower the lake level, but these largely failed because of the heavy rainfall and consequent large inflow of water. Creep reportedly reached a rate of 80 cm/day immediately before the slide which suddenly involved a mass of rock 1.8 km long, 1.6 km wide, with a volume of 250 Mm^3, moving across the valley in a minute or less. The most remarkable feature of the slide is that it crossed the 100 m wide gorge and pushed its front 140 m up the opposite side of the valley with the

entire front retaining its structural integrity. The valley and lake were filled by material moving behind the coherent slide front. The slide plane was probably stepped but mostly located along joints within clay-rich strata which had an inclination of 40–45° in the upper part of the slide mass, but were nearly horizontal towards the base.

It has been hypothesised by Müller (1964) that the catastrophic acceleration from a velocity of a few cm/day to at least 8 m/s was caused by a sudden decrease in internal friction of the slide mass so that a quasi-plastic creep behaviour was transformed into that of a viscous fluid. The enormous potential energy of the slide mass was transformed into kinetic energy in a few seconds. Such behaviour is analagous to that of some quick clays and snow avalanches in which particles lose contact almost instantaneously and flow but, as soon as a critically low velocity is reached again, bonding is reformed and the mass recovers frictional strength. Such behaviour in soil masses is said to be thixotropic but had, hitherto, never been recognised in rock masses. The behaviour was therefore unexpected and unpredictable.

The distinction between rock avalanches and rock slides is not always clear. It is based entirely on the presence of a well-defined failure plane, usually along pre-existing joint surfaces, in rock slides with less deformation of the slide mass. If the Vaiont thixotropic-type behaviour is not unique it may be that the viscosities of the moving masses may not be as different as was once thought.

One of the most dangerous features of many large rock avalanches and rock slides is that they may dam even large rivers for a while and then as the river cuts through the debris, the water behind the dam may be released as a catastrophic flood. Such dam bursts are unhappily common in the Andes and Himalayas, especially in the headwaters of the Indus.

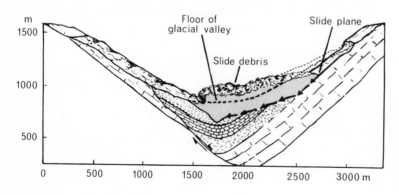

Fig. 7.25 Geological factors influencing the Vaiont rock slide (after Kiersch, 1965).

Wedge failures are usually relatively small features produced by failure along two intersecting joint planes. Such failures are important in the formation of serrated ridges and peaks of high mountains. The wedges may fail as a unit or be subdivided internally by joints radiating from the base of the wedge (Plate 7.7).

Rock mass creep

Creep of rock involves long-term slow deformation in which rock behaves plastically. A zone of creep may extend to depths as great as 300 m below the ground surface, and involve rates of movement ranging from 1 mm/year to 10 m/year (Ter-Stepanian, 1977). It is distinguished from soil creep by its great depth and isolation from daily and seasonal climatic conditions, and from landsliding by the lack of a single clearly defined failure plane and slow rate of deformation.

The existence of rock mass creep may be recognised by the presence of one or more forms of surface disturbance including the exposure of shear planes at rock steps (which may be up to 40 m high), and ridges or furrows, up to 100 m wide and hundreds of metres long, all running across a slope parallel with the crest. The scarps may face up or down the slope, and furrows and crestal sags may disturb drainage and hold shallow ponds. Very plastic deformation may give rise to intense surface rippling, or if it occurs beneath beds of more brittle rocks these may be rafted, tilted, rotated, depressed, or raised in a chaotic manner. The bases of slopes may bulge and cause diversion of streams, or rivers may undercut the creeping slope and cause increases in the velocity of movement.

Some of the geological conditions under which deep creep may occur are illustrated in Fig. 7.26 and detailed accounts are given by Mahr and Nemčok (1977), Mahr (1977), and Radbruch-Hall (1978). Common conditions include: (1) deep-seated bending, folding, and plastic flow of rocks on slopes (Plate 7.8); (2) bulging, spreading, and fracturing of steep-sided ridges in mountains; (3) incremental movement along a dipping rough-surfaced plane, especially where foliated rocks like schists dip towards a valley and continued creep causes progressive decrease of frictional strength towards a residual condition (Huder, 1976); (4) movement distributed over a thick zone in relatively uniform rock, as in mudstones or in schists and gneisses dipping into a mountain; (5) distortion and buckling of dipping interbedded strong and weak rocks, or

7.7 A small wedge failure, with radiating fractures, in sand-stone, Antarctica.

by creeping of rigid over soft rocks without buckling; and (6) valleyward squeezing-out of weak ductile rocks overlain by, or interbedded with, more rigid rocks, causing tensional fracturing and outward movement of the more rigid rocks, sometimes with upward bulging in the centres of valleys.

The last type of movement has occurred in the iron-ore open-cast mines of central England where valleys are cut through Jurassic limestones into underlying Lias clays. The bedding is nearly horizontal but the deformation of the clays, as they bulged towards the valleys, has caused fracturing of the limestone and, either cambering of jointed rock masses towards the valley (Hollingworth *et al.*, 1944), or reverse tilting of the limestones. Deformation of some limestones has opened fissures, called gulls, parallel to the contours. In the Bath area disturbance of the strata occurs to depths of 30–40 m (Chandler *et al.*, 1976). While the deformations can be interpreted as the squeezing out of plastic clays from a loaded zone to an unloaded one, the process may have been aided by saturation of the clay by water in colder climatic conditions and

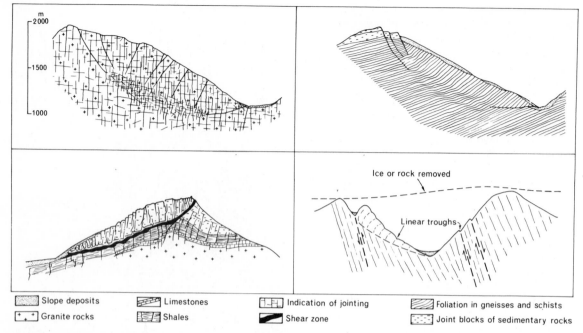

Slope deposits	Limestones	Indication of jointing	Foliation in gneisses and schists
Granite rocks	Shales	Shear zone	Joint blocks of sedimentary rocks

Fig. 7.26 Geological conditions and effects of rock mass creep.

possibly by freeze-thaw effects (Horswill and Horton, 1976).

Water on rock slopes

The importance of water within rocks has already been discussed. Water also, of course, washes over the ground surface, as overland flow, in stream channels, or as ice and snow. The main discussion of these effects is given in Chapter 5 but the importance of water in debris flows and snow or slush avalanches is so great on rock slopes, compared with soil-covered slopes, that these two phenomena are discussed here.

Debris flows

Debris flows are slurries of viscous soil and boulders which can move at high velocities over rough surfaces. They are most common in high mountains and in semi-arid areas which have little or no vegetation. Several conditions are necessary before they can form, of which the following are the most important (a number of these factors have to be active at the same time):

(1) Availability of water from intense rainfall, snowmelt, snow thaw, ground water rise, or volcanic eruptions.

(2) Deposits of unconsolidated soil such as alluvium.

(3) Existence of clay minerals in the soil.

(4) A mechanism for mixing soil and water, e.g. a landslide.

(5) A channel for the debris flow to follow.

(6) A lack of vegetation cover to provide bare erodible soil.

(7) Deposits of transportable weathered or shattered bedrock.

(8) Unstable slopes in the source area.

(9) Undercutting of a slope toe by a stream.

(10) Earthquake or other form of vibration.

Debris flows range in size from a few metres to over 1000 m in width and may be tens of metres thick in places; more commonly they are 1 to 5 m thick. They are often channelled down valleys or gullies and where they emerge into wider areas they become thin and lose water laterally or to underlying porous material, with the result that they become increasingly immobile and finally halt. The loss of velocity and water frequently causes banks (called levees) of angular material to be formed at the margin of the flow and a lobate mound at the terminus (Plate 7.9).

Debris flows have a seemingly random distribution of boulders in a muddy or sandy matrix

7.8 Creep in shales has caused bending of the strata. (Photo: by permission of The Institute of Geological Sciences.)

although there may be some sorting with the largest particles forming the levees or lobes. Particles may be graded during deposition from wide flows which thin and lose their transporting power, with large blocks being dropped first.

The strength of a flow is largely controlled by the clay fraction. Where the clay fraction is small or the water content very high the capacity to carry boulders is reduced, but clay-rich (i.e. greater than 30 per cent clay) flows have a density of 2.1 to 2.4 Mg/m^3 compared with a density of 2.5 to 2.6 Mg/m^3 for their boulders, so the boulders can virtually float and are thus readily transported (Johnson, 1970). A clay-water matrix of lesser density may support large grains where the weight of the large grains is transferred to the water, with a consequent increase in pore water pressure in an undrained condition (Hampton, 1979).

Material in a dense flow apparently moves as a plug of non-turbulent material sliding over a slurry basal layer. As a result there is no shearing within the mass and virtually no abrasion of the boulders. Flows may continue to move over slopes of low angles because their frontal lobes are very steep and thrusting may occur within them.

It is estimated that debris flows occur with a frequency of once in 30 to 100 years in Californian desert canyons. They are more frequent in many mountains, such as the Pamirs, where severe convective summer rainstorms occur at high altitudes, and send floods of debris into the arid valleys, each year.

Snow avalanches

Snow avalanches can entrain boulders, trees, and other debris as well as destroying stands of trees. They also contribute this debris to the accumulations at the base of slopes. Four types can be distinguished: powder, mass, mixed, and slush avalanches (Plate 7.10).

7.9 Channels with levees left by debris flows, Austrian Alps.

7.10 A mixed slab-powder avalanche disintegrating as it falls, Mount Cook, New Zealand Alps.

In powder avalanches most of the snow swirls through the air as a snow cloud. In flowing mass avalanches the snow moves in a turbulent tumbling motion. Most avalanches are of mixed powder and mass varieties. Large blocks and slabs of snow bounce over the ground, breaking up as they do so, and a cloud of snow dust travels above them. Avalanches may run over the snow ground layer or penetrate through the snow pack and run on the ground, entraining rock debris as they do so. Slush avalanches are wet, dense masses of snow which behave rather like small debris flows and they entrain much fine rock material and boulders. They commonly occur in spring melts or during rainstorms (Leaf and Martinelli, 1977; Perla and Martinelli, 1976).

Large avalanches usually occur on slopes of 30–50°, small avalanches on slopes of 50–65°, and minor shedding of small snow accumulations on steeper slopes where accumulation is most difficult. On slopes of less than about 30° snow avalanches

are not common although dry snow can avalanche at angles as low as 25° and wet snow at angles as low as 10° (Mellor, 1978). The optimum conditions for geomorphically significant avalanche snow accumulation are on slopes of 30–50° with an even surface, and hence above the tree line, with well-established tracks, below the accumulation area, covered by a mantle of loose debris (Luckman, 1977).

The pressure (p) exerted on obstacles by snow avalanches depends upon the density of the snow mass (ρ) and its velocity (v):

$$p = \tfrac{1}{2}\rho v^2.$$

Maximum bed shear stresses may range as high as 1 to 10 kN/m^2, but impact stresses can be far higher and the effect of an avalanche is, therefore, potentially greater on obstacles than upon even ground surfaces. Typical impact stresses and velocities are given below:

	Stress (kN/m^2)	Velocity (m/s)
small dust avalanches	1–10	20–40
dust cloud of large avalanches	up to 100	30–70
small, dry, surface avalanches	10–100	10–30
dense, fast, confined surface avalanches	100–600	20–40
extreme cases	up to 1000	up to 125.

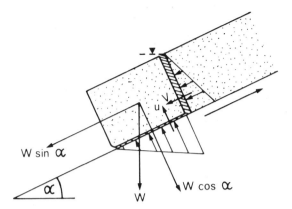

Fig. 7.27 Stresses acting on a joint block on a slope.

The major landforms produced by snow avalanches are both erosional and depositional: gullies and tracks are eroded; talus slopes, boulder tongues, and irregular basal slope deposits indicate the accumulation areas.

Stability analyses

Stability analyses are usually more difficult to carry out, and the results are less reliable, for rock than for soil slopes. Complex geological conditions are the cause of this situation. It is the discontinuities in rock masses which largely control the stability and many features of such masses are difficult to measure. Cleft water pressures are almost impossible to estimate unless a dense array of piezometers is in place and it is usually impossible to determine the cohesion along the joints forming a potential slide plane. As a result it is common practice to undertake a back analysis (i.e. a post-mortem) in which $F = 1$, cleft water pressures are assumed to be at the worst case condition, ϕ_j is determined from shear box tests of the rock (to give ϕ), and i is estimated from joint surveys. A stability equation is then solved to determine possible values of c, and this value is used to estimate the stability of other slopes, in the area of a landslide, which have similar geological conditions.

Where a large rock slide occurs with complex shearing patterns there is seldom much hope of accurately modelling all the parameters. A simple infinite slope analysis is then appropriate. Such an analysis was applied to the largest historical rockslide and debris flow, which occurred in the Peruvian Andes in April 1974. The Mayunmarca slide had an estimated volume of $1000 \, \text{Mm}^3$, a length of the

debris mass of 8 km, a vertical difference of 1.9 km, and occurred on valley slopes in the range of 35 to 9° (Kojan and Hutchinson, 1978).

Deep rotational failures occur in soft rocks, highly shattered bedrock, and in strongly weathered rock. Where there is no control on the slide plane exerted by a discontinuity, then the methods of analysing rotational failures, discussed in Chapter 6, for soil slopes are appropriate.

Planar slides

The simplest case for stability analysis is the condition of a block of rock on a potential slide plane (Fig. 7.27). The block will be on the point of sliding, or in a condition of limiting equilibrium, when the disturbing force acting down the plane is exactly equal to the resisting force on the basal area A of the block:

$$W\sin\alpha = c_jA + W\cos\alpha\tan(\phi+i).$$

If $c = 0$ then the equation simplifies to $\alpha = \phi_j$, where $\phi_j = (\phi+i)$.

The influence of water pressure in a tension crack is indicated in Fig. 7.27 where it is shown that the water pressure increases linearly with depth in the crack and produces a total force V acting on the rear face of the block, and down the inclined shear plane. Assuming that the water seeps along the slide plane and flows out of the base of the block at atmospheric pressure, then the water pressure distribution results in an uplift force u which reduces the normal force acting across the plane at the base of the block.

The condition for limiting equilibrium is then defined by:

$$W\sin\alpha + V = c_jA + (W\cos\alpha-u)\tan\phi_j.$$

Not all planar slides are rectangular in section and tension cracks vary in their location. It is common for a failing block to have a triangular or trapezoidal section (Fig. 7.28).

For sliding to occur on a single plane the plane must strike parallel or nearly parallel (within ±20°) to the slope face; the failure plane must outcrop in the slope face or at the toe; the dip of the failure plane must be greater than the angle of friction of this plane, i.e. $\alpha>\phi_j$; the lateral boundaries of the slide must provide negligible resistance.

Assuming that the tension cracks are vertical and filled with water to a depth z_w, that there is no rotation of the block, and that we are considering a slice of unit thickness then:

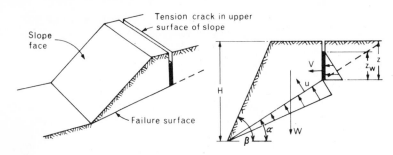

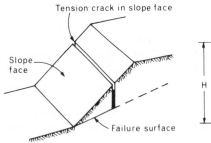

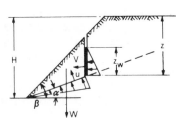

Fig. 7.28 Geometry of a rock slope with a planar slide with tension cracks in upper and lower parts of a sliding wedge.

$$F = \frac{c_j A + (W\cos\alpha - u - V\sin\alpha)\tan\phi_j}{W\sin\alpha + V\cos\alpha}$$

where from Fig. 7.28

$$A = (H-z)\operatorname{cosec}\alpha$$
$$u = \tfrac{1}{2}\gamma_w \cdot z_w (H-z)\operatorname{cosec}\alpha$$
$$V = \tfrac{1}{2}\gamma_w \cdot z_w^2.$$

For a tension crack in the upper slope surface

$$W = \tfrac{1}{2}\gamma H^2 \left[(1-(z/H)^2)\cot\alpha - \cot\beta\right]$$

and for a tension crack in the slope face

$$W = \tfrac{1}{2}\gamma H^2 \left[(1-(z/H)^2 \cot\alpha(\cot\alpha \cdot \tan\beta - 1)\right].$$

Where the location, depth, and water height in a tension crack are known, solution of the equation for a factor of safety is relatively simple, but if a variety of ground water conditions have to be considered then repeatedly solving the equation becomes tedious. Graphical plots, or stability charts, may then be used to solve components of these equations. Stability charts are given by Hoek and Bray (1977) and alternative forms of the analysis for various ground water conditions by Stimpson (1979).

Limitations of stability analyses

It has been stated already that determining some of the parameters used in a stability analysis may be difficult. In mathematical analyses simplifying assumptions are fundamental, but if the assumptions are wrong, or the selection of parameters is wrong, then precise computation is in vain. It has to be accepted that detailed field observations may be more valuable in interpreting or predicting the causes or possibility of landslides in rock. Careful mapping of geological structures and hillslope forms may be a better guide than much study in the laboratory or much calculation. Unusual contour changes and breaks in slope, signs of large-scale creep, the opening of joints, frontal or toe bulges, lateral shearing or dislocation, remnant head scarps, disturbed drainage, anomalous vegetation changes, tilted or displaced structures, or a history of instability are indicators which have to be sought and considered. It also has to be recognised that the larger the potential instability the more difficult it may be to detect. High altitude aerial photographs or satellite photographs may then be the most useful tools with which to start an investigation.

Deposits below rock slopes

The debris which has fallen from a mountain wall or a cliff may be carried away by streams, glaciers, or waves, or it may accumulate to form a talus deposit

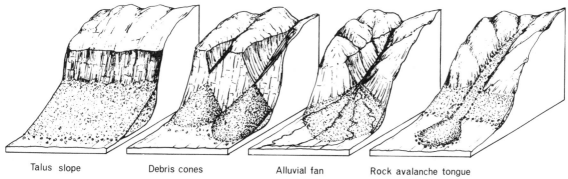

| Talus slope | Debris cones | Alluvial fan | Rock avalanche tongue |

Fig. 7.29 Debris accumulation below rock slopes.

(also known as a scree). Talus may take the form of a sheet of debris, known as a talus slope, where it has accumulated below a cliff which has weathered more or less uniformly along its face and shed its debris on to the slope below; or a talus cone may form below a chute or gully through which descending debris is funnelled (Fig. 7.29). Rare but high energy events such as avalanches can produce long tongues of debris.

Talus deposits may form any combination of slope and cone forms depending upon the nature and distribution of weathering and erosion processes along a cliff face, and upon whether the joint pattern permits the development of chutes. Because streams may also be active on mountain slopes there is no simple distinction between cones formed by rockfalls, landslides, snow avalanches, or debris flows and alluvial fans produced by stream deposition and debris flows. Thus a continuum may exist from steep-angled rock debris cones to lower-angle alluvial fans. Cones, however, are virtually always steeper than about 11° and more commonly have angles steeper than 28° and up to 46°; they may also lodge against a rock slope and be discontinuous with the chute which feeds them. Alluvial fans are nearly always less steep than 11° and are always continuous in long profile with the mountain valley which supplies their deposits.

Talus slopes

Talus forms are so varied that exceptions can be found to most statements made about them. In long profile they may be straight; they may be steep at the top with a straight segment below; but most commonly they are straight in their upper segments and have a concave profile at the base. Slope angles may vary from about 28 to 46° but most talus slopes have angles between about 30 and 38° (Plates

7.11 and 7.12). Steeper upper slopes are usually associated with debris from large rockfalls, especially where the debris has a slab-like shape and cannot roll down the slope. More commonly upper talus slopes are formed of smaller fragments and particle sizes increase down slope. Basal concavities are particularly associated with large debris which has rolled, bounced, or slid over the upper talus. On some slopes there is no evidence of particle sorting.

Relatively smooth talus slopes are usually associated with single-grain accumulations or small rockfalls which cause repeated movements of material, already on the slope, as a result of impacts. Avalanches, debris flows, and landslides are usually less regular in their occurrence and are more likely to produce irregular lobes, small slide scarps and hollows, miniature cones or, in the case of some debris flows, levees.

7.11 A debris cone formed of schist debris with a wide range of particle sizes.

Talus deposits as seen in section show a variety of fabrics and degrees of organisation and bedding of the material. Many of the bedding forms can be associated with particular processes and are good indicators of the significance of those processes. Four main types of fabric are commonly recognised (Fig. 7.30):

(1) An open-work fabric, in which there are few small clasts, is supported on the point contacts between grains and usually results from the fall of individual grains or from small rockfalls. It may have imbricate structure in which elongated particles pack against each other and dip back into the slope, or it may show a majority of particles dipping down the slope.

(2) A partly open-work fabric in which some voids are filled with fine materials.

(3) A closed clast-supported fabric in which all of the voids are filled with fine materials, resulting from the falling and washing of fine materials into an originally open-work fabric and indicative of water washing activity on the talus surface.

(4) A matrix-supported fabric, in which the coarse fragments are totally enclosed within a fine matrix, resulting from deposition by a debris flow rich in mud and other fines.

Talus processes are extremely varied and contribute to both the formation and reshaping of talus slopes; the actual combination at a site is dependent upon the local climate. As many active talus slopes are in alpine, polar, or subpolar regions, freeze–thaw, saturated flow of debris over frozen ground (known as solifluction), and avalanching of snow and slush

7.12 A talus slope with the typically lobate debris left by slush avalanches.

are important contributors to those talus slopes (Gardner, 1979), but more important in the accumulations of arid zones are rockfalls and debris flows during rare rainstorms. Mixing and burial may be involved in downslope movement of surface material, which can occur at rates of up to a metre per year.

Debris from granular disintegration or rockfalls is probably the most universal contributor to talus slopes. Laboratory experiments indicate that individual grains falling on to a talus slope roll, bounce and slide downwards until they reach a zone where particles of similar size to their own form the surface (Kirkby and Statham, 1975). It can be imagined that large particles will readily roll over smaller ones and that small particles will fill gaps between large particles, so it is not surprising that sorting takes place. It must also follow that in a sorting system the smallest particles will dominate the upper slopes because they cannot roll beyond the highest gaps on the slope. The sieving action of a sorted talus slope is also supplemented by the greater kinetic energy of large particles. This can be appreciated from the relationship:

$$e = \tfrac{1}{2}mv^2$$

where e is kinetic energy, m is the mass of the solid, and v is its velocity.

The concavity of the lower part of the talus profile is not a function of the internal friction of the talus materials for the large particles forming the talus would be expected to have a high angle of friction, and be able to support a steep angle of slope. The lower concavity may arise from the probability that some particles will travel more than the average distance downslope to produce a progressively thinning apron round the base. It is also to be expected that debris falling from a high cliff face will have the greatest kinetic energy and should therefore be able to form an extensive thinning 'tail' of large fragments at the lowest part of the talus slope (Statham, 1973). As the cliff face recedes as a result of erosion, the height of free fall is reduced and the extent of the tail should also be reduced so that on slopes from which the cliff face has been nearly eliminated we might expect a straight talus with no basal concavity. This theoretical conclusion is verified by both laboratory studies and by many field observations, although exceptions do occur (Kirkby and Statham, 1975).

Falling particles knock against other talus materials and push them down the slope, but they may have far less influence on particle distributions and slope changes than the processes involving water and snow. Debris flows and snow or slush avalanches carry fine and coarse materials to positions on the slope controlled by the energy of the flow and by its water content. They may thus come to rest at almost any point on the talus slope or below it. 'Wet' processes frequently form lobate or levee deposits where the mobilising water drains away and the flowing mass comes to rest.

As fine-grained particles accumulate in the void space of the talus they add frictional strength to it. The presence of fines, however, and the movement of water within the slope permit pore-water pressures to develop, especially where drainage is impeded by soil ice or lenses of matrix-supported material. The possibility of landsliding within the talus materials thus increases with the age and degree of weathering of the deposit and landslides may contribute to extensions of the basal concavity (Brunner and Scheidegger, 1974). Thus vegetated relict talus slopes formed under colder climates may be subject to landsliding under more temperate environmental conditions (Rouse, 1975).

Slope angles on talus materials is a topic about which there has been much discussion and confusion as a result of the variety of terms used and the different definitions applied to the same term. If grains are poured on to a pile from a low height a cone develops with sides at an angle with the horizontal. If the platform upon which the pile stands is then tilted a steeper angle is attained, up to as much as $10°$ steeper, before failure occurs. The angle at which the failed material comes to rest is usually lower than either of the other angles already mentioned. Unfortunately all three angles have at various times been called the angle of repose.

The difference in the angles of slope of the pile are partly related to the differences between the dynamic friction angles which are relevant to the grains falling on the pile and the static friction angles which are relevant to the pile at rest. The lowest angle achieved by grains after failure has occurred is related to the residual strength of the materials. The angles of slope of a talus are seldom, perhaps never, as steep as the static friction angle achieved by the tilted pile. This static friction angle is the same as, or close to, that measured in a shear box test. The angle of slope achieved by grains in their loose state after a slide is often closely related to the angle of residual shear (ϕ_r) and provided that particles moving over a talus do not have a high kinetic energy this angle will approximate to the angle of the talus slope. Where particles do have high kinetic energy they are likely to come to rest at lower angles than those of ϕ_r.

In summary it is usually true to say that the maximum angle of a tilted pile is related to the static friction angle which approximates to ϕ; the maximum angle of rest of slipped grains is related to ϕ_r; and the angle of slope achieved by grains with

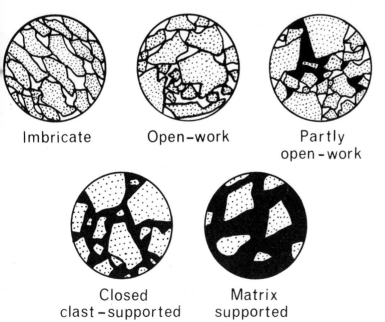

Imbricate   Open-work   Partly open-work

Closed clast-supported   Matrix supported

Fig. 7.30 Fabrics of talus materials.

high energy is related to dynamic friction and is lower than ϕ or ϕ_r. Thus although ϕ for clean sand is about 44°, a pile of loose clean sand will form a talus slope of about 33°.

Talus slopes are seldom composed of materials of a single-grain size and they may be affected by the shape of their particles. In general, massive, densely packed, coarsely grained, angular, or rough fragments form steeper slopes, while loosely packed, rounded, slaty, schistose, fine-grained, or smooth rounded particles form gentler slopes. High pore water pressures may also reduce stable talus angles; for clean sands the angle of rest of saturated material with internal seepage is about half that of dry material. Processes such as slush avalanches or surface wash also have the effect of reducing talus slope angles to below those attributable to the angle of rest or the strength of talus materials alone.

Friction angles ($\phi°$) of some common talus materials are:

Rounded gravels 41–44
Sandy gravel 45
Mixed talus 37
Sandy clay (with 45 per cent sand) 29
Sandy clay (with 35 per cent sand) 38

Alluvial fans

An alluvial fan is a fluvial or debris flow deposit whose surface forms a segment of a cone that radiates downslope from an apex where the depositing stream leaves its upland source area (Plate 7.13). The source area is usually in a mountain and the depositional area is in a valley, a basin, or at the margin of a plain. The long profiles of the stream channel are concave and are continuous from the mountain reaches onto the alluvial fan. Alluvial fan slope angles overlap with those of steeper debris cones and erosional pediments which are cut across the edge of an upland. Pediments nearly always have lower angles of slope than fans and their debris

7.13 An alluvial fan formed within a glaciated valley since the retreat of the valley glacier, New Zealand Alps. An attempt has been made to control the stream channels by the planting of trees.

forms a thin veneer across a cut surface, while fans are thick depositional features (Fig. 7.31).

The lateral coalescence of many alluvial fans along a mountain front is commonly called a bajada in North America. Along bajadas many small fans lose their fan shape where they are restricted by adjacent larger fans. An alluvial lowland that lacks the form of coalescing alluvial fans is best called an alluvial slope.

Fans vary greatly in size from a few metres in length to more than 200 km and the thickness of their deposits is likewise variable. Where deposition is occurring in a sinking basin, such as a graben, fan deposits may be hundreds or even thousands of metres thick. As a general guideline fans may be distinguished from thinner deposits, such as those of pediments, by the thickness of their deposits exceeding 1/100th of the length of the landform from apex to toe.

Alluvial fans have a great diversity of sizes, slopes, types of deposit, and source-area characteristics. They are most commonly seen in arid and semi-arid parts of the world although they do occur also in humid areas. However, their dominant areas of occurrence are tectonically active mountains where the upland source area is being raised by

contrast with the depositional zone. Such tectonic environments permit the deposition of very thick fans. Fans are thus common in mountains of humid regions which have been glaciated and, because a ready supply of debris is necessary for fan formation, they develop in terrains of readily eroded rock such as mudstones or highly fractured hard rocks. Massive resistant rocks seldom provide suitable environments for fan formation because trunk streams in valleys can remove the small quantities of debris carried by tributaries.

Fan morphologies and processes are largely controlled by the stream which transports debris from the eroding source area to the depositional zone. Deposition is caused by decreases in stream depth and water velocity. This may occur at the apex of the fan or, where the stream is entrenched in the apex deposits, some distance down the fan. Here the channel opens from its confined single trench and spreads out on the fan to form numerous distributaries as water infiltrates into permeable fan sediments. The loss of discharge causes streams to drop their sediment load. Inevitably the largest-calibre debris is dropped first and the size of material progressively decreases down the fan. This simple relationship, however, is often disturbed because

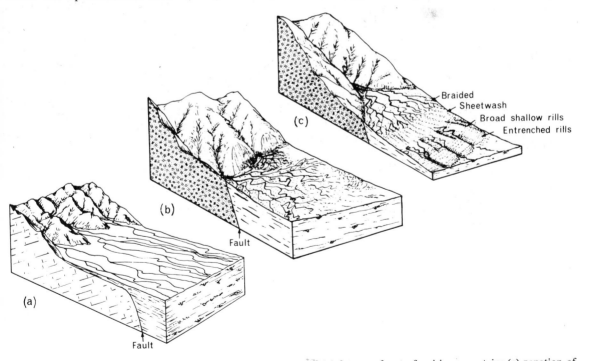

Fig. 7.31 (a) pediment forming at a mountain front; (b) an alluvial fan at a front of a rising mountain; (c) zonation of processes on a fan (based on Packard, 1974).

rare large floods carry coarse material beyond the apex and deposit it among the finer sediments characteristic of the middle and toe sections of the fan.

The deposition of coarse debris near the head of the fan causes the stream to braid, and continuing infiltration may prevent it from maintaining even a braided system, so a channel network may be replaced by sheetflow in the middle part of the fan. Then towards the toe the more impermeable deposits may prevent further infiltration and the sheetflow may become reconcentrated into broad rills which may cut headwards, especially where they receive additional flow from springs. Thus a sequence of types of flow from the apex to the toe may be: entrenched channel → braided channel → sheetflow → rills → headcut channels (Packard, 1974) (Fig. 7.31c).

The location of each type of flow on a fan will change both along the length of the fan in response to the magnitude of the discharge, and across the fan as one part of it receives the flow and builds up until it is higher than segments on either side of it, when discharge will be diverted to a lower segment and build that up. Repeated changes of course thus change the locus of stream activity and are the primary cause of the radiating system of channels and deposits which give the fan its shape. On very large fans segments may be inactive for hundreds of years, during which a soil develops, before being buried by deposits of a renewed phase of deposition.

The proportion of a fan that is flooded by a rare storm flow varies with the size of the source and fan area. Fans larger than 100 km^2 may have only 5 per cent of their area affected by the flood, but smaller fans of 1 km^2 may have more than 20 per cent of their area covered by a flood of the same occurrence interval.

Four main types of deposits are common to alluvial fans, although any one fan may have only one or any combination of these kinds. The most common type is the sediments of the braided channels and distributary channels. The second type consists of the fillings of channels which were temporarily entrenched. The third type has been called sieve deposits (Hooke, 1967) which are formed when the fan surface is so coarse and permeable that infiltration is extremely rapid and a hummocky deposit of coarse detritus is left on the surface. The fourth type is the detritus of debris flows and sheetflows. These may be broad spreads

of fine-grained materials or a range of increasingly coarse material associated with leveed channels terminating in lobes.

The size and materials of alluvial fans have a tendency towards mutual adjustment among a complex set of controlling conditions which include the area, lithology, the relief and vegetation cover of the source area, slope of the stream channel, the water and sediment discharge, the climatic and tectonic environment, the geometry of the mountain front, adjacent fans, and the basin of deposition. Changes in one or more of these variables will tend to cause a readjustment of the fan morphology.

In general fan slopes decrease as fan sizes and drainage basin sizes become larger, as flow discharges become greater and more regular, and as particle sizes become smaller. Thus small fans and those with debris flow, sieve deposits, and coarse detritus tend to be steepest.

Climate is an important variable affecting fans. Periods of accumulation of debris often coincide with accelerated erosion in the source area, perhaps as the result of a decrease in vegetation cover, a lowering of the tree line, an increase in storminess, or more severe physical weathering. Alternatively, accumulation may occur because of decreased competence of transport processes at the mountain foot. This may result from decreased flow or a reduction in flood flows.

Thick alluvial fans are essentially tectonic in origin because it is only in mountainous areas that a great enough relative relief can exist for thick deposits to form. Optimum conditions for thick accumulations occur where the rate of uplift ($\delta u/\delta t$) exceeds the rate of downcutting ($\delta w/\delta t$) of the stream channel and deposition ($\delta s/\delta t$) at the mountain front. Where erosion rates ($\delta e/\delta t$) of fan deposits are greater than rates of uplift along the mountain front either fans will become inactive or eroded, and pediments may be formed (Bull, 1977). Thus fans form where:

$$\frac{\delta u}{\delta t} > \frac{\delta w}{\delta t} + \frac{\delta s}{\delta t}$$

and pediments form, fans fossilise or become entrenched where:

$$\frac{\delta e}{\delta t} > \frac{\delta u}{\delta t}$$

Deep entrenchment of a fan may shift the locus of deposition to near the toe where a minor fan may form at the base of the larger one. Erosion of the toe by a trunk steam may also cause entrenchment and the formation of secondary fans. The formation of a fan on one side of a valley may push the trunk river to the opposite side, so causing undercutting of the slope and instability of the rock.

Fans are of considerable economic importance in arid areas and in mountain regions. They provide flat sites with fertile soils and accessible ground or surface water, but their shifting channels, liability to flood, locations along faulted mountain fronts, and liability to subsidence after removal of ground water,

also produce many problems for settlement and maintenance of roads and railways.

Further reading
General texts on rock slope stability and processes include Young (1972) and Carson and Kirkby (1972). Landslides are discussed by Zaruba and Mencl (1969), in Schuster and Krizek (1979) and stability problems by Attewell and Farmer (1976) and by Chowdhury (1978). Rapp (1960a, b) has given detailed accounts of rock slopes in two small areas.

8

Tors and bornhardts

Upstanding bodies of rock on flat or sloping surfaces have attracted much attention and have been given special names. Small residual rock masses are called tors, and large domed residuals are called bornhardts. Bornhardts are part of a general class of large residual hills which usually occur surmounting an eroded plain; such features are called inselbergs. Inselbergs may be left as prominent landforms because of their superior resistance to erosion compared with their surrounds (inselbergs of resistance), or because they are on divides farthest from lines of active erosion (inselbergs of position). The general classes of inselbergs are best discussed in relation to the erosional plains they surmount, and will not be given further treatment in this book. Tors and bornhardts, however, are of particular interest because of the information studies of them have provided on the influence of joints upon landforms.

Tors

A tor is a residual mass of bare bedrock rising conspicuously above its surroundings from a basal rock platform which may be buried by the regolith. It is isolated by steep faces on all sides and is the result of differential weathering of joint blocks and mass slope wasting. A tor is seldom less than 3 m, or more than 50 m, in height. Tors may occur on summits, valley sides, valley floors, or on extensive plains.

A number of attempts have been made to confine the word 'tor' to rock residuals produced in specific ways, but recent research has shown that they may be produced in a variety of ways on a variety of rock types. The common feature in hypotheses of origin of tors is that tors are formed in areas of widely spaced joints and, therefore, are more resistant to weathering and erosion than surrounding areas with more closely spaced joints.

Tors have been formed in granite, gneiss, schists, quartzite, dolerite, and sandstones. They may have tower-like forms which, in Africa, are often called koppies (kopjes); they may be tabular or composed of spherical boulders (Plates 8.1, 8.2 and 8.3). The hypotheses which attempt to account for their origin usually fall into one of four classes: (1) an origin by deep chemical weathering along joints, followed by stripping of the regolith to leave upstanding tors; (2) retreat of scarps across bedrock to leave residuals; (3) scarp retreat under severe cold climates characterised by freeze–thaw activity; (4) downwearing of surrounding rock and removal of the debris. These hypotheses may involve two stages of production in which weathering precedes a stage of regolith stripping, or a one-stage process in which weathering and erosion of the weathering products are concurrent.

The first extensive discussion of tor origins was that of Linton (1955) who was primarily concerned with the tors of southwestern England and especially

8.1 A koppie in granite near Lake Victoria, Tanzania.

8.2 A tor of large corestones left by stripping of the regolith, Cameroon.

8.3 A tabular tor in granite, Dartmoor.

with those of Dartmoor. Linton believed that a tor is a residual mass of bedrock produced by deep chemical weathering acting along joints so that core stones are formed. Once the weathered regolith is stripped away core stones are left surmounting massive joint-bounded blocks. Linton also distinguished tors from valleyside rock masses. Subsequent work has shown that similar forms are produced in a variety of ways and that there is little to be gained by trying to distinguish the residuals formed on various slopes.

The deep weathering followed by stripping

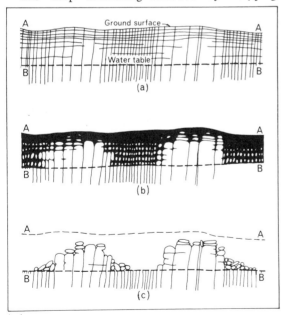

Fig. 8.1 Linton's hypothesis of the formation of Dartmoor tors by deep weathering followed by stripping of the regolith.

hypothesis for tor formation is illustrated in Fig. 8.1, and the reasons for its significance can be appreciated by consideration of Plate 8.4. In this view of a granite cliff the great range of joint block sizes is readily seen. Linton recognised that the deep weathering needed to isolate rock masses in the regolith may be very ancient, and in the case of Dartmoor probably Tertiary in age. His hypothesis that the effective depth of weathering is controlled by a water table has not met general support because it is recognised that water tables vary seasonally, and that they are certainly lowered as streams incise into the landscape. Stream incision itself may be most effective when it acts along zones of preferential weathering and so isolates hills formed upon more massive rocks. The existence of zones of advanced and deep weathering in places like Dartmoor is evident from exposures in quarries, but the weathering which produces a granitic grus of sandy quartz-rich material has to be distinguished from the deep rotting and kaolinisation of granite caused by hydrothermal activity. Hydrothermal alteration of Dartmoor granite gives a clay fraction of 30–70 per cent but the grus (also known as growan) of weathered granite has only 13-28 per cent clay and a far higher proportion of unaltered feldspar (Thomas, 1974). There is no evidence that hydrothermal activity is responsible for the development of subsurface tors (Eden and Green, 1971).

Once a tor has been produced within the regolith its exposure is dependent upon stripping of the regolith. In Europe this frequently may be attributed to the glacial stages when severe climates destroyed much of the vegetation and permitted vigorous erosion of the regolith. Phases of active erosion in cold climates may also have been responsible for the separation of rock masses from retreat-

8.4 Variable spacing of joints in granite, Antarctica.

ing cliff faces with little influence of deep weathering, as in parts of the Czechoslovakian highlands
(Demek, 1964). This process can be seen today
forming valleyside and slope crest tors in a one-
stage process in Antarctica and in areas of the
Arctic where freeze–thaw activities are very active
(Plate 8.5).

In the humid tropics, where phases of cold
climate were not experienced, exposure of the tor
may be by backwearing of a scarp (e.g. Ollier,

1960; Fig. 8.6) or by general downwearing of the
land surface as regolith is stripped away by stream
action (Fig. 8.7). Once rocks are exposed at the
surface the rate of weathering they suffer is probably far slower than that in the constantly moist
soil. Hence the core stones in the regolith may be
destroyed before those which are exposed, and this
type of tor may then cease to be rooted in bedrock
(Thomas, 1965) (Fig. 8.2).

A one-stage process of tor formation and de-

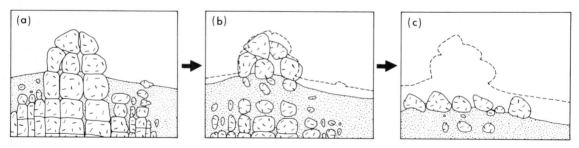

Fig. 8.2 Tors may become separated from their bedrock roots by weathering below them (after Thomas, 1965).

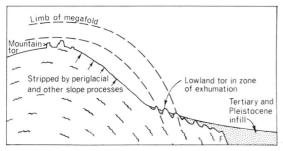

Fig. 8.3 Upland and lowland tors in the ranges of Central Otago showing their possible association with structural trends in the schist. The structure shown here is greatly simplified.

8.5 Valley side tors in dolerite are gradually diminished in size, Antarctica.

struction, in granite, by incision and enlargement of solution pans has been described from southeast Mongolia (Dzulynski and Kotarba, 1979). It is postulated that pans form on flat surfaces of granite, and etch high-level surfaces, by enlarging channels forming along major joints. Micropediments are formed at the bases of isolated joint blocks by pan formation which exploits the nearly horizontal sheet jointing. Micropediments enlarge as the pans undercut scarps and joint blocks which are being separated from the scarps as tors. Consequently a landscape of coalesced pediments is formed, surmounted by tors. The tors are eventually destroyed by the processes which created them as pans undercut them at the base and lower their summits, to leave a gently sloping broad pediment of barely covered rock pavements, adjusted to the base level formed by the local streams. It is presumed that rejuvenation will lead to another sequence of pan incision, micropediment, and tor formation, and then another pediment at a lower elevation.

Examples of tor formation by both single and two-stage processes acting in proximity occur in Central Otago, New Zealand. Here block faulted schist mountains were subjected to deep weathering in Tertiary times (McCraw, 1959) and at the base of the slopes tors were formed within the deep regolith (Fig. 8.3). During the late Quaternary many of these tors have been exposed by the stripping of the regolith. On the crests of the mountains upland tors have been formed in a climate in which severe frost action has shattered areas of the bedrock to leave only the most massive joint blocks standing, and a variety of processes — wind, solifluction, and sheet wash — have carried away the shattered schist in a one-stage process; these tors are probably of late Quaternary age. It has been suggested by Turner (1952) that the location and shape of the upland

tors is strongly influenced by the planes of schistosity so that tors are aligned along the crest of the slopes. Wood (1969) has further suggested that the trend of folds in the schist has an influence on the shape of the tors, with strong vertical lineation in the lowland tors, and strong horizontal trends in the upland tors making them more blocky. Where the dip of the schistosity is unfavourable, and on midslopes where erosion is most severe, tors are not formed.

One-stage tors are also found in Antarctica (Selby, 1972) where both frost action and salt weathering are active in preferentially shattering areas of closely jointed bedrock and wind blows away the residuum. The upper surfaces of boulders are frequently coated with iron oxides released from the micas of the granite and this forms a protective crust. Salt weathering produces many tafoni on the sides of joint blocks and eventually reduces them to small fragments. The process of tor formation is thus followed by tor destruction and ceases where either the zone of massive jointing terminates at depth in a zone of closely spaced jointing, or where the ground is covered by a residuum of fragments permanently frozen in place and no longer subject to weathering (Fig. 8.4).

Once tors have been exposed they may survive for a great length of time because they are no longer as subject to chemical weathering as the rock around them which is still covered with a regolith. With one-stage tors, however, the processes which exposed the tor may also gradually destroy it. This

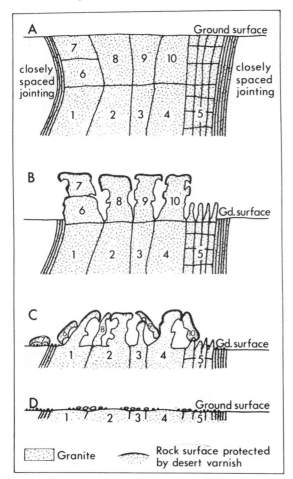

Fig. 8.4 The relationship between rock joint densities and tors in Antarctic granite (after Selby, 1972).

8.6 Upland tors, Central Otago, have a horizontal foliation. They have basal saucer-shaped depressions and the side of each which faces the wind (i.e. the side of the front tor nearest to the man) is being etched by wind abrasion.

clearly happens with Antarctic and upland Otago tors. Around the bases of many upland tors is a saucer-shaped depression containing shattered schist. McCraw (1965) has hypothesised that these tors are being undermined by accelerated frost weathering which takes place there because winter snows melt most readily, on sunny days, in contact with the rock and each night refreezing occurs. The remainder of the inter-tor surface is protected from repeated frost action by its continuous snow cover. A second destructive process is wind action which, being dominantly from the west, is preferentially abrading the western faces of the tors (Plate 8.6). Tors are thus temporary parts of the landscape even though their life-span may be considerable.

Bornhardts

Bornhardts are dome-shaped rock masses of con-siderable size — nearly always more than 30 m high and more in diameter, but often with dimensions of several hundred metres (Thomas, 1978) (Plate 8.7). They are usually regarded as a special form of inselberg and therefore as residual hills standing above a plain, but domed hills also occur within mountains and the genesis of all domed hill masses is best considered in one discussion. Many bornhardts are formed in granites and gneisses, but they may also be formed from sandstone and conglomerates. The two best-known domed inselbergs of central Australia — Ayers Rock and Mount Olga — are formed respectively in coarse arkose grit with rare bands of conglomerate, and conglomerates (Ollier and Tuddenham, 1962) (Plate 8.8).

Hypotheses of the origins of bornhardts fall into one of three classes: (1) that they are produced by scarp retreat across bedrock (Jessen, 1936; King, 1949); (2) that they are the result of scarp retreat across deeply weathered rocks (Ollier, 1960; Twidale, 1964; Mabbutt, 1961, 1965); or (3) that they result from differential weathering followed by down-wearing of the weathering mantle (Büdel, 1957; Thomas, 1965) (Figs. 8.5, 8.6, and 8.7). There also exist several special case hypotheses which attribute individual bornhardts to local effects such as up-warping and upfaulting, but these cannot have general applications.

An early hypothesis that bornhardts are the result of deep weathering under humid tropical climates can be discounted as a general hypothesis because they are now known to be widely distrib-

8.7 A large bornhardt (300 m high) in granite showing sheet structure near the top and the formation of orthogonal joints below, central Sahara.

uted. This climatic hypothesis probably gained adherents because many bornhardts are found on the great stable shields of Brazil, Africa, and Australia where deep erosion has laid bare granitic terrains (Fig. 8.8). Bornhardts are, presumably, not found on the shields of Asia and North America because

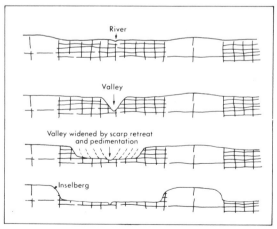

Fig. 8.5 L.C. King's (1949) hypothesis of bornhardt formation by scarp retreat across bedrock.

these have been glaciated and any inselbergs once standing above such plains would have been removed by ice sheets. In the young tectonic belts, cores of dome-shape are found, as in Yosemite National Park, but because they are not isolated residuals they are not called bornhardts and their close relationship with true bornhardts is not always recognised.

The hypothesis of scarp retreat across bedrock attracted many early workers and still finds support from L.C. King as being a generally applicable hypothesis, but it appears to most geomorphologists to be applicable to restricted localities only. The essence of the hypothesis is that lateral planation by streams is most effective in areas of closely spaced jointing and that erosional scarps gradually retreat until cores of massive rock limit further retreat. Resistant masses are, to some extent, self-perpetuating in that once exposed they shed runoff on to the surrounding area which is then preferentially weathered and eroded. It also follows that, if the land mass is uplifted and a new cycle of erosion cuts down, bornhardts may continue to exist on the one site, or they may be destroyed or newly

8.8 Mount Olga. The highest peak rises 450 m above the plain. (Photo: Australian News and Information Bureau.)

created on the lower land surface, depending upon the nature of the jointing at depth (Fig. 8.9).

Exposure of bornhardts by scarp retreat appears to be occurring in the Namib desert at the present time (Selby, 1977a). Long spurs protruding from the great escarpment of the Khomas Highlands, which form the inland boundary of the central Namib, are formed in schists but the cores of these spurs consist of granite intrusions which have a dome shape (Plate 8.9). Currently erosion is stripping the schist away from the domes and exposing them as bornhardts. These bornhardts are thus revealed as having their dome shape from their earliest exposure. They are manifestly more resistant than the schist and many domes are found on the main plains of the Namib. Because they diminish in size seawards from the escarpment it appears that they have been exposed and then reduced in size as the escarpment has

retreated inland, and weathering has had its greatest effect on the oldest bornhardts which are nearest to the sea. Nowhere in the Namib is there any evidence of deep chemical weathering and nearly everywhere the bedrock schist is within a metre of the ground surface. Any hypothesis which invokes deep weathering in the Namib relies upon analogy with other areas of Africa, and is essentially untestable (Ollier, 1978). The evidence that granite has been intruded in dome-shaped masses into the schist and then been revealed by scarp retreat across bedrock appears overwhelming.

Deep weathering hypotheses are well supported by field evidence from the inland plateaux of Africa. In Cameroon (West Africa) deep cutting for a quarry to provide ballast for a new railway line from the coast to the capital, Ngaoundere, revealed a low dome 50 m high and 200 m wide at the base. The

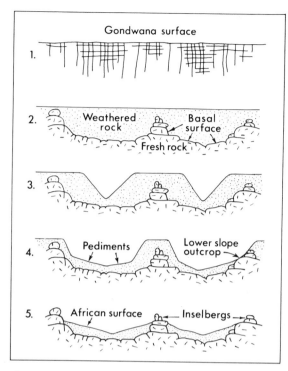

Fig. 8.6 Ollier's (1960) hypothesis of bornhardt exposure by scarp retreat across a regolith.

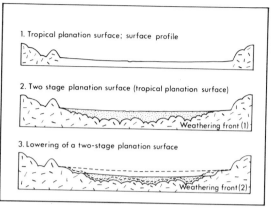

Fig. 8.7 Büdel's (1957) hypothesis of bornhardt exposure by downwearing of a deep regolith.

deeply weathered regolith was *in situ* above the massive dome of quartz-diorite, and the dome was surrounded by zones of more closely spaced jointing. Also on the buried dome tafoni-like hollows and platy exfoliation forms had formed within the regolith (Boyé and Fritsch, 1973).

The question of whether scarp retreat or downwearing of a deeply weathered regolith is responsible for exposure of domes is not of primary importance to the argument about their origin. It may, however,

8.9 Joint patterns in granites which are being stripped of the surrounding and overlying schist, central Namibia.

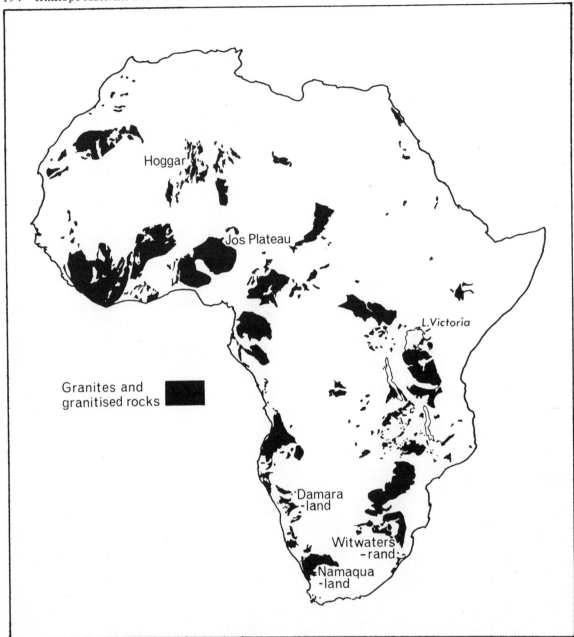

Fig. 8.8 The major outcrops of granitic rocks in Africa.

be of importance to the development of secondary features on the domes. It has been suggested by Pugh (1966) that the multiple-dome forms seen on some bornhardts are produced by successive phases of surface lowering around the bornhardt, with each phase of stability leaving an indentation or platform around the dome, so that a large dome may eventually have the form of a large basal dome surmounted by progressively smaller domes.

A similar argument has been used by Twidale and Bourne (1975a) who suggest that around a bornhardt chemical decomposition within the regolith, accelerated by runoff from the bare dome, will produce flared slopes to the vertical edges of the bornhardts, and nearly horizontal surfaces around them at the depth of the weathering front

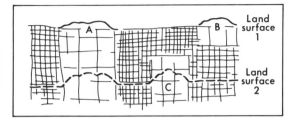

Fig. 8.9 Land surface lowering may destroy or create born-hardts depending upon joint frequency (according to the theory of L. C. King, 1949).

8.10 The road cutting has bisected a granite dome within a deeply weathered regolith, Witwatersrand, South Africa.

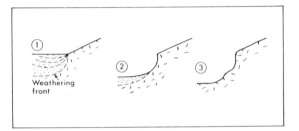

Fig. 8.10 Weathering around a rock outcrop may develop a subsurface flare in the rock with bevels produced by stages of regolith removal (after Twidale, 1971).

(Fig. 8.10). Episodic stripping of the regolith will then reveal successive steps cut into the hard rock (Fig. 8.11). Where the platforms of inselbergs have a widespread uniformity of altitude these may be correlated with ancient erosion surfaces, but this would clearly be a hazardous practice where correlation is extended over considerable distances, and where independent evidence for the erosion surfaces is not available.

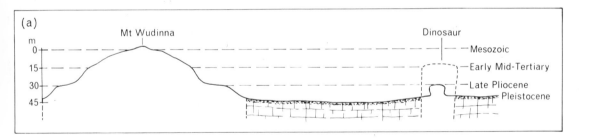

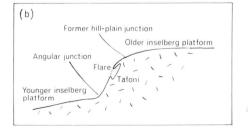

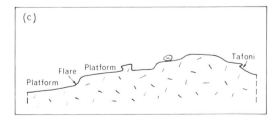

Fig. 8.11 Stages of downwearing of the landscape may be recorded by platforms cut at a weathering front. Platforms on Mount Wudinna, South Australia, are correlated with erosion surfaces (after Twidale and Bourne, 1975a).

The form of bornhardts is perhaps more controversial than the origins which now seem established as being of several kinds. The form characteristic which has attracted most discussion is the dome shape. Most bornhardts are characterised by sheet joints — that is, planar or gently curved joints conformable with the dome — and the central argument is related to whether the joints produce the dome form, or the dome form existed before the joints developed. At least four hypotheses have to be considered: (1) that dome forms and sheet joints develop in response to unloading and stress release; (2) that they result from intrusions into the crust and the formation of stretching planes; (3) that they result from faulting and the development of secondary shears; and (4) that lateral compression within the crust is the major cause of domes and sheeting. In considering these arguments it is essential to remember that bornhardts are always massive rock features with few joints, and that these are nearly always closed. Bornhardts exist because of their resistance to weathering and erosion.

The unloading or stress release hypothesis was expressed in its most persuasive form by G. K. Gilbert (1904). He, like many geologists since, was impressed by the evidence of rock bursts in mines and tunnels, of the springing up of slabs from valley floors after glaciers have retreated from over them, and by the common occurrence of thin slabs and slab failures on the walls of deglaciated valleys. The most obvious cause of such failures thus appeared to be the radial expansion of the rock following the removal of a confining stress. Further evidence in favour of this hypothesis comes from the observations of Dale (1923), Jahns (1943), and Johnson (1970) who noted that sheet structures are commonly flat; the spacing between them increases with depth, ranging from a metre or less at the surface to sheets 10 m or more in thickness at depth (Fig. 8.12); sheet joints transect structures or even dykes in the rock; sheet structures terminate laterally by gradually thinning or where they are transected by other structures such as vertical joints; they occur in a variety of lithologies such as granite, gneiss, quartzite, and massive sandstones and limestones; and where the relief is deeply indented they follow the topography, thus being essentially anticlinal on ridges and synclinal in valleys. These observations all seem to indicate that sheet joints form and progressively open as the confining stress is removed and that even when thick sheeting structures have formed at 50 m or

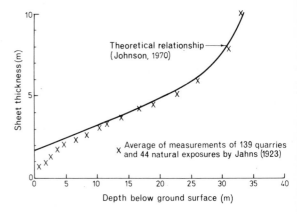

Fig. 8.12 The increase of thickness of sheets in granite with depth (after Johnson, 1970).

more below the surface they can be further subdivided by the development of more sheeting joints.

The main objections to the release of confining stress hypothesis have been put by Twidale (1973). He has pointed out that (a) there is no reason why radial swelling of the rock should be accompanied by the opening of joints and (b) bornhardts are always zones of massive rock which exist because of their lack of joints or because their joint systems are closed, and it is therefore contradictory to say that they owe their form to the development of a joint system.

The idea that domed rock masses are produced by the intrusion into the crust of granitic masses, or by the advance of a metasomatic front was proposed by Holmes and Wray (1912) to explain the bornhardts of Mozambique (Fig. 8.13). This hypothesis would certainly fit the evidence of plutons within the spurs reaching into the Namib desert, but it cannot be used to explain the domes formed in sedimentary rocks, nor does it fit all granitic intrusions, for satellite photographs of plutons outcropping in the Sahara show that, while the outline of the pluton may be circular, the relief within it is often very irregular. At best then, the intrusion hypothesis is appropriate to restricted cases.

The idea that faulting and the formation of secondary shears may be responsible for bornhardts has been suggested for explaining the Rio de Janiero 'sugar loaves' but there is little evidence for the faulting and, if it is valid, it can only be so for limited cases.

That closed joint systems require rock masses to be in a state of compressive stress has already

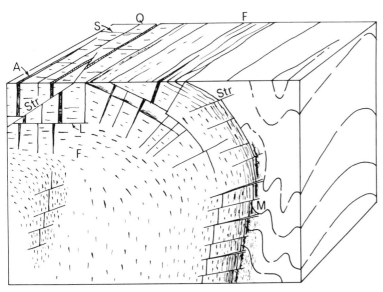

Fig. 8.13 The theoretical development of joints at the margins of an intruded granite pluton, emphasising the trends parallel to the margin of the pluton.

Q Cross joints   A Aplite dykes   S Longitudinal joints

F Flow layers and foliation   Str Planes of stretching

L Flat lying joints   M Marginal thrusts

been pointed out. It may also be noted that many rocks are in a far greater state of compressive stress than would be expected from vertical compression alone (Isaacson, 1957). Compressive stresses, however, may also be caused by (a) original compression of the rock induced during emplacement of a pluton; (b) recrystallisation during metamorphism; and (c) tectonism. Field evidence in favour of compression includes the presence of sheared rock, slickensides, discolouration of feldspar crystals by internal shearing, but most obviously it is the evidence for closed joints within the bornhardt. Thus the idea that excessive compressive stress during past orogenies is responsible for massive rock outcrops, and that release of this stress permits the opening of joints which may, or may not, follow the ground surface, appears to have much support.

In conclusion it may be felt that the arguments are still not entirely resolved. Special cases for which one or more hypotheses are most appropriate seem to be common. The most generally applicable hypothesis that accounts for both the bornhardts in the first place and then sheet structures controlling their outer form is that of tectonic stress and its subsequent release. In this hypothesis the joint system essentially controls the landforms, and as the joints are progressively opened so a bornhardt progressively loses its form and may even pass

through stages for which the terms koppie, or castellated tor, or tower tor may be most appropriate (Thomas, 1965). Finally a flat rock surface level with the general landsurface, called a ruware, may be formed. Until detailed investigations are carried out, however, it is impossible to tell if ruwares are the end-product of bornhardt or tor destruction, or the emerging upper surface of a bornhardt which is still buried within a rock mass or regolith (Fig. 8.14).

The hypothesis that joint spacing and joint curvature largely control the origin and form of bornhardts and tors has yet to be tested thoroughly for four main reasons.

(1) It cannot be known, only assumed, that rock is weathered and eroded preferentially because it is weaker – once the erosion has occurred the evidence has been destroyed.

(2) Closely spaced jointing may be either a cause or an effect of weathering at depth in the regolith – it is only assumed to be a cause.

(3) Much of the evidence required to test hypotheses of joint spacing is always hidden beneath the regolith and it is often assumed that there is structural continuity beneath the regolith, and a lateral variation in jointing or rock compression – cuttings through appropriate sites are rare.

(4) Residuals may be of great age and have

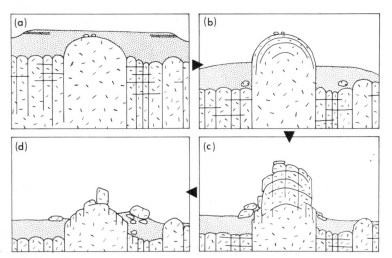

Fig. 8.14 The exposure and subsequent decay of a bornhardt in Nigeria as joints open (after Thomas, 1965).

undergone numerous changes of climatic, and therefore weathering and erosional, environments. The formation of a regolith 30–50 m deep in crystalline rocks may take anything from several hundred thousand years to ten million years, although the saprolite may be stripped in a few thousand years. Large bornhardts may only be produced from the adjacent bedrock or regolith in a period of tens of millions of years, and even small tors could only be produced over tens to hundreds of thousands of years. It is inevitable, therefore, that much of the evidence for the formation and evolution of such landforms should be unclear or absent.

Many of the arguments about origins may also be confused by assumptions that only one, or very few, origins are possible for the same landform.

The very clear evidence that both tors and bornhardts may be formed in a great variety of rock, in a variety of climates, in both single and multiple phase events, and by weathering and removal of surrounding saprolite, and of unaltered bedrock, indicates quite clearly that one end-form may be produced by a variety of processes, and that it is impossible to elucidate from a simple examination of shape the origin of a landform. The concept of multiple modes of origin being able to produce one type of end-form has been called the principle of equifinality (Bertallanffy, 1950). It is a salutary reminder to all earth scientists of the need for an open mind, for detailed study of all the evidence, and for the development of multiple working hypotheses.

9

Slope profiles and models of slope evolution

Hillslope profiles may be controlled by weathering, by transport, or by accumulation processes. *Weathering-controlled slopes* occur where the actual rate at which regolith can be produced is less than the potential rate at which it can be removed. Consequently the slope profile reflects the relative resistance of the rock of which it is composed. *Transport-limited slopes* are those on which the rate of regolith production is greater than the capacity of transport processes to remove it, so that regolith accumulates, and the slope profile is then controlled by the properties of the regolith and the nature of the processes acting on it. Where there is an equilibrium between rate of weathering and rate of removal a transportational slope will occur, with neither net gain nor net loss of material. A transportational slope is thus an intermediate case separating denudation slopes from accumulation slopes.

The definitions given above should be applied strictly only to points on a slope and there may be considerable variation in the control along a profile. The control will vary with climate and with lithology. The absence of a regolith implies a control by weathering. The existence of a thick regolith implies control by transport, especially where the regolith is weathered rock *in situ*. A thin regolith may be present on a slope subject to control by weathering, especially if the rate of weathering at the soil–rock interface is very low, and this is probably the case on many stepped scarps which have forms expressing the influence of their hidden structure.

Rock slope profiles

Profiles and rock mass strength

The profiles of rock slopes may be controlled by:

(1) the strength of the rock on which they are formed;

(2) processes or structural controls which operate to undercut or oversteepen the slope with respect to the rock strength available for maintaining long-term stability;

(3) processes and structural controls which operate to reduce the profile angle below that which could theoretically be supported by rock resistance.

Slopes which reflect rock resistance have units controlled by the operation of the parameters which were recognised in the rock mass strength classification (Chapter 4). By applying the appropriate rating (see Table 4.7) for each parameter the total strength rating for each unit may be calculated (Fig. 9.1a, b, c). If the underlying assumption is correct there should be a close correspondence between the total rating and the hillslope unit angle. Ratings have been calculated for the rock units on a number of slope profiles in both Antarctica and New Zealand (Selby, 1980), and more recently in many parts of southern Africa. Figure 9.2 shows that mass strength and rock unit angle of slope are highly correlated ($r = 0.88$) (Plate 9.1). There is, of course, some scatter in the data caused by the coarseness of a five-class method of classification, by observational error, by variations in intensity of processes across the slope, and by failure of the classification to include all of the effective parameters.

All available data indicate that slope angles that reflect rock resistance are widespread in dry climatic zones and mountains where soil formation is limited. They are recognisable in the field by the lack of undercutting, a lack of talus or regolith material, and by the absence of an overwhelming structural control on the landscape. The envelope which could be drawn to contain all the data points of Fig. 9.2 would also contain all other available data from slopes whose angles are in equilibrium with the mass strength of their rocks. This envelope has, therefore,

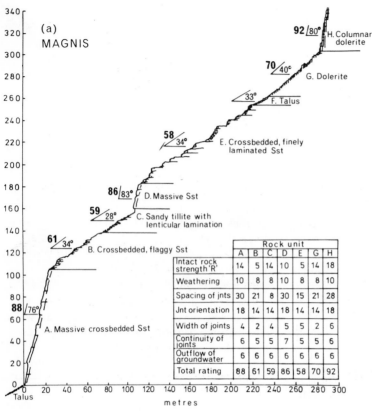

	Rock unit						
	A	B	C	D	E	G	H
Intact rock strength 'R'	14	5	14	10	5	14	18
Weathering	10	8	8	10	8	8	10
Spacing of jnts	30	21	8	30	15	21	28
Jnt orientation	18	14	14	18	14	14	18
Width of joints	4	2	4	5	5	2	6
Continuity of joints	6	5	5	7	5	5	6
Outflow of groundwater	6	6	6	6	6	6	6
Total rating	88	61	59	86	58	70	92

Fig. 9.1 Scaled profiles of rock slopes: (a) A slope in Magnis Valley, Transantarctic Mountains, showing the strength ratings for each rock unit and the slope angle; (b, c) Profiles of two granite bornhardts in the Namib desert showing slope angles and mass strength ratings.

(a)   (b)

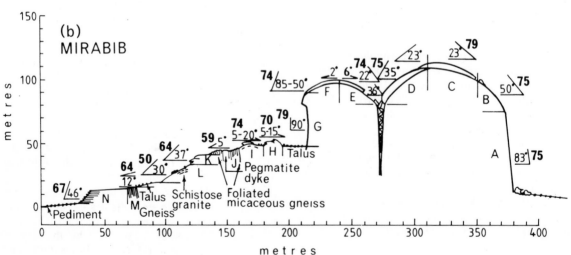

been called the strength equilibrium envelope and the slopes are known as strength equilibrium slopes.

Recognition of the existence of equilibrium slopes has important theoretical and practical implications.

One of the major problems of geomorphological theory has been the development of criteria, and accumulation of evidence, by which it is possible to determine the pattern of long-term slope evolution — especially to decide whether slopes decline in

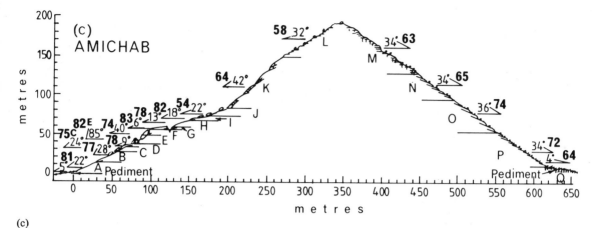

(c)

angle or retreat parallel to themselves as they evolve. The recognition that many rock slopes forming scarps and faces of inselbergs are in equilibrium implies that such slopes will maintain a constant angle as long as rock strength is maintained; these slopes will retreat parallel to themselves. If strength increases or decreases into the outcrop then the hillslope angle will increase or decrease in conformity as the slope retreats. This theory has, so far, been tested only on exposures in southern

Africa where all measured rock slopes, except some in massive granite bornhardts, were found to be strength equilibrium slopes. The lithologies of these equilibrium slopes included sandstones, tillites, shales, dolomite, marble, dolerite, gneiss, schist, basalt, pegmatite, and some granites; there was much variety in bedding and joint patterns.

The practical implications of recognising strength equilibrium slopes arise because, by using the regression equation relating mass strength and

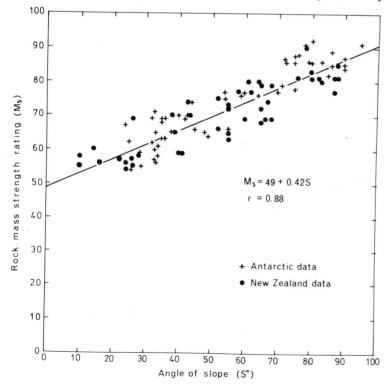

$$M_s = 49 + 0.42S$$

$$r = 0.88$$

+ Antarctic data

• New Zealand data

Fig. 9.2 The relationship between rock mass strength and slope angle for all measured Antarctic and New Zealand slopes (after Selby, 1980).

9.1 Rock slopes for which the relationship between rock mass strength and slope angle has been determined, Trans-antarctic Mountains.

outcrop slope angle, it is possible to estimate, from a knowledge of mass strength, the angle of slope with an error of ± 5°. This relationship may be of value in preliminary applied surveys in which it is required to estimate the angle of a cut face for long-term stability. In most cases, of course, more detailed geotechnical studies would be needed to check the reliability of this estimate.

Structural controls operating to form slopes which are either steeper or less steep than their equilibrium angle are recognisable in some intrusive granite domes (Fig. 9.1b, c). The flanks of domes such as Mirabib are very steep and either exceed equilibrium angles, or are in equilibrium, because the lack of joints gives a very high strength rating. The broad tops of many domes, however, and the low-angled flanks of some domes are at much lower angles than could be supported by the rock mass strength. As joints open and sheet structures are subdivided by joints of an orthogonal set the very steep slopes gradually come into equilibrium with the reduced mass strength. On the bornhardt

Amichab (Fig. 9.1c) slope unit K has sheeting structures breaking down as an orthogonal joint set develops; units A to J inclusive are below the equilibrium angle and have some talus accumulating on the exposures; units L to P are in equilibrium.

In areas of anticlinal fold structures a range of slope angles may exist with slopes steeper than equilibrium being formed where flanks of folds are temporarily oversteepened by folding or stream undercutting, and slopes below equilibrium angles occur on gently dipping rock units. Erosional rock scarps which are not being undercut will usually be found to be at equilibrium.

Application of the mass strength classification shows that major changes in a slope profile may result from variability in just one of the parameters. In Fig. 9.3 it can be seen that in a dolerite of uniform intact strength changes in the dip of joint blocks from about 35° into the slope to vertical columnar jointing can change slope angles from 60 to 90°, and a change from columnar to closely spaced jointing can cause slope angles to decline from 90 to 59-

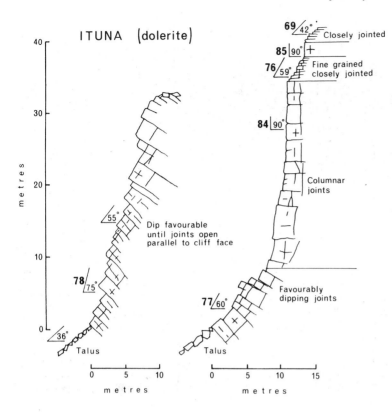

Fig. 9.3 Profiles of slopes in dolerite showing joint spacing and inclination and their effect upon slope angles, Ituna Valley, Antarctica.

42°. Similar changes are seen in the dolerites shown in Plate 9.2. The changes in jointing dip are minor but strongly influence slope angle, as does the incidence of closely spaced jointing at the large step in the profile.

Benches or steps in rock slope profiles may be either direct responses to resistance or they may be a reflection of the time available for planation of the bench. In the walls of the Grand Canyon of the Colorado, for example, benches are cut across certain strata but these may represent hiatuses in erosion or uplift, not a marked change in rock resistance. Another condition, in which factors other than rock unit resistance alone may operate, is that where strata of very variable strength are interbedded. A weak stratum may then fail or be cut back so that an overlying stratum will fail and appear to have a profile less steep than its inherent mass strength would indicate.

Cliff recession with talus accumulation

Where mass strength is the control, retreating slopes will retain profiles which reflect variations in strength, but on cliff faces which are retreating through uniform weathering a uniformly thick layer

of rock will be removed from the cliff, which will then retreat parallel to itself. The fallen debris accumulates at the base of the slope as a talus. It has been shown by Fisher (1866) and Lehmann (1933, 1934) that the rock slope buried by the talus will theoretically develop a convex shape (Fig. 9.4). The actual curvature of the convexity will depend upon the ratio of the volume of rock removed from the cliff to the volume which accumulates at the cliff base. Where all falling debris accumulates, the volume of debris will exceed the volume of solid rock that is removed because of the higher void space in the talus. Solution and removal by streams may, however, reduce the volume of the resulting talus to less than the volume of the intact rock which is removed.

Few exposures through talus into bedrock exist, but most reported profiles do not appear to support the theory that a convex rock core slope will develop beneath talus below a retreating cliff, and the value of this model is still uncertain.

A special case of the Fisher-Lehmann theory has been stated by Bakker and Le Heux (1952). In this model all the talus is steadily moved by rolling and sliding and the basal rock slope evolves at the

9.2 Dolerite cliffs showing the effect upon slope profiles of subtle changes in joint orientation and spacing, Britannia Range, Antarctica.

angle of rest of the talus material. Thus rock slopes with thin veneers of debris should be formed. It has been noted that such slopes are found quite widely in alpine and polar areas, and in the extreme environment of Antarctica they appear to be relatively common (Selby, 1974a; Chardon, 1976) (Plate 9.3). The analysis of Bakker and Le Heux appears to be generally valid even though it assumes (falsely) that talus slopes are usually at the angle of rest of their materials (see Chapter 7).

We can visualise the development of this type of slope – which is known as a Richter slope – by imagining particles falling from a cliff on to the top of a talus. If the newly fallen material just covers a little of the base of the cliff the next fall will be over the new talus and hence the base of the cliff will now be higher, so the cliff will recede by a series of minute steps at the angle of the talus. These steps may then be removed by weathering or the abrasion of sliding talus – or they may remain. In either case an essentially straight rock slope (i.e. a rectilinear slope) will be formed below the cliff. Eventually the free face should be eliminated and a smooth rock slope of uniform angle will be produced (Plate 9.4). Suggested successive stages in this development are shown in Fig. 9.5. Once the stage has been reached of either a uniform bare rock slope, or a uniform talus-covered slope, further evolution will depend upon the nature of the processes operating. Where uniform weathering occurs over a rock slope and the debris is blown away then it is likely that the rock slope will continue to get smaller but retain the same angle (Fig. 9.6); where wash processes occur the slope is likely to decline in angle as progressive weathering reduces the size of particles and basal regoliths thicken (Fig. 9.7).

Models of soil-covered slopes

The advantage of a model is that it generalises widely recognised features or processes, arranges them in a meaningful pattern, and simplifies the components so that they may be readily understood.

Models of landforms may be of several kinds. Analogue models represent the components of a landscape and, perhaps, associate each component with the dominant process acting on it; evolutionary models identify a sequence of changes in a landscape which are usually recognised from field evidence; mathematical models are expressions of landforms

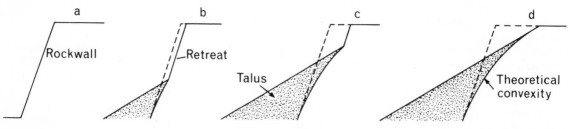

Fig. 9.4 A theoretical cliff, basal convex rock slope, and talus sequences in which it is assumed that there is no removal of talus from the base.

9.3 Richter denudation slopes, with a thin veneer of talus, extending headwards into sandstone cliffs, Transantarctic Mountains.

and processes as equations which can be repeatedly solved, with each subsequent solution being modified by the solution of its predecessor so that step-by-step changes are calculated.

Analogue models

Hillslope forms have traditionally been considered from the point of view of their long profiles from ridge tops to valley floors. This approach tends to ignore the complexity and variation of forms and processes across the slope, that is, along the contours. The divergence of water and debris from convexities and their convergence upon depressions and drainage lines is, of course, just as important as any process acting directly down a slope. The fact that, on many hillslopes, the heads of the drainage network are old landslide scars emphasises both the variation of forms and processes across a slope and the continuity of hillslopes with valley floors.

Discrete slope units, characterised by distinct inclinations and processes, may be portrayed upon block diagrams or specially prepared maps (Fig. 9.8). Before these are prepared, however, it is necessary to recognise the major types of unit which recur in the landscape. This may be done either by distinguishing flat, convex, concave, cliff, or inclined units or, more usefully, by recognising the relationship between pedological and geomorphic processes and the characteristic slope units on which they occur. Using this approach Dalrymple *et al.* (1968) and Conacher and Dalrymple (1977) have developed a nine-unit land surface model (Fig. 9.9). This model regards the hillslope system as a three-dimensional complex extending from the drainage divide to the centre of the channel bed, and from the ground surface to the uppermost boundary of weathered rock. Each of the nine slope units is defined in terms of form and the dominant processes currently acting on it. In reality it is unusual to find all nine units occurring on one slope profile;

9.4 Straight and even slopes cut across granite are interpreted as the final stages of Richter slope formation, Koettlitz Valley, Antarctica.

they do not necessarily occur in the order shown in Fig. 9.9 and individual units may recur in a single profile.

Concave-convex hillslopes are relatively common in many temperate environments so the order of units may be 1, 2, 3, 6, 8, 9; on steep faces with repeated rock outcrops the order might be, for example, 1, 2, 3, 4, 5, 4, 5, 4, 5, 6, 8, 9. A cliff above a river might have only units 1, 2, 3, and 4. The model thus provides a means of describing, and a means of mapping, slopes to show how they vary along the contours; it also relates processes to slope forms.

One advantage of the nine-unit model is that it can form the basis of a mathematical model of slope change in which the mode and rate of operation of processes characteristic of each slope segment are expressed in the form of equations, and repeated operations of the process are simulated.

Slope evolution models

Much of the geomorphological thinking in the English-speaking world, in the early part of the twentieth century, was devoted to study of the implications of the concept of the cycle of erosion, as expounded by W. M. Davis (1909) and his disciples. Later attempts to provide a framework for thought were those of W. Penck (1924) who sought to interpret the rate of crustal movement from slope morphology, and of L. C. King (1967) who was concerned with the influence of structure upon the major relief forms of the earth, with the widespread phenomenon (as he believed) of the parallel retreat of free faces on slopes and of semi-arid climatic influences.

Davis, Penck, and King sought to derive general and universal conclusions from widespread field observations. Unfortunately all these early evolutionary models of landscape change are essentially

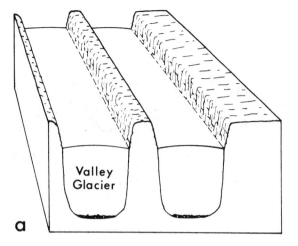

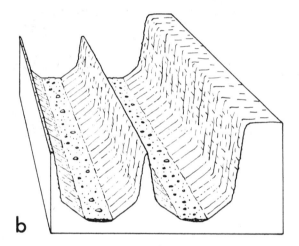

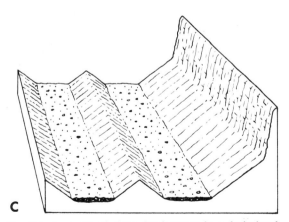

Fig. 9.5 A model of slope development in a deglaciated valley as Richter slopes extend. The Richter slopes are covered with a thin veneer of talus (after Selby, 1971b).

descriptive and untestable. They were formulated before the complexity, universality, and duration of climatic changes of the Late Cenozoic was appreciated, and before many data on the rates, incidence, and mechanisms of geomorphic processes were available.

Evolutionary models, usually represented as a sequence of slope profile changes or a sequence of block diagrams, still do have a place in geomorphology where they are founded upon detailed field work. Most such models employ the ergodic hypothesis which suggests that, under certain circumstances, sampling in space can be equivalent to sampling through time, and that space–time transformations are permissible (Chorley and Kennedy, 1971). Thus the slope profiles developed upon two adjacent till sheets of different ages, but similar composition, may be taken as representing two stages of development of one set of slopes, or slope profiles measured in the lower, middle, and upper reaches of a valley may be regarded as being part of a sequence with the headwater slope profiles being young and the lower reach profiles being older. In this latter case extraneous influences such as variations in geology, drainage diversion, antecedence, or base level change must be absent before the model could be regarded as acceptable.

Examples of applications of the ergodic hypothesis are the models of slope development in ice-free areas of Antarctica (Fig. 9.5) and in the Grand Canyon (Cunningham and Griba, 1973). In Antarctica progressive, and approximately dated, decay of glaciers in valleys through the Transantarctic Mountains has allowed Richter slopes to develop up valley walls across various lithologies. The Richter slopes in small areas are of varying lengths and have free faces above them of varying heights. The substitution of time for space permits these profiles to be arranged in a sequential order of development. In the Grand Canyon slope development and valley widening may be traced downvalley from first-order to progressively higher-order valley segments, and the assumption may be made that this represents an evolutionary sequence.

Mathematical models

The uncertainty always associated with evolutionary models, and usually with applications of the ergodic hypothesis, has led to a change in the focus of attention in modern research work. It is now primarily concerned with studies of the resistance of material and with processes of change and their

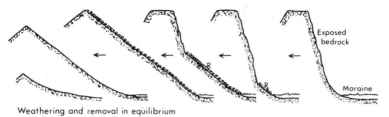

Fig. 9.6 The extension of a talus-mantled Richter slope (R) until the cliff face is destroyed. Thereafter the talus is removed by weathering and wind action, and the slope weathers uniformly, maintaining its angle but reducing in size.

Weathering and removal in equilibrium

results. Attempts are also being made to incorporate data on processes, and knowledge of mechanisms, into mathematical models which can be used to predict the way in which landforms will change under specified conditions of structure, climate, relief, and time for the operation of the processes. Once reliable models are produced they may have a number of applications including prediction of the effects of land-use changes upon slope forms and rates of evolution; the formulation of research programmes to provide data for, and tests of, models; the study of long-term effects of a single process, or group of processes – a condition normally impossible under natural conditions; and the explanation of geomorphological 'laws' such as the evolution of Richter slopes.

There are two main approaches to the development of mathematical models: analytical solutions based upon an assumed manner of action of processes, and simulation models based on calculating successively the effects of processes, whose rate and mode of action is assumed but from a manner indicated by field studies.

Analytical models

The necessary basis for any process–response model is the continuity equation, which is a statement that, if more material is brought into a slope section than is taken out then the difference must be represented by accumulation. Conversely, if less material is brought into a slope section than is removed the difference must come from net erosion of the section. The rate of debris transport is thus a major term in the continuity equation, and the variation of the rate of transport with relief largely controls the slope form and rate of change. For a satisfactory statement of the continuity equation we also need to specify the initial form of the profile, the conditions at the crest (usually regarded as fixed) and at the base of the slope (where constant removal of material is the simplest condition).

The equation has the form:

$$\frac{\delta y}{\delta t} = \frac{\delta S}{\delta x},$$

where y is the elevation of a point;

t is the time elapsed;

S is the sediment transport rate;

x is the horizontal distance of a point from the crest.

Where there is no limitation on the supply of material then:

$$S = f(x) \cdot \left(\frac{\delta y}{\delta x}\right)^n,$$

where n is an exponent describing the influence

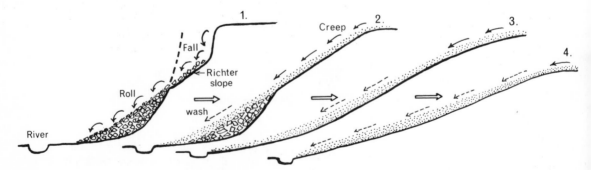

Fig. 9.7 A model of slope evolution by the development of a talus and Richter slope. Weathering produces a veneer of fine-grained debris and wash processes reduce the angle of the slope.

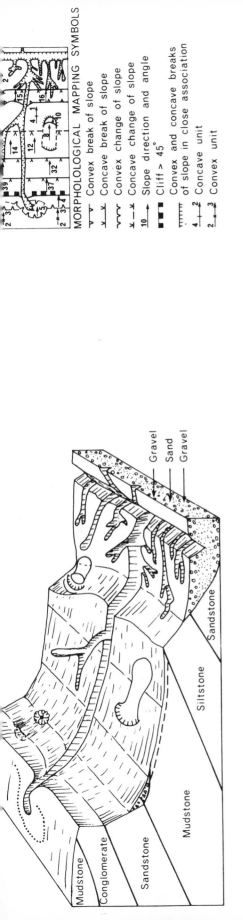
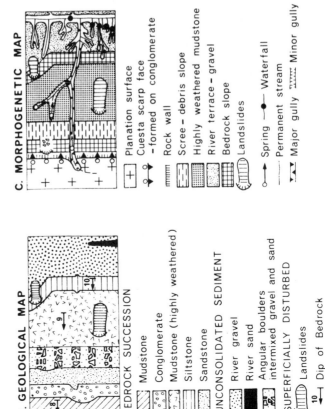
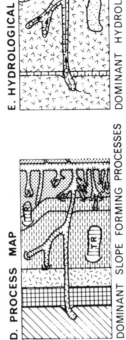

Fig. 9.8 Maps are some of the most common landscape models. These maps show various features and interpretations of one landscape represented in a block diagram. Such maps are very useful for recording field observations (modified and extended from Brunsden *et al.*, 1975).

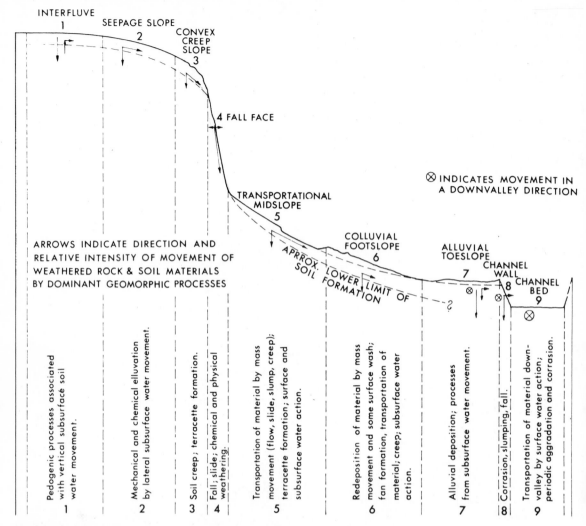

Fig. 9.9 The hypothetical nine-unit land surface model (after Dalrymple *et al.*, 1968).

of increasing gradient; its value will vary depending upon the controlling process.

As a first-order approximation for slopes of less than 30°:

$$S \propto \sin\beta \propto \tan\beta \propto \left(\frac{\delta y}{\delta x}\right)^n.$$

This model of transport rate as a function of slope angle (β) is confirmed from experimental data on soil erosion (see Chapter 5) and is used in analytical models such as those of Kirkby (1971) and Gossman (1976). By using values assigned to an initial condition (y, x, β) it is then possible repeatedly to solve the equation adjusting β to conform with the value of S determined in the previous solution. Changes

in slope profile can therefore be determined using a series of points along that profile.

Simulation models

Simulation models have been constructed by a number of workers including Young (1963a), Ahnert (1976b), and Armstrong (1976). Because it is relatively simple the model of Armstrong will be described here.

The model has a land surface in three dimensions, represented as a matrix of unit cells, each of which has two important properties — a height and a soil depth. The matrix of heights represents the form of the basin at any one time, and the direction of mass transfer of material is determined by the gradient at any point, so that the form becomes an important variable which modifies the action of slope processes. The soil depth at any point represents the total amount of material which is potentially mobile. An initial form of the landscape is specified as a map of heights and depths.

The processes operating in the model are selected, their mode of operation specified as an equation, and their magnitude is assigned a value which represents a natural rate of operation. It is thus possible to specify that each iteration of the model represents a set period of time. Armstrong ran his model for 20 000 iterations with each iteration representing one year, so that he assumed a period of denudation of 20 000 years. By using a high-speed computer the evolution of the slopes over that period can be calculated in a matter of minutes or a few hours.

The continuity equation for the slope system can be represented as a budget, so that over a unit of time (one iteration) can be calculated:

$$\delta H = I - O$$
$$\delta D = I - O + W$$

where H is the height of the ground surface above a datum;

 I is the inflow of material into the cell;

 O is the outflow of material from the cell;

 D is the soil depth;

 W is the amount of weathering.

Thus in each iteration is computed the weathering component and the outflow from each cell (representing the magnitude of the transport process), which is then added to the inflow of the cell next downslope.

In the model only three processes are considered:

(1) *Weathering* is represented by:

$$W_a = W_p e^{-K_w D}$$

where W_a is the actual weathering rate;

 W_p is the potential weathering rate at a bare rock surface;

 K_w is a constant;

 D is the soil depth;

 e is the base of natural logarithms.

Values assigned are $W_p = 1 \times 10^{-4}$ m/year, $K_w = 2.0, D = 1$ m.

(2) *Slope transport* is represented in the form of soil creep at a rate calculated from:

$$C_s = K_s \sin\beta$$

where C_s is the rate of soil movement;

 K_s is a constant;

 β is the slope angle.

Values assigned are $K_s = 10 \text{cm}^3/\text{cm/year}$ and slope angle was set by the initial landform and then modified according to the result of each iteration.

(3) *Fluvial transport* rate is given by:

$$C_r = K_r SQE$$

where C_r is the volumetric transport rate:

 K_r is a constant;

 S is the river slope;

 Q is the discharge;

 E is the river efficiency.

Values of S and Q are supplied by the model at each point, K_r is given a value of 4.13, and $E = 0.4$.

The results of the simulation are presented in Fig. 9.10. Starting from the initial form, block diagrams represent the result of the action of the three processes after the stated number of iterations. The main features of the landscape are the overall convexity of the landforms, their smoothness, and their stability. Individual slopes appear to be maintained in an equilibrium form, after first attaining convexity, with little change in shape, but a change in dimensions.

The simulation model of Ahnert (1976b) is more complex than that of Armstrong and includes weathering, structural effects, base level change, waste transport by splash, overland flow and wash, plastic flow, viscous flow, and debris slides. As the capacity of computers grows we may expect to see further components and complexities being included in models.

Specific results of some published simulations which are well related to natural conditions suggest the following conclusions:

(1) processes involving downslope soil transportation tend to cause slope decline, and the slope is transport-limited;

(2) processes involving direct removal of material from the slope tend to cause parallel retreat, and the slope is weathering-limited;

(3) downslope transportation at a rate which varies only with $\sin\beta$ produces a smooth slope convexity;

(4) stream incision at a rate in excess of the rate

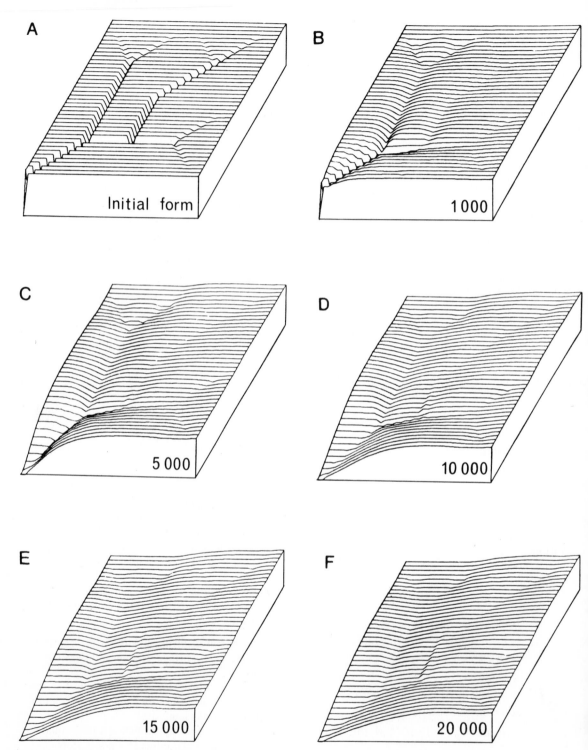

Fig. 9.10 Armstrong's (1976) computer-generated block drawings of a sequence of landforms. The numerals represent the number of iterations.

of transportation on the slope produces steep basal slopes which may then fail by landsliding.

The simplest process–response models are those which assume that only one process is operating upon a slope. Creep processes produce an expanding upper convexity on a slope, wash produces an increasing lower concavity, uniform solution produces a parallel downwearing, and shallow landsliding on a transportational midslope produces parallel retreat of that slope unit and a lower concavity where deposition occurs.

Results of modelling selected processes with assumed modes of action are shown in Fig. 9.11. Six theoretical cases are depicted: (1) the gradual reduction of slopes to increasingly gentle inclinations as a result of creep on upper convexities with slope wash on middle and lower slopes; (2) the parallel recession of slopes which undergo uniform rates of weathering and transport across the main slope units, with basal wash slopes; (3) the gradual elimination of steep slope units and the joining of upper convexities and lower concavities; (4) the undercutting of a cliff by waves or streams; (5) accumulations at the base of a slope as a result of a rise in baselevel; and (6) downcutting at the base of a slope as a result of a fall in baselevel.

Conclusions

The power of mathematical models is restricted by two main limitations: the ability of the modeller to select correctly and represent the significant variables in the evolution of a landscape, and the accuracy with which an equation describes a particular process. We have seen already how simple models are confined to a very restricted number of variables. Descriptive equations may also be limited: for example, the equation used to describe the rate of soil creep is commonly of the form:

$$C = K\sin\beta.$$

There are good theoretical grounds for this assumption as the resultant of the force of gravity acting at the ground surface varies in this way, but until far more long-term measurements are made of creep there can be no confidence that this equation is an adequate descriptor. If Young's (1978) measurements, indicating that creep is largely the result of solution and hence is largely an inwards directed process, are correct and universal, then the direct slope angle function may be incorrect.

Improved accuracy in modelling is thus very dependent upon long-term and detailed field measurement of processes, and correct representation of process mechanisms in descriptive equations. There has been considerable progress in mathematical modelling of processes since 1960, and detailed field and laboratory studies (e.g. Moeyersons and De Ploey, 1976) are providing more reliable data on the action of individual slope processes. Another

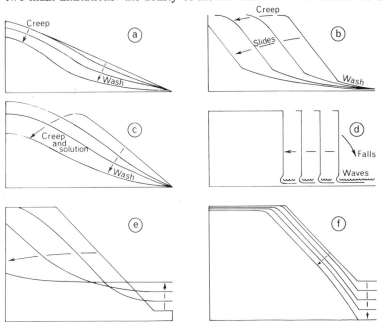

Fig. 9.11 Sequential slope profiles developed from mathematical models of slope change assuming given processes.

common limitation of models is that they are concerned with average rates of processes and do not consider catastrophic processes, even though these may account for a large part of total denudation in some environments.

The overriding problem with all models of landscape evolution is the ability to test them against natural conditions. Landforms, even in small drainage basins, evolve over thousands, or even millions of years, in which the intensity and type of dominant processes may change in a direction which cannot be known. It is very rare, therefore, for conditions to exist in which any model can be tested against any landscape. An attempt has been made by Parsons (1976) to test one mathematical model against the hillslope profiles measured by Savigear (1952) along an abandoned cliff on the coast of South Wales, where the spatial sequence of profiles from east to west was taken as an approximation to a developmental sequence through time. In this case reasonable agreement was found between the sequence of slope profiles predicted by the model and the actual slopes.

Models are probably most useful for predicting future changes where the dominant processes can be specified with some certainty. They are of less value for hindcasting because of the operation of the principle of equifinality, and because of our inability to define an initial landform to which the model must work back.

10

Magnitudes and frequencies of erosional events

The geomorphic importance of an erosional event is governed by the magnitude of the energy it expends upon the landscape, by the frequency with which it recurs, and by the work performed by processes operating in the period between severe erosional events. The greater the magnitude of an event the lower is the probability of its recurrence.

Recurrence intervals are expressed as a probability that an event will occur in a stated number of years (Table 10.1) A 10-year return period event has a 10 per cent chance of occurring in any one year, and a 100-year event a one per cent chance. An event of such magnitude that it has a probability of returning every ten years will not necessarily occur in every ten-year period, but it has a 99.9 per cent chance of occurring in every 50-year period. A statement of a return period is, consequently, not a forecast.

For reliable calculation of probabilities of occurrence the length of record should be at least as long as the recurrence interval. Calculation is from the relationship:

$$\text{return period} = \frac{N+1}{M}$$

where N is the number of years of record and M is the rank of an individual event in an array of annual maximum events of a similar class, such as floods or rainstorms (Dalrymple, 1960). The data for an individual site are usually plotted on logarithmic probability paper and a straight-line relationship describes the recurrence interval of events of given magnitudes (Fig. 10.1).

Records for longer than a few tens of years are not available for many parts of the world, and it is not possible to estimate the magnitude of very long return period events with confidence. A second source of uncertainty occurs with climatic events for, as the

record gets longer, it becomes increasingly subject to climatic changes, so that the calculated recurrence interval may no longer represent the probability of return of an event under modern conditions. Estimates of the magnitude of past events may be obtained from geological deposits, such as

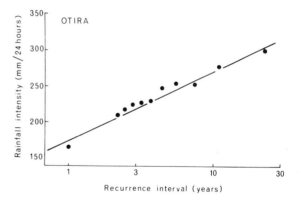

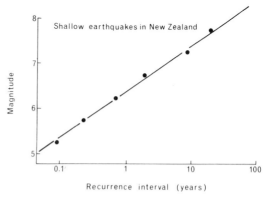

Fig. 10.1 Recurrence intervals for rainfall of given intensity for Otira Township in the New Zealand Alps, and for all shallow earthquakes of given magnitudes in New Zealand (data compiled by J. Adams).

Table 10.1 Probabilities of events occurring in set periods.

One hundred years	Fifty years	Twenty-five years	Ten years	Any one year	Return period years
					2
				50	
				40	
				30	
				25	
				20	5
		99	80	15	
	99.9	94	65	10	10
	90.5	71	40	5	20
86	61	40	18	2	50
64	39	22	9.6	1	100
40	22	12	5	0.5	200
18	9.5	5	2	0.2	500
10	4.8	2.5	1	0.1	1000
5	2.3	1.2	0.5	0.05	2000
2	1.0	0.5	0.2	0.02	5000
1	0.5	0.25	0.1	0.01	10 000

(Header of the first five columns: "Percent chance of getting one or more such or bigger floods in this many years")

events (Plate 10.1). Even where these deposits can be dated, however, they can seldom be fitted into a data array but only used as indicators.

The magnitude of an event usually influences the area it will effect. Very severe storms or earthquakes are usually experienced over large areas, but the area of the greatest intensity, and hence return period, is confined and lesser intensities occur away from the centre. This can be seen particularly well in the isoseismals for earthquakes (Fig. 10.2). Earthquakes which cause severe modification of landforms have intensities, that is degrees of shaking, of VII or greater on the Modified Mercalli Scale. At the centre of the earthquake shaking may be severe, but it declines away from the source. Where earthquakes are common isolines for the return period of events of given intensity can be mapped and seismically active and quiet zones delineated. It will be evident from Fig. 10.2 that there has been considerable extrapolation from recorded data by using geological evidence as a guide to earthquake occurrence and intensity.

Some storms have some of their effect through their prolonged duration (Fig. 10.3) compared with low-magnitude events, but this is not usually the case with earthquakes as, no matter what their intensity, the latter are all of similar duration.

Equilibrium

The concepts of magnitude and frequency of landscape-forming events implies that energy inputs

alluvial terraces which are known to have formed in a short period of time and may be related to single extreme storms, landslide deposits, or suddenly raised beaches which may be related to seismic

10.1 Boulder lines in a colluvial deposit indicating at least two periods of mobilisation of large slope debris separated by a period of lower energy soil deposition, Inland Karroo, South Africa.

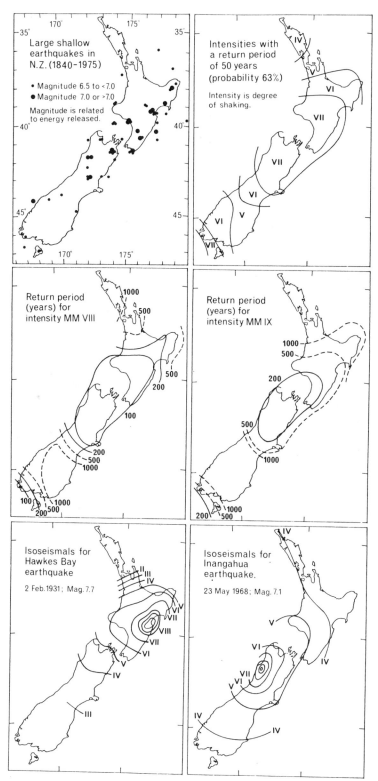

Fig. 10.2 Magnitudes, intensities, return periods, and isoseismals of New Zealand earthquakes (data from Smith, 1978a, b).

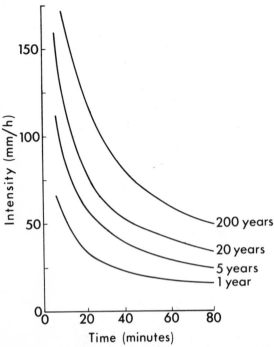

Fig. 10.3 Intensity-duration relationship of rainfall for various periods (after Wiltshire, 1960).

and change are not steadily progressive, unless viewed from the very long-term scale of the geological record in which indications of all discrete events are eradicated, or subsumed in the end-product of landform evolution.

In the shorter time scale of a few hundreds, or thousands, of years there appears to be an approximate relationship between landforms and the processes acting to modify them, such that hillslope processes which are active frequently, and are normally effective every year, do not disturb the approximate balance which usually exists between the rate of weathering and the removal of regolith. Except in deserts, therefore, there is an approximate equilibrium between weathering, soil development, the vegetation cover, and the rate of erosion on hillslopes. Extreme events destroy this equilibrium, breaking the vegetation cover, stripping away regolith along the track of gullies and landslides, and producing large influxes of sediment into valley floors. As it is uncommon for several extreme events to occur in a short interval, the scars and the deposits may then be slowly modified by lesser intensity processes such as creep, solution, and wash, while weathering, soil formation, and vegetation gradually re-establish a surface which is in approxi-

mate equilibrium with the energy of the processes usually acting on the slope. After a period the equilibrium may be broken by another extreme storm. This type of episodic development of the land surface can be visualised as occurring in a step-like manner (Fig. 10.4) in which storm events are followed by gradual periods of adjustment to the normally active processes and then, when an approximate equilibrium is re-established, by a period of relative stability.

Step-like functions are also evident from studies of seismic energy release (Fig. 10.5) (Robinson, 1979). There are clearly periods of seismic quiescence followed by the release of accumulated strain energy. At no time does the rate of energy release appear to have a value close to the long-term average for any length of time. It is still uncertain if the pattern of energy release can be extrapolated to provide a reliable prediction of earthquake occurrence. Even if it could, the earthquakes would not necessarily cause landform change, as their effect depends upon their location and depth as well as their intensity.

The term 'dynamic equilibrium' has been defined in many ways, but is frequently used to describe the balanced fluctuations about a constantly changing condition which is characterised by a sequence of unrepeated average states through time. A hillslope in a humid climate, undergoing modification only by creep and solution processes, appears to be in a steady state but is, presumably, experiencing some fluctuation in the intensity of those processes. These minor fluctuations appearing to be in a steady state define the dynamic equilibrium for that hillslope.

A second hillslope, also in a humid climate, but with higher relief and steeper slopes, may evolve episodically by landsliding followed by a period of adjustment to the landsliding event. This adjustment may involve gradual modification of the landslide debris and revegetating of the scar and debris lobe. For such a hillslope the dynamic equilibrium includes the severe events. Some writers (e.g. Chorley and Kennedy, 1971) have used the term 'dynamic metastable equilibrium' to describe this condition, but the distinction becomes arbitrary, and unhelpful, as the return period of the dominant event becomes shorter.

Thresholds

Any change in the landscape depends upon a threshold being passed in which the strength of a

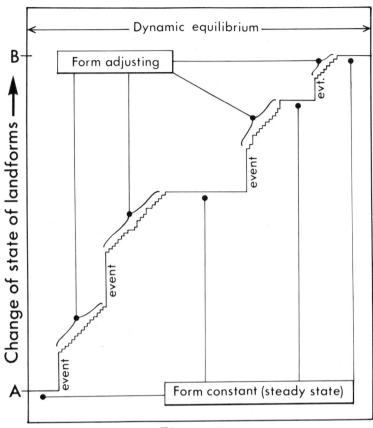

Fig. 10.4 Within the term dynamic equilibrium there are subsumed the three states of: (i) a landforming event; (ii) the adjustment of form which follows after that event; and (iii) a period of steady state in which there is virtually no adjustment of form. The curve which represents the change of landforms with time may, therefore, rise very steeply, gradually, or hardly at all, depending upon the magnitude and frequency of the dominant process (after Selby, 1974b).

rock or soil material is exceeded by an applied stress. For a sand grain to be moved by surface wash the stress applied by the moving water to initiate transport need be only very low ($<1N/m^2$), but the stress required to initiate a landslide is thousands, or millions, of times larger.

The exceeding of a threshold stress causes a step-like change in landforms, but the nature of the threshold may be one of three kinds.

(1) An increase in external stress produces a sudden change — as when a storm causes an increase in flow velocity and depth over bare soil. If the soil is vegetated, however, the soil particles may remain immobile until a shear stress large enough to cut through the ground cover is applied. Then, a threshold will be crossed suddenly and rill or gully erosion of the underlying soil will be rapid and, perhaps, severe.

(2) A reduction in internal resistance by progressive weathering may operate less obviously by lowering the shear stress that is required to initiate instability. Internal changes can consequently give

rise to threshold conditions without the operation of large external stresses.

(3) Another type of threshold is that resulting from gradual landform changes until a condition of

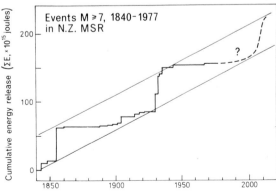

Fig. 10.5 Cumulative energy release for the main seismic region of New Zealand for earthquakes of magnitude greater or equal to 7. The step-like pattern is evident and a possible extrapolation is shown by the dashed line (after Robinson, 1979).

potential instability is reached. Where a clay stratum is overlain by a resistant sandstone stratum progressive weakening of the clay by seeping water may cause it to fail, and the sandstone will collapse once a critical support has been removed.

Periods of form adjustment

It has been shown that the energy required to reach a threshold stress is related to gradient, lithology, soil cover, vegetation, and climatic or seismic events. Similarly the period of recovery from such an event is controlled not only by the magnitude of the event but by many external and internal factors. A severe storm producing say 300 mm of rain in 24 hours may cause landsliding on steep, but forested, hillslopes in a humid climatic zone. Within five years the scars may be covered by grasses and herbs, and 50 years later by maturing trees: virtually no trace of the storm may then survive in the landscape. The same storm in a semi-arid zone may cause so much erosion, by wash, gullying, and debris flows, that its effect will still be noticeable hundreds of years later, because of debris flow levees on the hillslopes and alluvial infills in the headwater channels.

The period of adjustment after an event is consequently a guide to the work done by that event in modifying the landscape. Its *effectiveness* is, therefore, best understood by scaling it as a ratio to the mean annual erosion (Wolman and Gerson, 1978). Recovery times from one event of a certain magnitude may thus vary from one to hundreds of years depending upon the local environment (Fig. 10.6). For such data to be accumulated for different parts of the world it is essential that all events be reported with a statement of the recurrence interval and recovery period from the event.

It must also be recognised that the dominant geomorphic event on hillslopes may vary with the slope angle so that, as Simonett (1970) reported in New Guinea, the threshold angles of slope for various processes may be: mudflows − 2°, rotational slumps − 8°, debris slides and complex landslides − 15°, debris avalanches − 25 to 30°.

Extreme events in slope evolution

The importance of extreme events in hillslope change has been a neglected topic until recently because of the lack of data, and inability to date

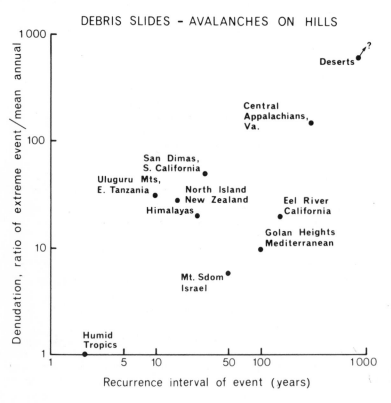

Fig. 10.6 Denudation of hillslopes, as indicated by volume of debris removed, during storm events in various parts of the world. The data points indicate the importance of extreme, compared with common, events in different environments (modified from the compilation by Wolman and Gerson, 1978).

past large-magnitude events. The occurrence of a number of extreme events in the twentieth century has focused attention on extremes (e.g. Starkel, 1979).

Mountains are particularly subject to the influences of extreme events because of high precipitation at high altitudes caused by extreme windiness and large water vapour flux; their high relief energy; high drainage density; glacially oversteepened or deeply incised river valleys; the presence of unconsolidated glacial and colluvial debris; the presence of ice and snow; the lack of vegetation at high altitudes; and the occurrence of earthquakes in young fold mountains. As a result the gradual processes of solution and creep may be of far less relative significance in mountains than upon the lowlands of the world.

Large rockfall, rock avalanche, and rock slide events may leave their imprint on the landscape for thousands of years. In parts of the Rocky Mountains cliff falls involving 1 to 100 Mm^3 of rock have occurred with a frequency of slightly less than once in a thousand years, since deglaciation, with greatest frequency occurring during and immediately after deglaciation (Gardner, 1977). Rock avalanches of the magnitude of the Huascarán event are likely to occur in the Andes only once in a thousand years or more, and in individual valleys catastrophic events, which do enormous amounts of work, may have an even lower frequency. For example, a rainstorm, in the basin of the Guil River in the southern Alps of France, lasting only 48 hours, removed more talus from the mountain catchment than had been moved in the preceding 10 000 years (Tricart, 1962).

Individual sites in alpine areas may be affected by mudflows perhaps once in 10 or once in 100 years, but cliff faces may experience repeated or nearly continuous minor rockfalls in the course of one year.

As a result of nearly continuous processes cliff faces may recede at rates of 0.1 to 2.5 mm/year with common values being around 0.7 mm/year (Caine, 1974; Barsch, 1977). This is probably 10 to 1000 times the rate at which broad interfluves are lowered. The great difficulty is to assess the significance for total denudation of the rare events. It has been suggested by Whalley (1974) that very infrequent events can still have a major effect upon total erosion. For example, the Rhine from its Alpine reaches is depositing about 1 Mm^3/year of sediment in Lake Constance. The volume of the Flims rock slide was about 12 Gm^3: if such a slide occurs only once in 1000 years it will still be making a large contribution to denudation in the Alps and may exceed the significance in a large catchment of all minor processes together. Its debris, of course, still has to be removed by other processes.

In any part of the world mountains are far more subject to extreme events than lowlands, but even in mountains the effects of extreme denudation are frequently limited to quite small localities. Thus in a mountain range catastrophic events might occur nearly every year, but each time hitting a different area so that overall the frequency is much less than one event a year. Similarly the sediment produced by a storm may range from a few tonnes/km^2 to hundreds of thousands of tonnes/km^2 and the average downwearing of the affected area from 0.01 to 200 mm. This range occurs because the proportion of the ground, in one small valley, suffering extreme erosion during a storm may range from less than 1 per cent to more than 50 per cent.

Regions of extreme climatic events

The importance of extreme events in different regions of the world is very variable. Starkel (1976) has reviewed studies of extreme events and concluded that four classes of region may be distinguished.

(1) Regions with a frequency of extreme events of 5–10 per century and with events of such magnitude that in each the denudation greatly exceeds the denudation produced by all low-intensity processes during 100 years. Such regions are most common in tropical monsoon and Mediterranean climates, and in steep uplands and farming lands of the temperate zone where human action has removed or changed the vegetation cover.

(2) Regions in which extreme events are rare and in which such events do not exceed the total denudation of 100 years of low-intensity processes. In these areas intense rainfalls or snowmelts occur each year. Such regions are common in the semi-arid zone.

(3) Regions with very rare extreme events. Because of normally very low precipitation a storm may achieve much more denudation than usually occurs in 100 years. Arid zones and some parts of the boreal zone are in this category.

(4) Stables regions show little variation from normal denudation rates. Many Arctic, Antarctic, continental-boreal, and lowland zones of the temperate regions are in this group.

Data on the frequency of extreme events are not available for many parts of the world, but some

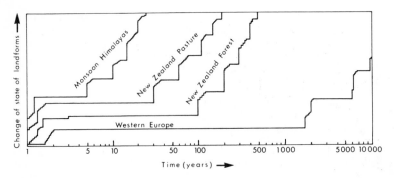

Fig. 10.7 The magnitude and frequency of landsliding in different environments.

comparisons are presented in Fig. 10.7. In the hills and lowlands of Western Europe periods of landsliding appear to be related to intervals of wet climate since the last glacial (Fig. 10.8). These periods have frequencies of about 1 in 2000 years, although each wet period lasted several hundred years and may have contained many extreme storms. At the other end of the scale the frontal ranges of the Himalayas north of the Ganges delta receive prolonged and intense rainfalls nearly every year and extreme landforming events probably have a frequency of about once in five years or less.

Less frequent are the storms causing landslides in the North Island of New Zealand. Under original forest the hills (steeper than 20°) are affected by shallow landsliding about once every 100 years, but in the last century many uplands have been cleared of forest and pastures now cover the hills. These are less protective than the forests, and storms of low intensity, which recur about once in 30 years, now cause slope instability (Selby, 1967a, b, 1976; Pain, 1969; Starkel, 1972a, b; Jones, 1973).

Effect on valley floors

The debris from a destructive storm may be carried directly into a large river and removed, but in the small catchments of many uplands it is stored, at least temporarily. The debris may come to rest as

fans, talus slopes, or debris cones, but frequently it becomes incorporated in a debris flow which causes rapid infilling of valley floors. The infill may later be dissected by streams to leave a terrace (Plates 10.2 and 10.3).

Periodic infilling of valley floors has been dated in North Island, New Zealand, by the oldest trees on a terrace and from the known age of volcanic ash deposits which have fallen on each terrace surface — the basic assumption being that the terrace cannot be younger than the trees or ashes on it. In the Hawke's Bay area of New Zealand terraces are dated as being A D pre-130, 1450, 1650, and 1840. The more recent terraces suggest that very large catastrophic events may occur every 200 years or so, but it is certain that more frequent severe storms also cause much erosion. The debris of lesser events, however, may be largely removed from the catchment either in a series of pulses as large floods carry waves of debris down the channels, or by more gradual processes occurring every few months (Grant, 1965). In forested areas fallen trees temporarily block stream channels creating infilled floodplains behind the dam. When the dam breaks the infill may again be incised and a low terrace formed (Pain, 1968). This type of infilling and terrace formation is clearly a temporary phase in mountain valley development, for the long-term trend is

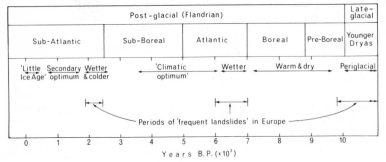

Fig. 10.8 Starkel's (1966) periods of frequent landslides in Europe.

10.2 A large alluvial fan formed by deposition of landslide debris during intense storms, Northern Ruahine Range, New Zealand. Subsequent stream incision has created large terraces in less than a year.

towards stream incision and debris removal. It is however very common in upland areas (e.g Machida, 1966) and emphasises the great importance of extreme events, not only for slope evolution but also for valley floor changes.

Accelerated, induced, and normal erosion

Enough has been said to demonstrate that the rate of change of most landscapes is extremely variable, yet terms such as 'accelerated' and 'normal' erosion are in common use. Without very long-term records of climate and landform change it is not possible to say confidently what is a 'normal' rate, and once the frequency of landform changing events is greater than a few hundred years the effect of climatic change has to be considered. It is probable therefore that 'normal' rates of erosion are inherently variable and that periods of faster or 'accelerated' erosion are part of the common sequence of events.

Landform change induced by human interference,

10.3 Deposition of gravels, originally derived from landslides, has buried the lower terraces on which trees were growing and left a record of the storm event. Northern Ruahine Range, New Zealand.

especially with the vegetation cover, has often accelerated erosion rates by 10 to 1000 times those of normal erosion beneath natural vegetation. To avoid confusion with climatically controlled accelerated erosion it is useful to call that resulting from human interference 'induced' erosion.

Conclusion

Extreme or catastrophic events are of greater significance, in the long-term development of hillslopes, than has been commonly accepted. Their significance can, however, be recognised only when their magnitude and frequency are known, and when the length of the recovery time is known, so that the effect of one event can be evaluated against the rate of change caused by slower, less obvious processes which modify the landscape.

It has been contended that in rivers most of the work of transportation is carried out by floods which are of such magnitude that they recur with a frequency of at least once in five years (Wolman and Miller, 1960). On many hillslopes subject to landsliding the dominant process of change may have a much lower frequency, but greater magnitude, and the valley floors of low order headwater channels in uplands may be strongly influenced by the landslide debris, so that the frequency of dominant events is far less than once in five years. More exact statements must await the accumulation of data over long periods, of at least 50 years, from a much greater range of environments.

11

Rates of denudation and their implications

Denudation is the result of weathering and the stripping of weathering products from the surface of the earth by the processes of erosion.

Four types of evidence are commonly used for estimating the rate of terrestrial denudation:

(1) The most common method, and the one most suitable for large-area studies, is to measure sediment and dissolved load discharged by rivers and then to convert this information to a rate of downwearing of the landscape.

(2) For small areas the sediment accumulated in reservoirs in a given period of time is a valuable, and sometimes very accurate, indicator of erosion and can be used to estimate the results of human interference with the landscape.

(3) Processes occurring on slopes may be measured directly; the rates of soil creep, surface wash, and landslides can be computed and, in a few cases, their contribution to total denudation assessed.

(4) Accurately dated land surfaces and the landform changes which have occurred on them can give indications of total areal denudation.

Pioneer studies of denudation rates fall into two main classes. The first reliable estimate of the overall rate of ground loss was made from a summary of river sediment loads in the United States by Dole and Stabler (1909). Sediment yields from drainage basins were also used in the three attempts to assess denudation on a world basis (Corbel, 1964; Fournier, 1960; Strakhov, 1967). The first attempts to record quantitatively all the slope processes within a stream catchment were those of Jäckli (1957) in Switzerland, Rapp (1960a) in Kärkevagge, Sweden, and Iveronova (1969) in the Kirghiz SSR. Many hundreds of reports of individual processes or catchment yield are now available; Young (1974) has summarised the data to 1973.

Methods of reporting data

Many different ways of reporting data on erosion rates are in use; consequently it is often difficult to compare information from different areas. A major problem is that most data are derived from discharge of material by streams and consequently are expressed as the mass, or volume of sediment, transported out of a catchment (in kg, t, m^3 or parts per million (ppm), with $1 \text{ g/m}^3 = 1$ ppm for sediment in suspension). This method has the advantage that masses or volumes may be averaged for a unit area of the catchment in a unit of time, to give a mean rate of lowering of the land surface. Data on transport rates of soil on hillslopes, however, are usually defined by the velocity of tracer material (m/year) or by the discharge of superficial soil through a unit contour length (m^3/m/year). The two classes of data are not readily converted and the results may be difficult to visualise.

A convenient measure is the Bubnoff Unit (B) with $1 \text{ B} = 1 \text{ mm}/1000$ years, which is equivalent to 1 m/M years. Average rates of ground lowering may be converted to volumes or masses of material removed, by the relationship: 1 mm of ground lowering = removal of $1000 \text{ m}^3/\text{km}^2$, which may be converted to masses by multiplying by the bulk density of soil or rock. Average densities (Mg/m^3) are:

silts	1.25	silty gravel	1.8
gravel	2.1	bedrock	2.65

Hence 1 mm/year of bedrock lowering removes 2650 t/km^2 per year, so $1 \text{ B} = 2.65 \text{ t/km}^2$ per year for removal of bedrock. The Bubnoff Unit for rate of ground lowering is convenient as it can be easily visualised either for rates of channel incision, surface lowering, or cliff retreat. It indicates a rate of change of landforms. It has the disadvantage of being an average which suggests a uniformity through time and space which is unreal.

A second difficulty in comparing data derives from the nature of storage within a geomorphic system. Many slope processes involve only transfers of material within a drainage basin and not its export through the channel. Slope debris is commonly stored in fans, talus slopes, on terraces, or on flood plains, so that even though hillslopes may be changing rapidly this change is not necessarily expressed in channel transport. The disparity is particularly noticeable with increasing area of the catchment, as hillslope debris in small catchments is commonly delivered directly into the channel, but in large catchments is held in a store (see Figs. 1.1 and 1.2).

A third difficulty is that the usual methods of expressing erosion are not comparable with the data for energy inputs, such as rainfall intensity or earthquake energy (both of which can be measured in joules or watts).

It has been suggested by Caine (1976), therefore, that erosion and sediment transport should be expressed as geomorphic work in power units (W). Erosion of the landscape involves the movement of sediment from higher to lower elevations and is a form of physical work which involves a reduction of the potential energy of the landscape. Potential energy (E) is given by:

$$E = mgh$$

where m is the mass of sediment moved, g is the gravitational acceleration, and h is the elevation. The change in potential energy through time is given by:

$$\Delta E = mg(h_1 - h_2)$$

with units of: Joules = kg $\times$ 9.81 m/s^2 $\times$ m. The erosion rate, or work accomplished, will then be defined as power in J/s = W; with power being an estimator of erosional intensity due to the movement of sediment, not to the transporting agent.

In fluvial studies the difference in elevation between the site of erosion and outlet from the catchment may be known with some precision, in many other cases it will be estimated from the areally weighted mean elevation of the catchment. For most stream channel situations the equation for ΔE becomes:

$$\Delta E = v\rho g(h_1 - h_2)$$

where v is the volume of sediment (m^3), and ρ is its density (Mg/m^3).

In the study of slope erosion where particles or tracer velocities are used, an estimate is required of the depth and area of the soil to which the measured velocity can be applied. Then a volume is defined and converted to mass by multiplying by the soil bulk density. Where sediment is discharged across a unit contour length the thickness of the moving mantle is usually observed directly. The quotient of discharge and thickness gives the mean velocity, and the product of planimetric area and thickness gives the volume of moving waste. In both cases the velocity or distance (d) of movement, is measured along the slope and must be reduced to its vertical component by the sine of the slope angle ($\sin\beta$). The equation then becomes:

$$\Delta E = v\rho g(d\sin\beta).$$

Within a catchment, this equation may be evaluated for different environments and the result summed to give an estimate of total erosional activity. In Table 11.1 are presented data on geomorphic activity in one small catchment in the San Juan Mountains, Colorado, and comparable data from many other areas. In studying this information it must be remembered that discrete events, such as rockfalls and landslides, may be infrequent and their contribution is corrected by an estimate of their frequency. For such processes frequency data are usually inadequate for a reliable estimate.

Volumes of material removed from catchments and rates of downwearing are best expressed in the equivalent units: 1 B = 1 mm/1000 year = removal of 1 m^3/km^2 per year. For comparison of the effectiveness of different processes, power units (W) are most appropriate.

World denudation studies

Some of the problems of deriving accurate estimates of denudation can be appreciated from a comparison of the work of Corbel, Fournier, and Strakhov.

Corbel summarised total denudation data for different climatic zones in three humidity categories and two relief categories. His results are summarised in Table 11.2. They indicate that erosion from mountainous areas is considerably higher than that from plains and that there is a general climatically controlled trend with low erosion rates in all tropical areas, except for humid mountain zones, and high rates in temperate and cold climates. Erosion increases with precipitation in all climatic zones. Compared with other estimates Corbel's figures

Table 11.1 Geomorphic Activity in a Small Mountain Catchment and Comparable Data from many sources

Process	Area (km²)	Thickness (mm)	Bulk Density (Mg/m³)	Annual Power (W)	Comparable data from other studies. Annual Power (W)
Surface wash	0.32	5	1.9	0.144	0.002–0.190
Soil creep and solifluction	0.77	200	1.9	0.107	0.075–2.271
Mudflows	0.01	–	1.9	0.355	0–51.078
Rockfall	0.20	–	2.6	0.130	0–0.102
Snow avalanche debris	0.02	–	2.6	0.001	0–0.015
Solute transport	0.98	–	2.6	0.102	4.040–20.104
Suspended load		Not measured			307.3–587.2
Total	0.98	–	–	0.841	

Source: modified from Caine, 1976.

Table 11.2 Corbel's Estimates of Rates of Regional Erosion (Expressed in Bubnoff Units)

Climate	Arid under 200 mm		Normal 200–1500 mm		Humid over 1500 mm	
	Mountain	Plain	Mountain	Plain	Mountain	Plain
Hot 15°N–15°S	1.0	0.5	25	10	30	15
Tropical 15°–23.5°N and S	1.0	0.5	30	15	40	20
Extra-tropical, Temperature over 15°C	4	1	100	20	100	30
Temperate, Temperature 0–13°C	50	10	100	30	150	40
Cold, Temperature under 0°C	50	15	100	30	180	–
Polar	50	15	100	30	150	–
Glaciated, polar	50	–	1000		2000	
Glaciated, non-polar	–	–	–	–	2000	

appear to underestimate the rates of erosion in the tropics, but it has to be recognised that he worked with few data.

Fournier studied suspended sediment yield in 78 drainage basins which vary in size from 2460 to 1 060 000 km². He correlated the sediment yield with a climatic parameter p^2/P, where p is the rainfall of the month with the greatest rainfall and P the mean annual rainfall (in mm). The results show three distinct groupings into: (Ia, Ib) basins with low relief; (II) basins with high relief and a humid climate; (III) basins with high relief and a semi-arid climate. There is clearly a relationship between sediment yield and increase in rainfall, but the relief factor is of major importance (Fig. 11.1). Fournier derived the empirical equation for predicting sediment yield when climate and relief are known:

$$\log E = 2.65\log(p^2/P) + 0.46\log\bar{H}.\tan\alpha - 1.56$$

where E is suspended sediment yield (t/km² per year), $\bar{H}$ is mean height (m), and α is the mean slope of the drainage basin. Where $\bar{H}.\tan\alpha$ is not known erosion may be calculated from the regression equations for each of the three groups (I, II, III above) of Table 11.3 in which Y is the parameter p^2/P, and X the sediment yield.

Using only data for p^2/P and sediment yield

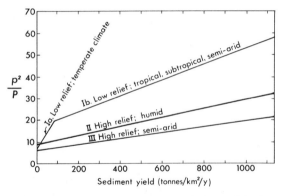

Fig. 11.1 Relationships between Fournier's precipitation parameter p^2/P and suspended sediment yield for drainage basins in different climatic regions (after Fournier, 1960).

Fournier was then able to map the world distribution of erosion rates (Fig. 11.2). He agrees with Corbel on the climatic and relief controls on erosion but, in contrast, he finds maximum erosion rates in the seasonally humid tropics, lower rates in the equatorial regions where the seasonal effect is lacking (and the vegetation more complete), and the lowest rates in the arid regions where the fluvial transport of sediment is virtually nil. The rate of erosion rises in the seasonally wet Mediterranean lands, but in temperate and cold areas it is low, except in mountain areas. In brief, Corbel and Fournier estimate opposite trends in the humid areas.

Strakhov (1967) has attempted to estimate world denudation rates by extrapolating from sediment yields for 60 river basins which vary in yield from 0.82 to 1000 Mt/y. His main conclusions are that denudation is strongly related to both climate and relief. Two broad parallel zones with distinct indices are recognised (Fig. 11.3):

(1) A temperate moist belt in the northern hemisphere with an annual precipitation of 150–600 mm. Its southern boundary is the +10 °C annual isotherm. In this zone the intensity of mechanical denudation is small and around 10 t/km² per year. In a few places it is up to 15 t/km² and only in one zone of North America does it reach 50–100 t/km² per year.

(2) The second zone includes the humid areas of the tropics and sub-tropics, and lies between the +10 °C isotherms of each hemisphere. Over most of this zone the mean annual temperature does not fall below 20 °C and rainfall is high at 1200–3000 mm a year. Mechanical denudation is markedly greater

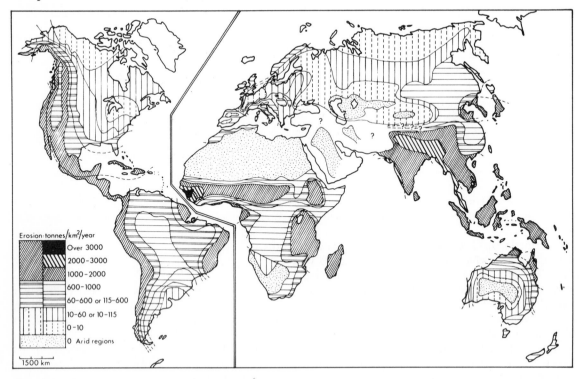

Fig. 11.2 World denundation rates (after Fournier, 1960).

Table 11.3 Fournier's Relationship between Climate and Sediment Yield

Curve	Regression equation	t	Degrees of freedom	p
Ia	$Y = 6.14X - 49.78$	9.16	46	<0.001
Ib	$Y = 27.12X - 475.4$	22.07	38	<0.001
II	$Y = 52.49X - 513.21$	18.1	29	<0.001
III	$Y = 91.78X - 737.62$	28.06	21	<0.001

than in the temperate zone of the northern hemisphere. In large areas it is 50-100 t/km², but in a number of places it rises to 100-240 t/km², and in southeastern Asia it averages 390 t/km². In the basins of the Indus, Ganges and Brahmaputra the value exceeds 1000 t/km² per year.

Strakhov's estimate of the significance of relief is illustrated in Fig. 11.4. It is evident that denudation is far more intense in areas of high relief — which are areas of greatest tectonic activity.

The effect of relief is particularly well shown in a comparison of the loads carried by major rivers (Gibbs, 1967; Grove, 1972). The Amazon has far higher total loads than the other tropical rivers (Table 11.4) and Gibbs estimates that 80 per cent of its load is derived from its Andean headwaters

even though they occupy only 12 per cent of the catchment area.

Figure 11.4 also shows that although there is a varying mathematical relationship between the values of mechanical and chemical denudation, it is clear that chemical denudation increases with the increase of mechanical denudation and vice versa. The correlation coefficient between the two types of erosion for all samples is 0.97. Both types of erosion therefore are subject to the same climatic controls.

In general the patterns of erosion deduced by Strakhov and Fournier are similar but the values obtained may differ by up to an order of magnitude, with Fournier's estimates being the higher.

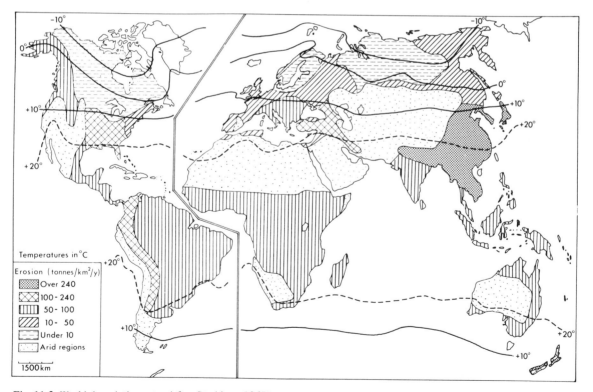

Fig. 11.3 World denudation rates (after Strakhov, 1967).

Table 11.4 Discharge and Denudation in Major Catchments

	Amazon	Congo	Mississippi	Niger	Benue	Niger–Benue
Discharge $(1 \times 10^{12}$ m^3/y)	5.5	1.2	0.5	0.10	0.11	0.21
Erosion rates Total basin (Mt/y)						
by solution	232	98.5	118	5.5	4.5	10
suspended solids	499	31.2	213	9	22	31
combined	731	129.7	231	14.5	25.5	41
Per cent by solution	32	76	36	38	17	24
Per km^2 (t/y)	116	37	100	19	77	37
downwearing (B)	46.4	14.8	40.2	8	36	15

Note: The data assume that all solutes are derived from the catchment (and none from dust or rain).
Rates of downwearing assume that all materials are derived from the slopes and that all material from the
slopes is carried away by the rivers.
Sources: R. J. Gibbs (1967) and A. T. Grove (1972).

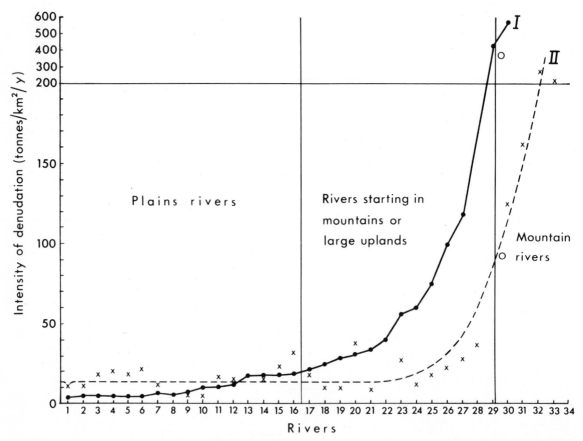

Fig. 11.4 Relation of mechanical (I) to chemical denudation (II). 1. Neva, 2. Yenisei, 3. Luga, 4. Narova, 5. Dnieper,
6. Onega, 7. Ob, 8. Western Dvina, 9. Kolyma, 10. Yana, 11. Mezen, 12. Southern Bug, 13. Northern Dvina, 14. Ural,
15. Don, 16. Volga, 17. Pechora, 18. Indigirka, 19. Amur, 20. Dniester, 21. Kuma, 22. Kalaus, 23. Syr-Darya, 24. Amazon,
25. La Plata, 26. Yukon, 27. Mississippi, 28. Kuban, 29. Kura, 30. Amu-Darya, 31. Terek, 32. Rion (Rioni), 33. Samur,
34. Sulak. The circles indicate the rivers of southeastern Asia: Indus, Ganges, Brahmaputra, Irrawaddy, Mekong, Yangtze
Kiang, and others (after Strakhov, 1967).

Sources of error

Data collection problems are the main cause of the different values, obtained by the three workers quoted, for world-wide erosion rates. A secondary contributing factor is the different methods of calculation.

Rates of denudation are derived from the rates of removal of earth material by a river from its catchment. Thus the annual rate of transport of earth material past a point in a stream, whose catchment area above that point is known, gives a rate in t/km^2 per year.

The areas of catchment can be measured accurately but the mass of material moved by the river is difficult to measure. Rivers transport material as bed load, in suspension, and in solution. Most early estimates of denudation were based on suspended load and solution load only. It is still impossible to measure bed load accurately and estimates of the proportion of bed load to total load vary from nil in the lower reaches of streams to 55 per cent or more in mountain reaches.

The accuracy of estimates of suspended and solute load transport depends upon the frequency of the measurements and the length of the period over which the measurements are made. The annual suspended load may vary by as much as a factor of five in successive years. Dole and Stabler (1909), with only one year's record of suspended sediment transport, estimated that the rate of denudation for the whole of the United States of America is 33 B, but Judson and Ritter (1964) with a longer record estimated the rate at 61 B.

For the few catchments in which daily suspended sediment and solute samplings have been carried out over an extended period, it is possible to construct sediment and solute hydrographs and annual discharges of load can then be calculated. Where the measurements of load are less complete, load and water discharges are correlated and total load is then estimated from the water discharge hydrograph. When sufficient correlations have been made the flow duration, or annual frequencies of discharge, can be converted to a suspended sediment or solute curve. With corrections for bed load this can be made into a denudation curve. Suspended sediment loads vary even with equal discharges so the reliability of the correlations depends upon the size of the sample of measurements.

Inaccuracies may result from inadequate measurement of water discharge; inaccurate laboratory analyses of sediment and solutes; poor sample collection in the field; and inadequate sampling times. From studies carried out on rivers in Devon, Walling (1978) has suggested that absolute errors associated with suspended sediments can be as high as +60 per cent for annual loads and between +400 and −80 per cent for monthly loads. For solution loads errors may be up to ±60 per cent for monthly loads.

Drainage basin discharge measurements can only estimate the debris which leaves the basin. They cannot take account of the colluvium distributed on slopes or alluvium left on flood plains. In a study of ten large river basins in southeastern USA Trimble (1977) found that while upland erosion was proceeding at about 9.5 B sediment yields at the mouths of catchments were only 0.53 B. The delivery ratio was thus only 6 per cent, and the difference is stored in valleys and channels. Storage times in some mountain valleys may reach thousands of years where late Pleistocene glacial and talus deposits are not being removed from the catchment.

The opposite situation may also occur with little erosion on fully vegetated slopes but channel bank erosion producing large quantities of debris. In such areas the denudation rate as measured at stream outlets is far greater than that from the slopes. Similarly if a river suddenly erodes unconsolidated glacial or talus deposits its load will increase far above the rate of slope denudation.

Human interference with natural vegetation may be the major source of error in estimates of denudation. Most stream load data collected by national agencies are obtained not for the purposes of geomorphological research but for water quality, river training, flood control, or some other purpose. Consequently they are most readily available from drainage basins with intense land use and often with considerable population. Deforestation and wasteful farming practices have been characteristic features of the opening up of North America and other new lands for settlement. The increase in runoff and acceleration of soil erosion which has resulted still continues in spite of soil conservation measures. Silted stream beds, buried floodplains, infilled reservoirs and estuaries are all witness to the results of human interference. It has been estimated that the conversion of forest to cropland in the middle Atlantic seaboard states of USA has increased sediment yield up to tenfold (Meade, 1969). It has also been estimated that accelerated erosion has stripped some three to four Gm^3 of sediment from the upland of the Piedmont in southeastern USA since AD 1700. Over 90 per cent of this sediment remains stored on hillslopes, in stream bottoms, and reservoirs. Only a small proportion of the sediment has so far reached the sea (Meade, 1976). Upland soil profiles have lost 150 mm. Since the decline of agricultural land use in the early 1900s small upland tributaries have adjusted to decreased sediment loads by entrenchment into, and erosion of, sediment deposited since the initiation of colonial agriculture (Costa, 1975).

The general pattern of relationships between land use, erosion rates, and channel conditions was demonstrated by Wolman (1967) for the area around Washington, D.C. (Fig. 11.5). Such changes are severe but locally they can be even more catastrophic. Pearce (1976) has shown that the destruction of vegetation around Sudbury, Ontario, over an area of 125 km² has increased local denudation rates by two orders of magnitude to about 37000 B. It has to be remembered that such rates cannot continue for long as they are caused by erosion of the soil and regolith which may have taken thousands of years to form. Denudation rates will fall when the regolith has been severely depleted.

Coal-mining, urbanisation, and highway construction have all increased sediment yields. By the end of the nineteenth century some of the rivers of Pennsylvania became so choked with anthracite coal debris that their bottom sediments could be dredged profitably for the coal they contained. Anthracite debris from the Susquehanna River basin is found in the modern bottom sediments of Chesapeake Bay as far as 40 km beyond the mouth of the river. Even though the Pennsylvania anthracite industry has declined since 1917, 10 per cent of the suspended matter measured in the Susquehanna

River in the spring flood of 1960 was coal. It has been shown that similar increases of loads of rivers in populated areas occur in many other parts of USA, in Australia, Malaysia and Europe (e.g. Douglas, 1967; Judson, 1968).

As a result of his studies of river discharge data for streams draining to the Atlantic seaboard of the USA Meade (1969) has concluded that

... as the present dissolved load of the Atlantic streams is nearly equivalent to the detrital load ... the errors in estimating each type of load are roughly equivalent. Considering the distribution and effects of different land uses in the Atlantic states, I estimate that the present sediment loads are four to five times what they were before the European settler arrived. On the basis of studies made in North Carolina and New Hampshire I estimate that about one-fourth of the dissolved loads of the streams represents material contributed by the atmosphere. Another one-tenth of the dissolved load may represent material added directly to the streams by human activity. Previous estimates of the natural rate of denudation of the Atlantic states therefore have probably been too large by at least a factor of two'.

Dissolved loads are particularly difficult to assess for many natural sources of solute may contribute to the total solute load. Janda (1971) has pointed out that dissolved gases, cyclic salt (i.e. salt from oceans or land areas, carried into the basin by wind), soil dust, volcanic ash, connate water, and soil organic matter may all greatly increase the solute load and give erroneously high estimates of denudation (Fig. 5.31). Janda has estimated that compu-

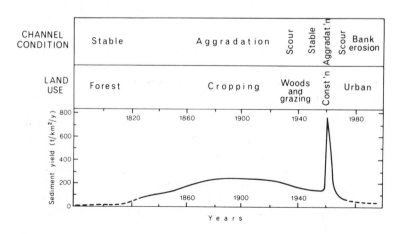

Fig. 11.5 The effect of land use changes on sediment yields near Washington, D.C. (after Wolman, 1967).

tations of chemical denudation rates which use total dissolved load – as most published computations have – exaggerate those rates by 1.4 to 2.4 times the real significance of this process. In areas of close human population solutes from sewerage, pesticides, fertilisers, and industrial wastes can also greatly increase solute loads. It has been pointed out that in 1958 the discharge of chloride ion alone into the Rhine River at Rees-Lobith was about 9.2 Mt (Durum *et al.*, 1960).

Not only man-made sources of solutes make calculations of denudation difficult. Natural sources of contamination are also a problem. Rainfall and snow carry low concentrations of soluble salts, some of which may find their way into the drainage waters. Coastal areas are particularly subject to precipitation containing marine salts and the decline in chloride content of rain away from the coast is often very distinct. In continental areas salts from endoreic drainage basins, soil dust, and, in a few places, dust derived from decaying vegetation are added to drainage waters.

Man-made sources of airfall material frequently obscure natural effects, as the following figures for sulphur concentrations in air over Britain readily show (Stevenson, 1968; Garnett, 1967):

Lerwick, Shetland Islands: $<5\ \mu g/m^3$
Leeds: $12\text{–}57\ \mu g/m^3$
Sheffield: $>12\,000\ \mu g/m^3$.

The contributions of domestic and industrial coal burning increase the concentration of sulphur in the air enormously, under specific weather conditions, in Sheffield. The effects of sulphur pollution from England and the Ruhr have even been recorded in Norway and Sweden. Pollution control schemes have reduced local concentrations but not over the earth as a whole.

Over England air pollution provides sulphate, chloride, calcium, magnesium, sodium, and ammonium ions as well as traces of antimony, arsenic, cobalt, lead, manganese, nickel, and zinc (Gorham, 1961) to the drainage water.

An estimate (Anon, 1972 in *Nature*, 240, p. 320-1) of the amounts of particulate matter taken up each year by the atmosphere over the northern hemisphere takes the quantity of natural particles as a steady 690 Mt (110 Mt from oceans and 580 Mt from land surfaces). Added to this is a man-made contribution put at 480 Mt today, compared with 120 Mt in 1880 and a predicted 760 Mt by AD 2000.

The composition of dissolved load varies season-

ally, and between rising and falling stages of stream discharge. A 'spring' burst of released solutes is a well-recognised phenomenon in many catchments. The non-seasonal release of geothermal waters into rivers also has a marked effect upon dissolved loads.

An, as yet unassessed, source of error in measurements of dissolved load is the quantity of solute which is transported as a gel and that which is adsorbed on to colloids. It may be expected that in drainage basins yielding sediment with colloids of high base exchange capacity – such as montmorillonitic clays – a considerable proportion of the cations may be adsorbed. It has been estimated by Pitty (1971) that in some catchments, during periods of high sediment concentration, cations adsorbed on suspended sediment may approach or even exceed the cations carried in solution. By ignoring this condition it is possible that the geomorphic importance of mechanical weathering in relation to chemical weathering could be greatly exaggerated.

It is clear that estimates of rates of natural denudation are liable to serious error. It is also clear that man may be a very effective geomorphic agent. These possible errors help to explain the considerable differences between the erosion patterns revealed by the studies of Corbel on the one hand and by Fournier and Strakhov on the other, and between the denudation rates of Fournier and Strakhov. Furthermore, attempts to extrapolate present denudation rates backwards into geological time must be regarded with extreme caution.

Local erosion patterns

Local erosion patterns are most easily studied by measuring the debris transported by streams or debris deposited in reservoirs in a known period of time. Langbein and Schumm (1958) have collated data from many observations throughout the United States. They found that, although stream and reservoir data indicate the same trends when sediment yield is plotted against mean annual precipitation, reservoirs capture about 1.8 times the amount of sediment measured in river discharges. This is partly because reservoirs capture the bed load which may be ignored in stream measurements, but also because reservoirs are seldom built in typical catchments as they are easier to place in areas with steep slopes and narrow channels, which usually have high rates of erosion. They are also usually built in small catchments which normally

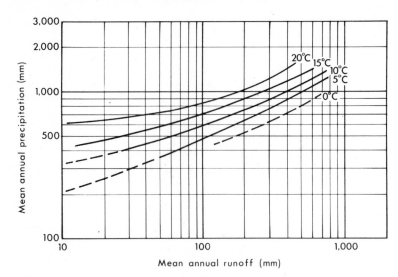

Fig. 11.6 The relationship between mean annual runoff and mean annual precipitation from United States data for given mean annual temperatures (after Langbein, 1949).

have a higher yield of sediment per unit area than larger catchments, because of the flatter slopes of the latter. Correction factors therefore have to be used in the data analysis.

From work done in the United States it has been possible to draw curves which represent the relationship between rainfall and runoff in several temperature zones (Langbein, 1949). These curves (Fig. 11.6) indicate the importance of evapotranspiration – which is largely controlled by temperature – as an indicator of runoff and hence as a major factor in determining the amount of sediment yield. They also show that runoff increases directly with precipitation. Sediment yield, however, is strongly influenced by a complex of factors which include precipitation, soils, vegetation, runoff, and land use. When the data used by Langbein and Schumm are plotted to show the relationship between vegetation and erosion (Fig. 11.7) it is seen that there is a distinct curve indicating low yield of sediment in arid areas, a peak yield in the semi-arid, a decline with increasingly close grass cover, and a fairly uniform yield from forests. This curve shows a different relationship from that derived by Fournier from world data (Fig. 11.8). Fournier's curve is a parabola with steppe and semi-arid basins yielding high sediment loads, and those with an evenly distributed rainfall, and hence close vegetation cover, yielding little sediment. These two parts of the curve are similar to that of Langbein and Schumm, but Fournier omits an arid zone where sediment yield should fall to zero, and he has a second zone of high yields from catchments with a

monsoon climate (this zone is not covered by the Langbein and Schumm data, but other studies (especially by Judson and Ritter, 1964) suggest that Fournier's curve may be more representative). The Fournier curve, then, suggests that on sediment yield alone it may be possible to distinguish four erosion regions – arid, semi-arid, moderate, and perhumid. Both Fournier and Schumm ignore glaciated regions in their studies.

It has been pointed out by Wilson (1973, 1977) that, although the Langbein–Schumm curve may be appropriate as a very generalised model of the relationship between rainfall and denudation rate in lowland USA, it obscures the extremely signifi-

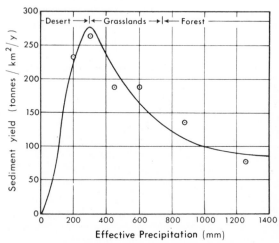

Fig. 11.7 Variation of sediment yield with effective precipitation (after Langbein and Schumm, 1958).

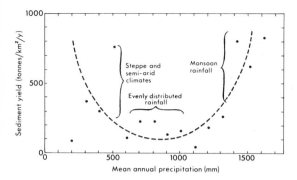

Fig. 11.8 Parabolic relationship between sediment yield and mean annual precipitation (after Fournier, 1960).

cant influence of drainage basin area, relief, and climatic regime. Wilson contends that the major control is not mean annual precipitation but seasonality of climate, with seasonally wet and dry climates producing maximum erosion, a contention which supports Fournier's approach. Douglas (1966) working on Australian catchments has concluded that rainfall totals and seasonality are the main controls on differences in denudation, rather than temperatures. Jansen and Painter (1974) came to broadly similar conclusions: their predictive

equations for different climates show that sediment yield increases with increasing runoff, altitude, relief, precipitation, temperature, and rock erodibility, but that it decreases with increasing area and vegetation cover.

It is clear that the data are, as yet, too unreliable and conflicting to draw any but the most generalised conclusions on the controls of denudation. The large number of variables and the complex manner in which they interact imply that a realistic model will itself be complex. With that reservation in mind, however, it must be noted that Fournier's approach has had considerable success where it can be used to fill in gaps in the data for a relatively large area.

It has been shown by McSaveney (personal communication, Figs. 11.9 and 11.10) that for the New Zealand Alps sediment yield closely follows the Fournier relationship. Using all available measurements this relationship has then been used to establish a general pattern for denudation of the whole country (Fig. 11.11). The correspondence between annual rainfall, relief, and denudation is consequently partly a product of the method. It obscures the probability that denudation caused by tectonic instability and seismic activity may account for up to half the total denudation in areas of greatest uplift of the New Zealand Alps.

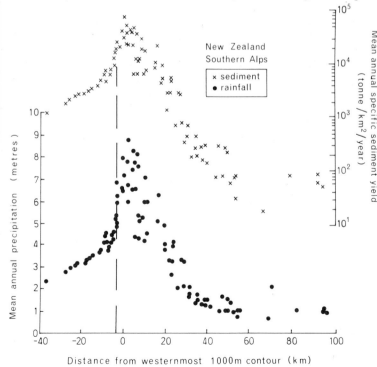

Fig. 11.9 The relationship between mean annual precipitation and sediment yield for a transect across the New Zealand Alps (supplied by M. McSaveney).

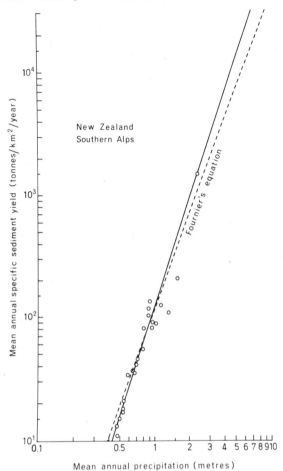

Fig. 11.10 The relationship between an estimate of denudation from Fournier's equation, and measured sediment yields for the New Zealand Alps (supplied by M. McSaveney).

Direct measurements

The number of studies using direct measurements of denudation on slopes is still very small. It has been pointed out by Young (1974) that if only four major variables are considered — climate, rock type, vegetation, and slope angle — some 500 environments would have to be sampled to give a minimum of studies for comparative purposes (ten climates each with its climax vegetation, ten rock types, and five classes of slope angle). There are probably fewer than 500 studies, using direct measurements, available and these have not been planned to give a statistically acceptable sample.

Soil creep on slopes, in humid climates, with angles of 15-30° produces movements downslope of 1-3 mm/year and volumetric movement of 0.1-10.0 cm^3/cm per year. Rates become higher in the cooler temperate zones, where frost heave increases creep, and in the semi-arid and humid tropics. The effect of solution on creep measurements is unclear.

Solifluction in cold climates involves both soil flow and frost heave mechanisms. Rates of solifluction are 10-100 times faster than creep with surface movements of 0.1 to 0.5 m/year and volumetric transfers of 50 cm^3/cm per year. Movement is faster in tundra than in polar desert areas and where sites are wet. As with other process studies there may be a strong bias in the data because there is always a tendency to take measurements where a process is obviously active.

Surface wash is studied with reference pins or poles, where the process is very active, and more commonly with sediment plots ranging in area from 4 m^2 to 1000 m^2. The range of values of slope recession obtained from plots is directly controlled by the amount of vegetation cover and ranges from 0.1 to over 1000 B, with the highest values being recorded in semi-arid climates, on badlands, and steep slopes.

Solution rates are usually calculated from dissolved matter in river water. Rates are generally assumed to be highest on limestones, but the effect of vegetation and soil on solution rates is not clear (see Table 5.5). Some studies suggest that solution of bare rock is only half that of rock covered by a soil and vegetation. It is probable that in many environments solution accounts for at least half of all denudation and leads directly to ground loss. Better understanding of solution rates, and controls on it, should be a major objective of process studies.

Landslides vary greatly in their importance, and in the ease with which that importance can be assessed. The biggest single problem is the lack of data on magnitude and frequency of landsliding. One major storm or earthquake, for example, may produce more slope erosion than all other processes in an area over a period of say 20 years. But in the absence of records it may not be possible to determine if this event has a return period of 50 or of 500 years. A second source of error is that measurements of landslide volumes are virtually confined to areas where they are particularly active, and their significance may consequently be exaggerated. Rates of ground lowering in areas of recorded active landsliding vary from 10 to 5000 B (see Selby, 1979, for examples).

Cliff retreat rates are, probably, most strongly

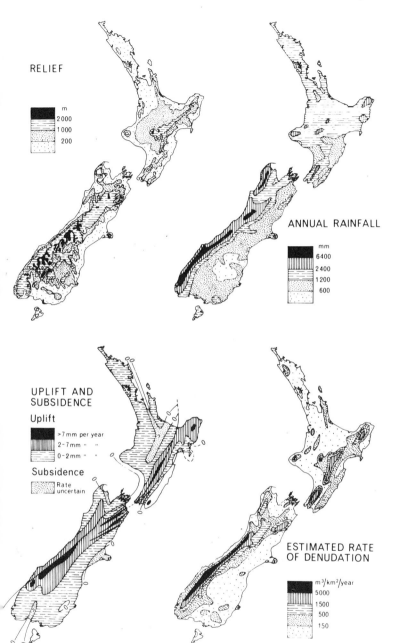

Fig. 11.11 The relationship between relief, annual rainfall, uplift and estimated denudation rates in New Zealand. Uplift rates are corrected from Wellman (1967).

controlled by rock mass resistance with the most resistant rocks having rates varying from 100–7000 B and the weakest rock rates up to 10 kB. As with landslides magnitude and frequency effects are important: it has been pointed out by Gordon *et al*. (1978), for example, that one rockfall on to the Lyell Glacier in South Georgia represented a minimum of 93 years of 'normal' subaerial erosion.

Slope retreat can seldom be measured directly over a period sufficiently long to give reliable data. Rates of 100–1000 B are probably of the right order of magnitude for most conditions, but may be biased towards the high rates because of interest in rapidly changing slopes.

Landform changes

Where a land surface can be dated accurately the modification of that surface since a known date can be expressed in quantitative terms. Since radiometric dating has been available several measurements have been made of the dissection of lava flows. Ruxton and McDougall (1967) working in the Hydrographers Range, Papua, on lavas ranging in age from 650 000 to 700 000 years BP (by potassium–argon dating) have found that calculated denudation rates are directly related to relief and range from 80 B at 60 m above sea level to 800 B at 760 m above sea level. These rates are similar to those calculated by other methods for hot moist hilly areas. Potassium–argon dating (Stipp and McDougall, 1968) has shown that Lyttelton volcano, New Zealand, was active between about 10 and 12 million years ago and that Akaroa volcano activity occurred between eight and nine million years ago. The central regions of both these volcanoes, which make up Banks Peninsula, have been very deeply eroded. Studies of dykes and valley fills have suggested that rates of erosion of the upper parts of Lyttelton volcano reached 28 kB. This very high rate of removal seems to have occurred mostly in late Miocene times and involved removal of pyroclastics.

In a study of denudation rates in the White Mountains, of eastern California, Marchand (1971) used the base of a basalt lava flow, dated at 10.8 million years, as his datum level and calculated the volume of material removed from below this level. Present rates of chemical denudation, corrected for contributions from the atmosphere and biosphere, he estimated as ranging from 1.4 to 21.0 B.

It has been estimated by L. C. King (1953) that free faces on major erosion scarps in many continents retreat at a nearly uniform rate of one to two kB.

Where the ages of trees are known it is sometimes possible to estimate rates of soil loss from around their roots. Eardley and Viavant (1967) have studied bristlecone pines, in southern Utah, which are easily dated by counting the tree rings in a core sample from the trunk. Measurements of soil removed from around the roots of the trees indicated mean soil losses of 290 B, but marked variations with slope such that:

$$\text{Denudation} = 53.1 \times \text{age of tree in } 1000 \text{ year units} \times \sin\beta.$$

On a 30° slope losses were 670 B and on a 5° slope 140 B. As the ages of the trees ranged from 170 to 2480 years this method is shown to be very useful in suitable areas.

Exposure of tree roots has been used, also, in Kenya (Dunne *et al.*, 1978) where it can be shown that modern rates of denudation are largely controlled by land use and in overgrazed areas many reach 10 kB. By contrast differences in the elevation of Cenozoic erosion surfaces suggest a rate of 8.4 B as characteristic of late Cenozoic time – a rate similar to that from Kenyan forest areas with an annual precipitation of about 750 mm at the present.

Denudation rates in the European Alps have been estimated by Clark and Jäger (1969) at between 400 and 1000 B. They have used dates of mineral assemblages derived from Rb/Sr and K/Ar ratios and assumed that the dates indicate the age at which the minerals were crystallised in metamorphism during Alpine folding. The metamorphic mineral assemblage indicates the range of temperatures to which the rock was subjected, and these temperatures must have occurred at pressures, or depths, that are compatible with the observed mineralogy. The rate of denudation can be deduced, since a rock collected at the surface at the present time was at a depth corresponding to the critical temperature of the mineral (in this case biotite) at the time in the past given by the age of that mineral. This rate contrasts with that of 40 B derived from an estimate of rock eroded from the Pelvoux Massif in the last 70 My (Montjuvent, 1973) but Montjuvent's study could not include the unknown general lowering of mountain peaks.

Another method which may be of wider use than has yet been attempted involves the dating of archaeological sites and then measuring the amount of soil removed from around buildings, etc. Judson (1968) measured soil loss from around a Roman cistern in central Italy, the footings of which are now exposed. He estimated average rates of erosion of 500 B. Cores from lake beds, or other sediments, overlying human artefacts may also be used for estimating denudation rates.

Tectonic uplift and denudation

The existence of large mountain chains for prolonged periods is clearly dependent upon continuing uplift. A mountain 4000 m high could, theoretically, be levelled in 0.5 My at the rate of denudation recorded for the central Taiwan Alps of 5.5 kB (Li, 1976). Rates of sediment transport in rivers in Taiwan, however, suggest that the denudation rate approximately equals the uplift rate.

High general rates of tectonic uplift are about 8 kB, and the highest rates of postglacial isostatic rebound are about 11 kB (for the Gulf of Bothnia (Walcott, 1972)). No known rates of denudation for large areas equal the maximum known rates of uplift. For small areas where rates are known it is possible to determine the actual increase in relief in a given time. In the Transverse Ranges of southern California, for example, local uplift rates reach 7.6 kB but the maximum rate of denudation, although high, is only 2.3 kB (Scott and Williams, 1978). The relationship between the two rates for the Japanese Alps has been demonstrated by Yoshikawa (1974) who has found that denudation rates may exceed uplift rates in the Outer Zone of southwest Japan and in the central mountains, but elsewhere modern rates of uplift exceed rates of denudation (Fig. 11.12). The western Caucasus also appears to be increasing in altitude with uplift rates apparently reaching 20-25 kB but denudation rates being close to 200 B (Gabrielyan, 1971).

The wide variations in rates of uplift and down-wearing make it difficult to generalise on how long it would take to erode a mountain chain, but through the operation of isostatic adjustments, as denudation occurs, it appears to be closer to a 100 My, for a high chain, than to 10 My. Oscillating rates appear to preclude the possibility of any long-term steady state relief (Ahnert, 1970).

Conclusions

The relationship between climate and rates of denudation appears to be close but is not necessarily uniform within one climatic zone. It operates primarily through the completeness of vegetation cover on potentially mobile soil and regolith materials, and is very strongly influenced by intensely seasonal precipitation.

The relationship between denudation and lithology is strong but not easy to quantify. Resistant, often ancient, massive rock has relatively low rates of denudation. Where rocks have a thin regolith even human interference may not greatly increase rates of denudation over large areas. It has been shown by Gordon (1979), for example, that in central New England, where soils are generally thin, most of the sediment removed has been trapped in Long Island Sound, and that over the last 8000 years the rate has been nearly uniform and low at around 50 B. Most of the sediment has come from

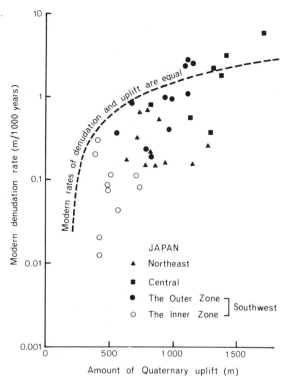

Fig. 11.12 The relationship between total uplift in the Quaternary and modern denudation rates for Japan (after Yoshikawa, 1974).

river bank collapse and not from soil erosion. Similar results were obtained by Pearce and Elson (1973) for areas in Quebec.

There appears to be rather broad agreement in the available data that for lowland areas in many parts of the world mean rates of 10 to 50 B are common, but there is also evidence that rates of downwearing of landscapes with soils upon ancient hard rocks, such as granite, may be limited by the rate of weathering to <5 B (e.g. Owens and Watson, 1979). For mountainous areas, and even lowland areas with highly erodible rocks such as bare shales, rates are commonly two to ten times higher than mean rates for lowlands. There is still considerable doubt about how generally applicable are higher rates, in excess of 100 B. Slaymaker (1977), for example, has found that for the Canadian Cordillera rates of 60 to 120 B are common. Young (1974), by contrast, suggests that mountain lands more usually have rates closer to 500 B. There is no doubt that differences in relief and slope angle are of paramount importance in controlling denudation rates, but whether the available data are biased by excessive sampling in rapidly eroding areas is not clear, and must await more measurements taken in a statistically acceptable manner.

Another major problem lies in uncertainty about how far modern rates can be extrapolated to explain rates of change of landforms. This is clearly a contentious area for relict landforms, such as bornhardts and duricrusts, may survive in the landscape for millions of years (Twidale, 1976b), but with what degree of alteration of form and volume is unknown: valley terraces and some slope deposits may last for many thousands, or even hundreds of thousands, of years. It is still not always possible to determine whether this survival is the result of slow process rates, or of selective preservation. The question of the degree to which landforms are relict from past environments must be answered, in the absence of direct dating, by a better knowledge of the variation of denudation rates in small areas and by better estimates of the variation in process intensities.

If measured rates are extrapolated they indicate that in two million years a mountain range is reduced in altitude by 1000 m (ignoring uplift) and a gently sloping landscape by 100 m. A possible estimate for the lowering of a major erosion surface is 50 to 200 m in ten million years. Such extrapolations ignore the effects of human interference with erosion which appears to exceed normal geological rates by at least a factor of two. The extrapolations also ignore the effects of climatic change.

The resulting time-scale for geomorphological change suggests that slopes are unlikely to have been greatly modified during the last 10 000 years, so studies of slope forms must take into account the results of late glacial and Holocene climatic changes. Minor valleys 20–50 m deep are reasonably attributed to the late Pleistocene but intermediate-sized features have to be considered in relation to the whole of Pleistocene time. The origin of erosion surfaces of continental extent, such as those of Africa or Australia, may reasonably be attributed to Tertiary or even earlier times. Even L. C. King's (1962) suggestion that the earliest 'Gondwana' surface of Africa originated in the Jurassic is not contradicted by the rates of denudation which have been measured.

Appendices

Appendix A: *The International System of Units*

The International System of Units, or SI system, has been increasingly used since 1970. It is a development of the metric system formulated in France in the 1880s.

The SI system is based on six primary units:

Quantity	Unit	Symbol
length	metre	m
mass	kilogramme	kg
time	second	s
electric current	ampere	A
temperature	degree Kelvin	°K (or the customary °C)
luminous intensity	candela	cd

It should be noted that the kilogramme is the unit of mass or quantity of matter. It is NOT the unit of force. One thousand kilogrammes is known as a tonne (1000 kg = 1 t). The mass of a body is independent of the gravitational force, but weight is a measure of the gravitational force acting upon a mass at a particular location, hence:

weight = mass × gravitational acceleration.

As the value of gravitational acceleration (*g*) varies somewhat over the surface of the earth the weight of an object will vary also. This variation is slight, and usually neglected, and the standard acceleration of 9.806 m/s^2 is used to convert between mass and weight. Weight is expressed in units of force, and the basic unit of force is the newton (N) thus:

$$1N = 1 \text{ kg} \times 1 \text{ m/s}^2.$$

An object with a mass of 1000 kg has a weight (at the surface of the earth) of 1000 kg × 9.806 m/s^2 = 9806 N = 9.806 kN.

The same distinction applies to mass density (*ρ*) and unit weight (*γ*) so that:

$$\gamma = \rho g.$$

Multiples of SI units

Prefix	Symbol	Power		Example
tera	T	10^{12}	1000000000000	
giga	G	10^9	1000000000	gigametre (Gm)
mega	M	10^6	1000000	megametre (Mm)
kilo	k	10^3	1000	kilometre (km)
hecto	h	10^2	100	
deca	da	10^1	10	
deci	d	10^{-1}	.1	
centi	c	10^{-2}	.01	
milli	m	10^{-3}	.001	millimetre (mm)
micro	μ	10^{-6}	.000001	micrometre (μm)
nano	n	10^{-9}	.000000001	nanometre (nm)
pico	p	10^{-12}	.000000000001	picometre (pm)
femto	f	10^{-15}	.000000000000001	
atto	a	10^{-18}	.000000000000000001	

The mass density of water is 1000 kg/m^3 and its unit weight (γ_w) therefore is 9806 N/m^3 = 9.806 kN/m^3 (under standard conditions).

Stress and pressure have units of force per unit area, consequently the basic unit of stress and pressure is N/m^2. Because of the size of the unit it is common to use kN/m^2 for soil and MN/m^2 for rock. Thus the pressure 1 m below the surface of a body of water is 9.8 kN/m^2. An equivalent of the N/m^2 is the pascal (Pa).

Work and energy are expressed in joules (J) and power in watts (W)

$$1 \text{ W} = 1 \text{ J/s} = 1 \text{ N} \times 1 \text{ m/s exactly.}$$

Appendix B: *Soil properties*

If the total volume of soil is represented by $V(\text{m}^3)$, its mass by $M(\text{kg})$, the weight by $W(\text{N})$, the bulk density by ρ, and the subscript s is used for solid, w for water, and v for voids (air plus water filled voids) then the ratios relating these phases are:

Porosity (n) $= V_v/V$
Voids ratio (e) $= V_v/V_s$

Moisture content $(m) = M_w/M_s$

Specific gravity $(G_s) = M_s/V_s\rho_w$ (mass of solid/mass of water occupying the same volume as the solid matter in the soil where $\rho_w = 1000 \text{ kg/m}^3$). (Note: G_s is commonly in the range of 2.5 to 2.8 for soils.)

Bulk density $(\rho) (\text{kg/m}^3) = M/V$
Unit weight $(\gamma) (\text{N/m}^3) = W/V$

References

Ahnert, F. (1970). 'Functional relationships between denudation, relief, and uplift in large mid-latitude drainage basins', *American Journal of Science*, 268, 243-63.

Ahnert, F. (1976a). 'Darstellung des Struktur-influsses auf die Oberflächenformen im theoretischen Modell', *Zeitschrift für Geomorphologie, Supplementband*, 24, 11-22.

Ahnert, F. (1976b). 'Brief description of a comprehensive three-dimensional process–response model of landform development', *Zeitschrift für Geomorphologie, Supplementband*, 25, 29-49.

Aires-Barros, L., Graça, R. C., and Velez, A. (1975). 'Dry and wet laboratory tests and thermal fatigue of rocks', *Engineering Geology*, 9, 249-65.

Aldridge, R. and Jackson, R. J. (1968). 'Interception of rainfall by Manuka (*Leptospermum scoparium*) at Taita, New Zealand', *New Zealand Journal of Science*, 11, 301-17.

Armstrong, A. (1976). 'A three-dimensional simulation of slope forms', *Zeitschrift für Geomorphologie, Supplementband*, 25, 20-8.

Attewell, P. B. and Farmer, I. W. (1976). *Principles of Engineering Geology* (London: Chapman and Hall).

Augusthitis, S. S. and Otteman, J. (1966). 'On diffusion rings and spheroidal weathering', *Chemical Geology*, 1, 201-9.

Bakker, J. P. and Le Heux, J. W. N. (1946). 'Projective geometric treatment of O. Lehmann's theory of the transformation of steep mountain slopes', *Koninklijke Nederlandsche Akademie van Wetenschappen*, B49, 533-47.

Bakker, J. P. and Le Heux, J. W. N. (1952). 'A remarkable new geomorphological law', *Koninklijke Nederlandsche Akademie van Wetenschappen*, B55, 399-410 and 554-71.

Barsch, D. (1977). 'Eine Abschätzung von Schuttproduktion und Schutttransport im Bereich aktiver blockgletscher der Schweizer Alpen', *Zeitschrift für Geomorphologie, Supplementband*, 28, 148-60.

Barton, N. (1973). 'Review of a new shear-strength criterion for rock joints', *Engineering Geology*, 7, 287-332.

Barton, N. and Choubey, V. (1977). 'The shear strength of rock joints in theory and practice', *Rock Mechanics*, 10, 1-54.

Battle, W. R. B. (1960). 'Temperature observation in bergschrunds and their relationship to frost shattering', pp. 83-95 in W. V. Lewis (ed.), *Norwegian Cirque Glaciers* (London: Royal Geographical Society).

Baynes, J. and Dearman, W. R. (1978). 'The microfabric of a chemically weathered granite', *Bulletin of the International Association of Engineering Geology*, 18, 91-100.

Baynes, F. J., Dearman, W. R., and Irfan, T. Y. (1978). 'Practical assessment of grade in a weathered granite', *Bulletin of the International Association of Engineering Geology*, 18, 101-9.

Bennett, H. H. (1939). *Soil Conservation* (New York: McGraw-Hill).

Bertallanffy, L. von (1950). 'An outline of general systems theory', *British Journal of the Philosophy of Science*, 1, 134-65.

Betson, R. P. (1964). 'What is watershed runoff?' *Journal of Geophysical Research*, 69, 1541-51.

Beven, K. (1978). 'The hydrological response of headwater and sideslope areas', *Hydrological Sciences Bulletin*, 23, 419-37.

Bieniawski, Z. T. (1973). 'Engineering classification of jointed rock masses' *Transactions of the South African Institution of Civil Engineers*, 15, 335-44.

Bieniawski, Z. T. (1975). 'The point-load test in geotechnical practice', *Engineering Geology*, 9, 1-11.

Bisdom, E. B. A. (1967). 'The role of microcrack systems in the spheroidal weathering of an intrusive granite in Galicia (N.W. Spain)', *Geologie en Mijnbouw*, 46, 333-40.

Bishop, A. W. (1955). 'The use of the slip circle in the stability analysis of earth slopes', *Géotechnique*, 5, 7-17.

Bishop, A.W. (1966). 'The strength of soils as engineering materials', *Géotechnique*, 16, 91–128.

Bishop, A.W. (1971). 'The influence of progressive failure on the method of stability analysis', *Géotechnique*, 21, 168–72.

Bishop, A.W. and Bjerrum, L. (1960). 'The relevance of the triaxial test to the solution of stability problems', *Proceedings American Society of Civil Engineers Research Conference on Shear Strength of Cohesive Soils*, 437–501.

Bishop, A.W., Green, G.E., Garga, V.K., Andresen, A., and Brown, J.D. (1971). 'A new ring shear apparatus and its application to the measurement of residual strength', *Géotechnique*, 21, 273–328.

Bishop, A.W. and Morgenstern, N.R. (1960). 'Stability coefficients for earth slopes', *Géotechnique*, 10, 129–47.

Bishop, D.M. and Stevens, M.E. (1964). 'Landslides in logged areas in S.E. Alaska', *U.S. Department of Agriculture, Forest Service Research Paper*, NOR-1, 1–18.

Bjerrum, L. (1967). 'Progressive failure in slopes of overconsolidated plastic clay and clay shales', *Journal of Soil Mechanics and Foundation Engineering Division, American Society of Civil Engineers*, 93, 1–49.

Bjerrum, L. and Jørstad, F. (1957). 'Rockfalls in Norway', *Norwegian Geotechnical Institute Publication*, 79, 1–11.

Blackwelder, E. (1925). 'Exfoliation as a phase of rock weathering', *Journal of Geology*, 33, 793–806.

Blackwelder, E. (1933). 'The insolation hypothesis of rock weathering', *American Journal of Science*, 26, 97–113.

Blong, R.J. (1973). 'A numerical classification of selected landslides of the debris slide-avalanche-flow type', *Engineering Geology*, 7, 99–114.

Bloom, A.L. (1978). *Geomorphology* (Englewood Cliffs, N.J.: Prentice-Hall).

Bouyoucos, G.J. (1935). 'The clay ratio as a criterion of susceptibility of soils to erosion', *Journal of the American Society of Agronomists*, 27, 738–41.

Boyé, M. and Fritsch, P. (1973). 'Dégagement artificiel d'un dôme cristallin au Sud-Cameroun', pp. 31–63 in 'Cinq études de geomorphologie et de palynologie', *Travaux et Documents de Géographie tropicale, CEGET-CNRS*, 8, 31–63.

Bradley, W.C. (1963). 'Large-scale exfoliation in massive sandstones of the Colorado Plateau', *Geological Society of America Bulletin*, 74, 519–28.

Brice, J.C. (1966). 'Erosion and deposition in the loess-mantled Great Plains, Medicine Creek drainage basin, Nebraska', *U.S. Geological Survey, Professional Paper*, 352-H, 255–339.

Bridgman, P.W. (1911). 'Water, in the liquid and five solid forms, under pressure', *American Academy of Arts and Science Proceedings (Daedalus)*, 47, 439–558.

Broch, E. (1979). 'Changes in rock strength caused by water', *Proceedings of the 4th Congress of the International Society for Rock Mechanics*, 1, 71–5.

Broch, E. and Franklin, J.A. (1972). 'The point-load strength test', *International Journal of Rock Mechanics and Mining Science*, 9, 669–97.

Brown, C.B. and Sheu, M.S. (1975). 'Effects of deforestation on slopes', *Journal of the Geotechnical Engineering Division, American Society of Civil Engineers*, GT2, 147–65.

Browning, J.M. (1973). 'Catastrophic rock slide, Mount Huascarán, north-central Peru, May 31, 1970', *American Association of Petroleum Geologists Bulletin*, 57, 1335–41.

Brunner, F.K. and Scheidegger, A.E. (1974). 'Kinematics of a scree slope', *Revista Italiana di Geofisica*, 23, 89–94.

Brunsden, D., Doornkamp, J.C., Fookes, P.G., Jones, D.K.C., and Kelley, J.M.H. (1975). 'Geomorphological mapping techniques in highway engineering', *Journal of the Institution of Highway Engineers*, 22, 35–41.

Brunsden, D. and Jones, D.K.C. (1976). 'The evolution of landslide slopes in Dorset', *Philosophical Transactions Royal Society of London*, A283, 605–631.

Büdel, J. (1957). 'Die "Doppelten Einebnungsflächen" in den feuchten Tropen', *Zeitschrift für Geomorphologie*, 1, 201–88.

Büdel, J. (1977). 'Hanggeschichte und Hangalter', *Zeitschrift für Geomorphologie, Supplementband*, 28, 14–29.

Bull, W.B. (1977). 'The alluvial fan environment', *Progress in Physical Geography*, 1, 222–70.

Bunting, B.T. (1964). 'Slope development and soil formation on some British sandstones', *Geographical Journal*, 130, 73–9.

Butler, R.M.J. (1972). 'Water as an unwanted commodity: some aspects of flood alleviation', *Journal of the Institute of Water Engineers*, 26, 311–22.

Cabrera, J.G. and Smalley, I.J. (1973). 'Quick clays as products of glacial action: a new approach to their nature, geology, distribution and geotechnical properties', *Engineering Geology*, 7, 115–33.

Caine, N. (1974). 'The geomorphic processes of the alpine environment, pp. 721–48 in J.D. Ives and R.G. Barry (eds.), *Arctic and Alpine Environments* (London: Methuen).

Caine, N. (1976). 'A uniform measure of subaerial erosion', *Geological Society of America Bulletin*, 87, 137–40.

Calkin, P. and Cailleux, A. (1962). 'A quantitative study of cavernous weathering (taffonis) and its application to glacial chronology in Victoria Valley, Antarctica', *Zeitschrift für Geomorphologie*, 6, 317–24.

Campbell, R. H. (1975). 'Soil slips, debris flows, and rainstorms in the Santa Monica Mountains and vicinity, southern California', *U.S. Geological Survey, Professional Paper*, 851, 1–51.

Carroll, D. (1970). *Rock weathering* (New York: Plenum Press).

Carson, M. J. (1969). 'Models of hillslope development under mass failure', *Geographical Analysis*, 1, 76–100.

Carson, M. A. (1971). *The Mechanics of Erosion* (London: Pion).

Carson, M. A. (1976). 'Mass-wasting, slope development and climate', pp. 101–36 in E. Derbyshire, (ed.), *Geomorphology and Climate* (London: Wiley).

Carson, M. A. and Kirkby, M. J. (1972). *Hillslope Form and Process* (London: Cambridge University Press).

Chandler, M. P., Parker, D. C., and Selby, M. J. (1981). 'An open-sided field direct shear box', *British Geomorphological Research Group, Technical Bulletin*, 27.

Chandler, R. J. (1970). 'A shallow slab slide in the Lias Clay near Uppingham, Rutland' *Géotechnique*, 20, 253–60.

Chandler, R. J. (1971). 'Landsliding on the Jurassic escarpment near Rockingham, Northamptonshire', *Institute of British Geographers, Special Publication*, 3, 111–28.

Chandler, R. J. (1972). 'Lias Clay: weathering processes and their effect on shear strength', *Géotechnique*, 22, 403–31.

Chandler, R. J. (1973). 'The inclination of talus. Arctic talus terraces and other slopes composed of granular material', *Journal of Geology*, 81, 1–14.

Chandler, R. J., Kellaway, G. A., Skempton, A. W., and Wyatt, R. J. (1976). 'Valley slope sections in Jurassic strata near Bath, Somerset', *Philosophical Transactions of the Royal Society London*, A 283, 527–56.

Chardon, M. (1976). 'Observations sur la formation des versants regularisés ou versants de Richter', *Actes du symposium sur les versants en pays méditerranéens, à Aix-en-Provence, France, 28–30 avril 1975, Centre d'Etudes Géographiques et de Recherches Méditerranéenes*, 5, 25–7.

Chorley, R. J. and Kennedy, B. A. (1971). *Physical Geography: A Systems Approach* (London: Prentice-Hall International).

Chowdhury, R. N. (1978). *Slope Analysis* (Amsterdam: Elsevier).

Claridge, G. G. C. (1960). 'Clay minerals, accelerated erosion, and sedimentation in the Waipaoa River catchment', *New Zealand Journal of Geology and Geophysics*, 3, 184–91.

Claridge, G. G. C. (1970). 'Studies in element balances in a small catchment at Taita, New Zealand', *International Association of Scientific Hydrology*, Publication 96, 523–40.

Clark, S. P. and Jäger, E. (1969). 'Denudation rate in the Alps from geochronologic and heat flow data', *American Journal of Science*, 267, 1143–60.

Cleaves, E. T., Godfrey, A. E., and Bricker, O. P. (1970). 'Geochemical balance of a small watershed and its geomorphic implications', *Geological Society of America Bulletin*, 81, 3015–32.

Coates, D. F. (1965). 'Rock Mechanics Principles', *Canadian Department of Energy, Mines and Resources, Mines Branch Monograph*, 874.

Coates, D. R. (ed.) (1977). *Landslides*. Geological Society of America, Reviews in Engineering Geology, 3, 1–278.

Conacher, A. J. and Dalrymple, J. B. (1977). 'The nine unit landsurface model: an approach to pedogeomorphic research', *Geoderma*, 18, 1–154.

Connell, D. C. and Tombs, J. M. C. (1971). 'The crystallisation pressure of ice – a simple experiment', *Journal of Glaciology*, 10, 312–15.

Cooke, R. U. and Smalley, I. J. (1968). 'Salt weathering in deserts', *Nature*, 220, 1226–7.

Corbel, J. (1964). 'L'érosion terrestre, étude quantitative (Méthodes-techniques-résultats)', *Annales de Géographie*, 73, 385–412.

Costa, J. E. (1975). 'Effects of agriculture on erosion and sedimentation in the Piedmont province, Maryland', *Geological Society of America Bulletin*, 86, 1281–6.

Cotecchia, V. and Melidoro, G. (1974). 'Some principal geological aspects of the landslides of southern Italy', *Bulletin of the International Association of Engineering Geology*, 9, 23–32.

Coulomb, C. A. (1776). 'Essais sur une application des règles des maximis et minimis à quelques problems de statique relatifs à l'architecture', *Mémoirs présentées par divers Savants*. (Paris: Academie des Sciences).

Cousins, B. F. (1978). 'Stability charts for simple earth slopes', *Journal of the Geotechnical Engineering Division, Proceedings of the American Society of Civil Engineers*, 104(GT2), 267–79.

Crawford, C. B. and Eden, W. J. (1969). 'Stability of natural slopes in sensitive clay', pp. 453–70 in *Stability and Performance of Slopes and Embankments* (Soil Mechanics and Foundation Division American Society of Civil Engineers).

Crouch, R. J. (1976). 'Field tunnel erosion – a review', *Journal of the Soil Conservation Service of New South Wales*, 32, 98–111.

Crozier, M. J. (1973). 'Techniques for the morphometric analysis of landslips', *Zeitschrift für Geomorphologie*, 17, 78–101.

Cruden, D. M. (1976). 'Major rock slides in the Rockies', *Canadian Geotechnical Journal*, 13, 8–20.

Cruden, D.M. and Krahn, J. (1973). 'A re-examination of the geology of the Frank Slide', *Canadian Geotechnical Journal*, 10, 581–91.

Culling, W. E. (1963). 'Soil creep and the development of hillside slopes', *Journal of Geology*, 71, 127–61.

Culmann, C. (1866). *Graphische Statik* (Zurich).

Cunningham, F. F. and Griba, W. (1973). 'A model of slope development, and its application to the Grand Canyon, Arizona', *Zeitschrift für Geomorphologie*, 17, 43–77.

Curtis, C. D. (1976). 'Chemistry of rock weathering: fundamental reactions and controls', pp. 25–57 in E. Derbyshire (ed.), *Geomorphology and Climate* (London: Wiley).

Dahl, R. (1966). 'Blockfields, weathering-pits and tor-like forms in the Narvik mountains, Nordland, Norway', *Geografiska Annaler*, 48A, 55–85.

Dale, T. N. (1923). 'The commercial granites of New England', *U.S. Geological Survey Bulletin*, 738.

Dalrymple, J. B., Blong, R. J., and Conacher, A. J. (1968). 'A hypothetical nine-unit landsurface model', *Zeitschrift für Geomorphologie*, 12, 60–76.

Dalrymple, T. (1960). 'Flood frequency analysis: Manual of Hydrology, part 3, flood flow techniques', *U.S. Geological Survey Water Supply Paper*, 1543A, 1–80.

D'Andrea, D. V., Fisher, R. L., and Fogelson, D. E. (1965). 'Prediction of compressive strength of rock from other rock properties', *U.S. Bureau of Mines, Report of Investigations*, 6702.

Davis, W. M. (1909). *Geographical Essays* (reprinted 1954) (New York: Dover).

Day, M. J. and Goudie, A. S. (1977). 'Field assessment of rock hardness using the Schmidt test hammer', *British Geomorphological Research Group Technical Bulletin*, 18, 19–29.

Dearman, W. R. (1974). 'Weathering classification in the characterisation of rock for engineering purposes in British practice', *Bulletin of the International Association of Engineering Geology*, 9, 33–42.

Dearman, W. R. (1976). 'Weathering classification in the characterisation of rock: a revision', *Bulletin of the International Association of Engineering Geology*, 13, 123–7.

Deere, D. U. (1968). 'Geological considerations', pp. 1–20 in O. C. Zienkiewicz and D. Stagg (eds.), *Rock Mechanics in Engineering Practice* (New York: Wiley).

Deere, D. U. and Miller, R. P. (1966). 'Engineering classification and index properties for intact rock', *Technical Report No. AFNL-TR-65-116, Air Force Weapons Laboratory, New Mexico*.

Deere, D. U. and Patton, F. D. (1971). 'Slope stability in residual soils', *Proceedings of the Fourth Pan American Conference on Soil Mechanics and Foundation Engineering, Puerto Rico*, 1, 87–170.

Demek, J. (1964). 'Castle koppies and tors in the Bohemian Highland (Czechoslovakia)', *Biuletyn Peryglacjalny*, 14, 195–216.

De Ploey, J. and Cruz, O. (1979). 'Landslides in the Serra do Mar, Brazil', *Catena*, 6, 111–22.

Dole, R. B. and Stabler, H. (1909). 'Denudation', pp. 78–93 in 'Papers in the conservation of water resources', *U.S. Geological Survey Water Supply Paper*, 234.

Douglas, I. (1966). 'Denudation rates and water chemistry of selected catchments in eastern Australia and their significance for tropical geomorphology', Australian National University, Ph.D. thesis.

Douglas, I. (1967). 'Man, vegetation and the sediment yields of rivers', *Nature*, 215, 925–8.

Douglas, I. (1968). 'The effects of precipitation chemistry and catchment area lithology on the quality of river water in selected catchments in eastern Australia', *Earth Science Journal*, 2, 126–42.

Dragovich, D. (1969). 'The origin of cavernous surfaces (tafoni) in granite rocks of southern Australia', *Zeitschrift für Geomorphologie*, 13, 163–81.

Dudley, J. H. (1970). 'Review of collapsing soils', *Journal of the Soil Mechanics and Foundation Division, American Society of Civil Engineers*, 96, 925–47.

Dunn, J. R. and Hudec, P. P. (1966). 'Water, clay and rock soundness', *Ohio Journal of Science*, 66, 153–67.

Dunne, T. (1970). 'Runoff production in a humid area', *U.S. Department of Agriculture, Agricultural Research Service*, Paper, ARS 41–160.

Dunne, T. (1978a). 'Rates of chemical denudation of silicate rocks in tropical catchments', *Nature*, 274, 244–46.

Dunne, T. (1978b). 'Field studies of hillslope flow processes', pp. 227–93 in M. J. Kirkby, (ed.), *Hillslope Hydrology* (Chichester: Wiley).

Dunne, T. and Black, R. G. (1970a). 'An experimental investigation of runoff production in permeable soils', *Water Resources Research*, 6, 478-90.

Dunne, T. and Black, R. G. (1970b). 'Partial area contributions to storm runoff in a small New England watershed', *Water Resources Research*, 6, 1296-311.

Dunne, T., Dietrich, W. E., and Brunengo, M. J. (1978). 'Recent and past erosion rates in semi-arid Kenya', *Zeitschrift für Geomorphologie, Supplementband*, 29, 130-40.

Dunne, T., Moore, T. R., and Taylor, C. H. (1975). 'Recognition and prediction of runoff producing zones in humid regions', *Hydrological Sciences Bulletin*, 20, 305-27.

Durum, W. H., Heidel, S. G., and Tison, L. J. (1960). 'World-wide runoff of dissolved solids', *International Association of Scientific Hydrology*, 51, 618-28.

Dury, G. H. (1969). 'Rational descriptive classification of duricrusts', *Earth Science Journal*, 3, 77-86.

Dzulynski, St. and Kotarba, A. (1979). 'Solution pans and their bearing on the development of pediments and tors in granite', *Zeitschrift für Geomorphologie*, 23, 172-91.

Eardley, A. J. and Viavant, W. (1967). 'Rates of denudation as measured by bristlecone pines, Cedar Breaks Utah', *Utah Geological and mineralogical Survey Special Studies*, 21, 1-13.

Eden, M. J. and Green, C. P. (1971). 'Some aspects of granite weathering and tor formation on Dartmoor, England', *Geografiska Annaler*, 53A, 92-9.

Emmett, W. W. (1970). 'The hydraulics of overland flow on hillslopes', *U.S. Geological Survey Professional Paper*, 662-A, 1-68.

Emmett, W. W. (1978). 'Overland flow', pp. 145-76 in M. J. Kirkby (ed.), *Hillslope Hydrology* (Chichester: Wiley).

Ericksen, G. E., Plafker, G., and Fernandez, J. C. (1970). 'Preliminary report on the geological events associated with the May 31, 1970, Peru earthquake', *U.S. Geological Survey Circular*, 639, 1-25.

Evans, I. S. (1970). 'Salt crystallisation and rock weathering: a review', *Revue de Géomorphologie Dynamique*, 19, 153-77.

Everett, A. G. (1979). 'Secondary permeability as a possible factor in the origin of debris avalanches associated with heavy rainfall', *Journal of Hydrology*, 43, 347-54.

Exon, N. F., Langford-Smith, T., and McDougall, I. (1970). 'The age and geomorphic correlations of deep weathering profiles, silcrete and basalt in the Roma-Amby District, Queensland', *Journal of the Geological Society of Australia*, 17, 21-30.

F.A.O. (1977). *Soil Conservation and Management in Developing Countries*. Food and Agricultural Organisation of the United Nations, Rome, Bulletin 33.

Farres, P. (1978). 'The role of time and aggregate size in the crusting process', *Earth Surface Processes*, 3, 243-54.

Fellenius, W. (1936). 'Calculation of the stability of earth dams', *Transactions, 2nd Congress on Large Dams*, Washington, D.C., 4, 445-65.

Fieldes, M. and Swindale, L. D. (1954). 'Chemical weathering of silicates in soil formation', *New Zealand Journal of Science and Technology*, 36B, 140-54.

Fisher, O. (1866). 'On the disintegration of a chalk cliff', *Geological Magazine*, 3, 354-6.

Fleming, R. W. and Johnson, A. M. (1975). 'Rates of seasonal creep of silty clay soil', *Quarterly Journal of Engineering Geology*, 8, 1-29.

Fookes, P. G., Dearman, W. R., and Franklin, J. A. (1971). 'Some engineering aspects of rock weathering with field examples from Dartmoor and elsewhere', *Quarterly Journal of Engineering Geology*, 4, 139-85.

Foster, G. R., Meyer, L. D., and Onstad, C. A. (1977a). 'A runoff erosivity factor and variable slope length exponents for soil loss estimates', *American Society of Agricultural Engineers, Transactions*, 20, 683-7.

Foster, G. R., Meyer, L. D., and Onstad, C. A. (1977b). 'An erosion equation derived from basic erosion principles', *American Society of Agricultural Engineers, Transactions*, 20, 678-82.

Fournier, F. (1960). *Climat et Erosion: la relation entre l'érosion du sol par l'eau et les précipitations atmosphériques* (Paris: Presses Universitaire de France), 1-201.

Franklin, J. and Chandra, R. (1972). 'The slake-durability test', *International Journal of Rock Mechanics and Mining Science*, 9, 325-41.

Franzle, O. (1971). 'Die Opferkessel im quartzitischen Sandstein von Fontainebleu', *Zeitschrift für Geomorphologie*, 15, 212-35.

Fredlund, D. G. and Krahn, J. (1977). 'Comparison of slope stability methods of analysis', *Canadian Geotechnical Journal*, 14, 429-39.

Frietas, M. H. de and Watters, R. J. (1973). 'Some field examples of toppling failures', *Géotechnique*, 23, 495-514.

Gabrielyan, H. (1971). 'Quantitative characteristics of the denudation of the Caucasus', *Acta Geographia Debrecina*, 10, 37-40.

Gadd, N. R. (1975). 'Geology of Leda Clay', pp. 137-51 in *Mass wasting* (Norwich: Geo Abstracts).

Gage, M. (1966). 'Franz Josef Glacier', *Ice*, 20, 26-7.

Gardner, J. S. (1977). 'High magnitude rockfall-rockslide: frequency and geomorphic significance in the Highwood Pass area, Alberta', *Great Plains-Rocky Mountain Geographical Journal*, 6, 228–39.

Gardner, J. S. (1979). 'The movement of material on debris slopes in the Canadian Rocky Mountains', *Zeitschrift für Geomorphologie*, 23, 45–57.

Garner, N. F. (1976). 'Insolation warmed over: comment', *Geology*, 4, 264.

Garnett, A. (1967). 'Some climatological problems in urban geography with reference to air pollution', *Institute of British Geographers, Transactions*, 42, 21–43.

Gerber, E. and Scheidegger, A. E. (1969). 'Stress-induced weathering of rock masses', *Ecologae Geologicae Helveticae*, 62, 401–15.

Gibbs, H. S. (1945). 'Tunnel-gully erosion on the Wither Hills, Marlborough', *New Zealand Journal of Science and Technology*, A27, 135–46.

Gibbs, R. J. (1967). 'The geochemistry of the Amazon river system, 1', *Geological Society of America Bulletin*, 78, 1203–31.

Gilbert, G. K. (1904). 'Domes and dome structure of the High Sierra', *Geological Society of America Bulletin*, 15, 29–36.

Goldstein, M. and Ter-Stepanian, G. (1957). 'The long-term strength of clays and deep creep of slopes', *Proceedings, 4th International Conference of Soil Mechanics and Foundation Engineering*, 2, 311–14.

Goodman, R. E. (1970). 'The deformability of joints' in *Determination of the in situ modulus of deformation of rock*. American Society for Testing and Materials, Special Publication, 477, 174–96.

Goodman, R. E. and Bray, J. W. (1976). 'Toppling of rock slopes' pp. 201–34 in *Rock Engineering for Foundations and Slopes* (New York: American Society of Civil Engineers).

Gordon, J. E., Birnie, R. V., and Timmis, R. (1978). 'A major rockfall and debris slide on the Lyell Glacier, South Georgia', *Arctic and Alpine Research*, 10, 49–60.

Gordon, R. B. (1979). 'Denudation rate of central New England determined from estuarine sedimentation', *American Journal of Science*, 279, 632–42.

Gorham, E. (1961). 'Factors influencing supply of major ions to inland waters, with special reference to the atmosphere', *Geological Society of America Bulletin*, 72, 795–840.

Gossman, H. (1970). 'Theorien zur Hangentwicklung in verschiedenen Klimazonen', *Würzburger Geographisches Arbeit*, 31.

Gossman, H. (1976). 'Slope modelling with changing boundary conditions – effects of climate and lithology', *Zeitschrift für Geomorphologie, Supplementband*, 25, 72–88.

Goudie, A. (1973). *Duricrusts in Tropical and Subtropical Landscapes* (Oxford: Clarendon Press).

Goudie, A. (1974). 'Further experimental investigation of rock weathering by salt crystallisation and other mechanical processes', *Zeitschrift für Geomorphologie, Supplementband*, 1–12.

Grant, P. J. (1965). 'Major regime changes of the Tukituki River, Hawke's Bay, since about 1650 AD', *Journal of Hydrology (N.Z.)*, 4, 17–30.

Grant-Taylor, T. L. (1964). 'Stable angles in Wellington greywacke', *New Zealand Engineering*, 19, 129–30.

Gray, D. H. (1970). 'Effects of forest clear-cutting on the stability of natural slopes', *Bulletin of the Association of Engineering Geologists*, 7, 45–66.

Gregory, K. J. and Walling, D. E. (1973). *Drainage Basin Form and Process* (London: Arnold).

Griffith, A. A. (1921). 'The phenomenon of rupture and flow in solids', *Philosophical Transactions of the Royal Society, London*, A221, 163–98.

Griggs, D. T. (1936). 'The factor of fatigue in rock exfoliation', *Journal of Geology*, 44, 783–96.

Grim, R. E. (1962). *Applied Clay Mineralogy* (New York: McGraw-Hill).

Grim, R. E. (1968). *Clay Mineralogy* (New York: McGraw-Hill).

Grove, A. T. (1972). 'The dissolved and solid load carried by some West African rivers: Senegal, Niger, Benue and Shari', *Journal of Hydrology*, 16, 277–300.

Grove, J. M. (1972). 'The incidence of landslides, avalanches and floods in western Norway during the Little Ice Age', *Arctic and Alpine Research*, 4, 131–8.

Gunn, R. and Kinzer, G. D. (1949). 'The terminal velocity of fall from water droplets in stagnant air', *Journal of Meteorology*, 6, 243–8.

Habib, P. (1975). 'Production of gaseous pore pressure during rock slides', *Rock Mechanics*, 7, 193–7.

Haefli, R. (1948). 'The stability of slopes acted upon by parallel seepage', *Proceedings 2nd International Conference on Soil Mechanics and Foundation Engineering*, 1, 134–48.

Haefli, R. (1965). 'Creep and progressive failure in snow, soil, rock and ice', *Proceedings 6th International Conference on Soil Mechanics and Foundation Engineering*, 3, 134–48.

Hampton, M. A. (1979). 'Buoyancy in debris flows', *Journal of Sedimentary Petrology*, 49, 753–8.

Hamrol, A. A. (1961). 'A quantitative classification of the weathering and weatherability of rocks',

Proceedings of the 5th International Conference on Soil Mechanics and Foundation Engineering, 2, 771-4.

Hansen, W. R. (1965). 'Effects of the earthquake of March 27, 1964, at Anchorage, Alaska', *U.S. Geological Survey Professional Paper*, 542-A, 1-68.

Harrison, J. V. and Falcon, N. L. (1937). 'The Saidmarreh landslip, southwest Iran', *Geographical Journal*, 89, 42-7.

Haughey, A. (1979). 'Man's effect on water quality', *Soil and Water*, 15, 9-12.

Hedges, J. (1969). 'Opferkessel', *Zeitschrift für Geomorphologie*, 13, 22-55.

Heede, B. H. (1976). 'Gully development and control', *USDA Forest Service Research Paper*, RM-169, 1-42.

Heede, B. H. (1977). 'Case study of a watershed rehabilitation project: Alkali Creek, Colorado', *U.S. Forest Service Research Paper*, RM189.

Heim, A. (1882). 'Der Bergsturz von Elm', *Zeitschrift der deutschen geologischen Gesellschaft*, 34, 74-115.

Heim, A. (1933). *Bergsturz und Menschenleben* (Zurich: Fretz und Wasmuth).

Hem, J. D. and Cropper, W. H. (1959). 'Survey of ferrous-ferric chemical equilibria and redox potentials', *U.S. Geological Survey Water Supply Paper*, 1459-A, 1-31.

Hewlett, J. D. (1961). 'Soil moisture as a source of base flow from steep mountain watersheds', *U.S. Department of Agriculture Forest Service, Southeastern Forest Experiment Station, Asheville, North Carolina, Station Paper*, 132, 1-11.

Hewlett, J. D. and Hibbert, A. R. (1967). 'Factors affecting the response of small watersheds to precipitation in humid areas', pp. 275-90 in *International Symposium on Forest Hydrology* (Oxford: Pergamon).

Hibbert, A. R. (1967). 'Forest treatment effects on water yield', pp. 527-43 in W. E. Sopper and H. W. Lull, (eds.), *International Symposium on Forest Hydrology* (Oxford: Pergamon).

Hodder, A. P. W. (1976). 'Cavitation-induced nucleation of ice: a possible mechanism for frost-cracking in rocks', *New Zealand Journal of Geology and Geophysics*, 19, 821-6.

Hoek, E. (1968). 'The brittle failure of rock', pp. 99-124 in O. C. Zienkiewicz and D. Stagg (eds.), *Rock Mechanics in Engineering Practice* (New York: Wiley).

Hoek, E. and Bray, J. (1977). *Rock Slope Engineering*, 2nd ed. (London: Institution of Mining and Metallurgy).

Hohberger, K. and Einsele, G. (1979). 'Die Bedeutung des Lösungsabtrags verschiedene Gesteine für die Landschaftsentwicklung in Mitteleuropa', *Zeitschrift für Geomorphologie*, 23, 361-82.

Hollingworth, S. E., Taylor, J. H., and Kellaway, G. A. (1944). Large-scale superficial structures in the Northampton ironstone field', *Quarterly Journal of the Geological Society*, 100, 1-44.

Holmes, A. and Wray, D. A. (1912). 'Outlines of the geology of Mozambique', *Geological Magazine*, 9, 412-17.

Hooke, R. L. B. (1967). 'Processes on arid-region alluvial fans', *Journal of Geology*, 75, 438-60.

Horn, H. M. and Deere, D. U. (1962). 'Frictional characteristics of minerals', *Géotechnique*, 12, 319-35.

Horswill, O. and Horton, A. (1976). 'Cambering and valley bulging in the Gwash Valley at Empingham, Rutland', *Philosophical Transactions of the Royal Society, London*, A283, 427-62.

Horton, R. E. (1933). 'The role of infiltration in the hydrological cycle', *American Geophysical Union, Transactions*, 14, 446-60.

Horton, R. E. (1945). 'Erosional development of streams and their drainage basins: hydrophysical approach to quantitative morphology', *Geological Society of America Bulletin*, 56, 275-370.

Hoskins, E. R., Jaeger, J. C., and Rosengren, K. J. (1967). 'A medium-scale direct friction experiment', *International Journal of Rock Mechanics and Mining Sciences*, 5, 143-54.

Hradek, M. (1977). 'Distribution of fine-grained products of Pleistocene frost weathering in the Ceská vysocina (Czech highlands) and their geomorphological importance' [in Czech], *Zpravy Geografickeho Ustavu CSAV*, 14, 92-6.

Hsi, G. and Nath, J. H. (1970). 'Wind drag within a simulated forest', *Journal of Applied Meteorology*, 9, 592-602.

Hsü, K. J. (1975). 'Catastrophic debris streams (Sturzstroms) generated by rockfalls', *Geological Society of America Bulletin*, 86, 129-40.

Hsü, K. J. (1978). 'Albert Heim: observations on landslides and relevance to modern interpretations', pp. 71-93 in B. Voight (ed.), *Rockslides and Avalanches* (Amsterdam: Elsevier).

Huder, J. (1976). 'Creep in Büdner Schist', pp. 125-53 in N. Janbu, F. Jørstad, and B. Kjaernsli (eds.), *Laurits Bjerrum Memorial Volume* (Oslo: Norwegian Geotechnical Institute).

Hudson, N. W. (1971). *Soil Conservation* (London: Batsford).

Hudson, N. W. and Jackson, D. C. (1959). 'Results achieved in the measurement of erosion and runoff in Southern Rhodesia', Paper presented to the *Third Inter-African Soil Conference, Dalaba, 1959*.

Hutchinson, J. N. (1961). 'A landslide on a thin layer of quick clay at Furre, central Norway', *Géotechnique*, 11, 69–94.

Hutchinson, J. N. (1968). 'Mass movement', pp. 688–95 in R. W. Fairbridge (ed.), *The Encyclopedia of Geomorphology* (New York: Reinhold).

Hutchinson, J. N. (1977). 'Assessment of the effectiveness of corrective measures in relation to geological conditions and types of slope movement', *Bulletin of the International Association of Engineering Geology*, 16, 131–55.

Hutchinson, J. N. and Bhandari, R. K. (1971). 'Undrained loading, a fundamental mechanism of mudflows and other mass movements', *Géotechnique*, 21, 353–8.

Hutchinson, J. N. and Gostelow, T. P. (1976). 'The development of an abandoned cliff in London Clay at Hadleigh, Essex', *Philosophical Transactions of the Royal Society, London*, A 283, 557–604.

Hurlbut, C. S. (1971). *Dana's Manual of Mineralogy*, 18th ed. (New York: Wiley).

International Society for Rock Mechanics (1973). 'Suggested methods for determining the point-load strength index', *I.S.R.M. Committee on Laboratory Tests*, Document No. 1, 8–12.

Irfan, T. Y. and Dearman, W. R. (1978). 'Engineering classification and index properties of a weathered granite, *Bulletin of the International Association of Engineering Geology*, 17, 79–90.

Isaacson, E. de St. Q. (1957). 'Research into the rock burst problem on the Kolar gold field', *Mine and Quarry Engineering*, 23, 520–6.

Iveronova, M. I. (1969). 'An attempt at the quantitative analyses of contemporary denudation processes', *National Lending Library Translation*, RTS7436 (from Russian language original).

Jäckli, H. (1957). 'Gegenwartsgeologie des bundnerischen Rheingebietes-ein Beitrag zur exogenen Dynamik Alpiner Gebirgslandschaften', *Beiträge zur Geologie der Schweiz, Geotechnische Serie*, 36, 1–126.

Jacks, G. V. and Whyte, R. O. (1939). *The Rape of the Earth* (London: Faber & Faber).

Jahns, R. H. (1943). 'Sheet structure in granites: its origin and use as a measure of glacial erosion in New England', *Journal of Geology*, 51, 71–98.

James, P. M. (1971). 'The role of progressive failure in clay slopes', *Proceedings of the First Australia-New Zealand Conference on Geomechanics*, 1, 344–8.

Jamiolkowski, M. and Pasqualini, E. (1976). 'Sulla scelta dei parametri geotechnici che intervengono nelle verifiche di stabilita dei pendii naturali ed artificiali', *Atti dell'Instituto di Scienze delle Construzioni, Politecnico di Torino*, 319, 1–53.

Janbu, N. (1954). 'Stability analysis of slopes with dimensionless parameters', *Harvard Soil Mechanics Series*, No. 46 (Cambridge, Mass.: Harvard University Press).

Janbu, N., Bjerrum, L., and Kjaernsli, B. (1956). 'Soil mechanics applied to some engineering problems' [in Norwegian with English summary], *Norwegian Geotechnical Institute Publication*, 16.

Janda, R. J. (1971). 'An evaluation of procedures used in computing chemical denudation rates', *Geological Society of America Bulletin*, 82, 67–80.

Jansen, J. M. L. and Painter, R. B. (1974). 'Predicting sediment yield from climate and topography', *Journal of Hydrology*, 21, 371–80.

Jennings, J. E. (1971). 'A mathematical theory for the calculation of the stability of slopes in opencast mines', *Proceedings of the Open Pit Mining Symposium, South African Institute of Mining and Metallurgy*, 6, 87–102.

Jessen, O. (1936). *Reisen und Forschungen in Angola* (Berlin: Reimer).

Jocelyn, J. (1972). 'Stress patterns and spheroidal weathering', *Nature Physical Science*, 240, 39–40.

Johnson, A. M. (1965). 'A model for debris flow', Ph.D. thesis, Department of Geology and Geophysics, Pennsylvania State University.

Johnson, A. M. (1970). *Physical Processes in Geology* (San Francisco: Freeman, Cooper & Co.).

Jones, F. O. (1973). 'Landslides of Rio de Janeiro and the Serra das Araras escarpment, Brazil', *U.S. Geological Survey Professional Paper*, 697, 1–42.

Judson, S. (1968). 'Erosion rates near Rome, Italy', *Science*, 160, 1444–6.

Judson, S. and Ritter, D. F. (1964). 'Rates of regional denudation in the United States', *Journal of Geophysical Research*, 69, 3395–401.

Keller, W. D. (1957). *The Principles of Chemical Weathering* (Columbia, Mississippi: Lucas).

Kenney, T. C. (1967). 'The influence of mineral composition on the residual strength of natural soils', *Proceedings Geotechnical Conference, Oslo*, 1, 123–9.

Kenney, T. C. (1977). 'Residual strengths of mineral mixtures', *Proceedings of the 9th International Conference on Soil Mechanics and Foundation Engineering, Tokyo*, 1, 155–60.

Kent, P. E. (1966). 'The transport mechanism in catastrophic rock falls', *Journal of Geology*, 74, 79–83.

Kerr, P. F. (1963). 'Quick clay', *Scientific American*, 209, 132–42.

Kezdi, A. (1974). *Handbook of Soil Mechanics*, Vol. 1, *Soil Physics* (Amsterdam: Elsevier).

Kiersch, G. A. (1965). 'Vaiont reservoir disaster', *Geotimes*, 9, 9-12.

Kieslinger, A. (1960). 'Residual stress and relaxation in rocks', *International Geological Congress, Copenhagen*, Session 21, 270-6.

King, L. C. (1949). 'A theory of bornhardts', *Geographical Journal*, 112, 83-7.

King, L. C. (1953). 'Canons of landscape evolution', *Geological Society of America Bulletin*, 64, 721-51.

King, L. C. (1962 and 1967). *The Morphology of the Earth* (Edinburgh: Oliver and Boyd).

Kingdom-Ward, F. (1955). 'Aftermath of the great Assam earthquake of 1950', *Geographical Journal*, 121, 290-303.

Kirkby, M. J. (1971). 'Hillslope process-response models based on the continuity equation', *Institute of British Geographers, Special Publication*, 3, 15-30.

Kirkby, M. J. (ed.) (1978). *Hillslope Hydrology* (Chichester: Wiley).

Kirkby, M. J. and Chorley, R. J. (1967). 'Throughflow, overland flow and erosion', *Bulletin of the International Association for Scientific Hydrology*, 12, 5-21.

Kirkby, M. J. and Statham, I. (1975). 'Surface stone movement and scree formation', *Journal of Geology*, 83, 349-62.

Kojan, E. (1967). 'Mechanics and rates of natural soil creep', *U.S. Forest Service Experiment Station (Berkeley, California) Report*, 233-53.

Kojan, E. and Hutchinson, J. N. (1978). 'Mayunmarca rockslide and debris flow, Peru', pp. 315-61 in B. Voight (ed.), *Rockslides and Avalanches* (Amsterdam: Elsevier).

Krammes, J. S. (1965). 'Seasonal debris movement from steep mountainside slopes in southern California', *U.S. Department of Agriculture, Miscellaneous Publication*, 970, 85-8.

Krinsley, D. H. and Smalley, I. J. (1972). 'Sand', *American Scientist*, 60, 286-91.

Kutter, J. K. and Rautenberg, A. (1979). 'The residual shear strength of filled joints in rock', *Proceedings of the 4th Congress of the International Society for Rock Mechanics*, 1, 221-7.

Lajtai, E. Z. and Alison, J. R. (1979). 'A study of residual stress effects in sandstone', *Canadian Journal of Earth Sciences*, 16, 1547-57.

Lambe, T. W. and Whitman, R. V. (1979). *Soil Mechanics: SI Version* (Chichester: Wiley).

Langbein, W. B. (1949). 'Annual runoff in the United States', *U.S. Geological Survey Circular*, 52, 1-11.

Langbein, W. B. and Schumm, S. A. (1958). 'Yield of sediment in relation to mean annual precipitation', *Transactions of the American Geophysical Union*, 39, 1076-84.

Langford-Smith, T. (1978). *Silcrete in Australia* (Armidale: Department of Geography, University of New England).

Laws, J. O. (1941). 'Measurements of fall-velocity of water-drops and raindrops', *Transactions of the American Geophysical Union*, 22, 709.

Laws, J. O. and Parsons, D. A. (1943). 'The relation of raindrop size to intensity', *Transactions of the American Geophysical Union*, 24, 452.

Leaf, C. F. and Martinelli, M. (1977). 'Avalanche Dynamics', *U.S. Department of Agriculture, Forest Service Research Paper*, RM-183, 1-51.

Lehmann, O. (1933). 'Morphologische Theorie der Verwitterung von steinschlag wänden', *Vierteljahsschrift der Naturforschende Gesellschaft in Zurich*, 87, 83-126.

Lehmann, O. (1934). 'Ueber die morphologischen Folgen der Wandwitterung', *Annals of Geomorphology*, 8, 93-9.

Lelong, F. (1966). 'Régime des nappes phréatiques contenues dans les formations d'altération tropicale. Conséquences pour la pédogenèse', *Science du Terre*, 11, 203-44.

Leopold, L. B., Wolman, M. G., and Miller, J. P. (1964). *Fluvial Processes in Geomorphology* (San Francisco: W. H. Freeman).

Li, Y-H. (1976). 'Denudation of Taiwan Island since the Pliocene Epoch', *Geology*, 4, 105-7.

Lindner, E. (1976). 'Swelling rock: a review', pp. 141-81 in *Rock Engineering for Foundations and Slopes*, 1 (New York: American Society of Civil Engineers).

Linton, D. L. (1955). 'The problem of tors', *Geographical Journal*, 121, 470-87.

Lo, K. Y. and Lee, C. F. (1975). 'Stress distributions in rock slopes under high *in situ* stresses', pp. 35-55 in *Mass Wasting*, 4th Guelph Symposium on Geomorphology, 1975 (Norwich: Geo Abstracts).

Lohnes, R. A. and Handy, R. L. (1968). 'Slope angles in friable loess', *Journal of Geology*, 76, 247-58.

Loughnan, F. C. (1969). *Chemical Weathering of the Silicate Minerals* (Amsterdam: Elsevier).

Luckman, B. H. (1977). 'The geomorphic activity of avalanches', *Geografiska Annaler*, 59A, 31-48.

Lull, H. W. (1964). 'Ecological and silvicultural aspects', Section 6 in Ven Te Chow (ed.), *Handbook of Applied Hydrology* (New York: McGraw-Hill).

Lutton, R. J. (1969). 'Fractures and failure mechanics in loess and applications to rock mechanics', *U.S. Army Engineer Waterways Experiment Station Research Report*, S-69-1, Vicksburg, Mississippi.

Mabbutt, J. A. (1961). 'A stripped land surface in Western Australia', *Institute of British Geographers, Transactions*, 29, 101-14.

Mabbutt, J. A. (1965). 'The weathered land surface in Central Australia', *Zeitschrift für Geomorphologie*, 9, 82-114.

McConnell, R. G. and Brock, R. W. (1904). 'The great landslide at Frank, Alberta', *Canada Department of the Interior Annual Report 1902-1903*, part 8, appendix.

McCraw, J. D. (1959). 'Periglacial and allied phenomena in Western Otago', *New Zealand Geographer*, 15, 61-8.

McCraw, J. D. (1965). 'Landscapes of Central Otago', pp. 30-45 in R. G. Lister, and R. P. Hargreaves (eds.), *Central Otago* (New Zealand Geographical Society).

MacDonald, D. F. (1913). 'Some engineering problems of the Panama Canal in their relation to geology and topography', *U.S. Department of the Interior, Bureau of Mines Bulletin*, 86, 1-88.

McFarlane, M. J. (1976). *Laterite and Landscape*. (London: Academic Press).

McGreal, W. S. (1979). 'Factors promoting coastal slope instability in southeast County Down, N. Ireland', *Zeitschrift für Geomorphologie*, 23, 76-90.

Machida, H. (1966). 'Rapid erosional development of mountain slopes and valleys caused by large landslides in Japan', *Geographical Reports of Tokyo Metropolitan University*, 1, 55-78.

McSaveney, M. J. (1978). 'Sherman Glacier rock avalanche, Alaska, U.S.A.', pp. 197-258 in B. Voight (ed.), *Rockslides and Avalanches* (Amsterdam: Elsevier).

Mahr, T. (1977). 'Deep-reaching gravitational deformation of high mountain slopes', *Bulletin of the International Association of Engineering Geology*, 16, 121-7.

Mahr, T. and Nemčok, A. (1977). 'Deep-seated creep deformations in the crystalline caves of the Tatry Mountains', *Bulletin of the International Association of Engineering Geology*, 16, 104-6.

Maignien, R. (1966). *A Review of Research on Laterite*. UNESCO, Natural Resources Research, 4.

Mainguet, M. (1972). *Le Modelé Des Grès*, 2 vols. (Paris: Institut Geographique National).

Malaurie, J. (1968). 'Effets relatifs de la gélifraction en haute latitude – observations en Terre d'Inglefield (nord-ouest Groenland) et études expérimentales', *Société Savantes Haute Normandie Revue de Science*, 50, 57-67.

Marchand, D. E. (1971). 'Rates and modes of denudation, White Mountains, eastern California', *American Journal of Science*, 270, 109-35.

Meade, R. H. (1969). 'Errors in using modern stream-load data to estimate natural rates of denudation', *Geological Society of American Bulletin*, 80, 1265-74.

Meade, R. H. (1976). 'Sediment problems in the Savannah river basin', pp. 105-29 in B. L. Dillman and J. M. Stepp (eds.), *The Future of the Savannah River*. Proceedings of a Symposium held at Hickory Knob State Park, South Carolina, October 14-15, 1975 (Water Resources Research Institute, Clemson University, South Carolina).

Mellor, M. (1978). 'Dynamics of snow avalanches', pp. 753-92 in B. Voight (ed.), *Rockslides and Avalanches* (Amsterdam: Elsevier).

Meyerhof, G. G. (1969). 'Safety factors in soil mechanics', *Proceedings 7th International Conference on Soil Mechanics and Foundation Engineering, Mexico*, 479-81.

Middleton, H. E. (1930). 'Properties of soils which influence soil erosion', *U.S. Department of Agriculture Technical Bulletin*, 178.

Mitchell, J. K. (1976). *Fundamentals of Soil Behaviour* (New York: Wiley).

Mitchell, R. J. (1976). 'Earthflow terrain evaluation in Ontario: Kingston Ontario', *Queen's University Department of Civil Engineering Final Report*, Project Q-53, 1-23.

Mitchell, R. J. and Klugman, M. A. (1979). 'Mass instabilities in sensitive Canadian soils', *Engineering Geology*, 14, 109-34.

Moeyersons, J. and De Ploey, J. (1976). 'Quantitative data on splash erosion, simulated on unvegetated slopes', *Zeitschrift für Geomorphologie, Supplementband*, 25, 120-31.

Mollard, J. D. (1977). 'Some regional landslide types in Canada, in D. R. Coates, (ed.), *Landslides*. Reviews in Engineering Geology, Vol. 3, Geological Society of America, 29-56.

Montjuvent, G. (1973). 'L'erosion sur les Alpes francaises d'apres l'exemple du massif de Pelvoux', *Revue Geographie Alpine*, 61, 107-20.

Morgenstern, N. R. and Nixon, J. F. (1971). 'One-dimensional consolidation of thawing soils', *Canadian Geotechnical Journal*, 8, 558-65.

Morisawa, M. (1968). *Streams: their dynamics and morphology* (New York: McGraw-Hill).

Morton, D. H. and Campbell, R. H. (1974). 'Spring mudflows at Wrightwood, southern California', *Quarterly Journal of Engineering Geology*, 7, 377-84.

Moss, A. J., Walker, P. H., and Hutka, J. (1979). 'Raindrop-simulated transportation in shallow water flows: an experimental study', *Sedimentary Geology*, 22, 165-84.

Moye, D. E. (1955). 'Engineering Geology for the Snowy Mountains scheme', *Journal of the Institution of Engineers, Australia*, 27, 281-99.

Müller, L. (1958). 'Geomechanische Auswertung

gefügekundlicher Details', *Geologie und Bauwesen*, 24, 4–21.

Müller, L. (1963). *Der Felsbau* (Stuttgart: Enke).

Müller, L. (1964). 'The rock slide in the Vajont Valley', *Rock Mechanics and Engineering Geology, Vienna*, 2, 148–212.

Musgrave, G. W. (1947). 'Quantitative evaluation of factors in water-erosion – a first approximation', *Journal of Soil and Water Conservation*, 2, 133–8.

Nankano, R. (1967). 'On weathering and change of properties of Tertiary mudstone related to landslides', *Soil and Foundations (Tokyo)*, 7, 1–14.

Nemčok, A. (1977). 'Geological/tectonical structures – an essential condition for genesis and evolution of slope movement', *Bulletin of the International Association of Engineering Geology*, 16, 127–30.

Nemčok, A., Pašek, J., and Rybar, J. (1972). 'Classification of landslides and other mass movements', *Rock Mechanics*, 4, 71–8.

Netterberg, F. (1967). 'Some roadmaking properties of South African calcretes', *Proceedings, 4th Regional Conference for Africa on Soil Mechanics and Foundation Engineering, Cape Town*, 1, 77–81.

Norton, S. A. (1973). 'Laterite and bauxite formation', *Economic Geology*, 68, 353–61.

Oberlander, T. M. (1972). 'Morphogenesis of granitic boulder slopes in the Mojave desert, California', *Journal of Geology*, 80, 1–20.

Ollier, C. D. (1960). 'The inselbergs of Uganda', *Zeitschrift für Geomorphologie*, 4, 43–52.

Ollier, C. D. (1969). *Weathering* (Edinburgh: Oliver and Boyd).

Ollier, C. D. (1978). 'Inselbergs of the Namib Desert, processes and history', *Zeitschrift für Geomorphologie, Supplementband*, 31, 161–76.

Ollier, C. D. and Tuddenham, W. G. (1962). 'Inselbergs of Central Australia', *Zeitschrift für Geomorphologie*, 5, 257–76.

O'Loughlin, C. (1974). 'The effect of timber removal on the stability of forest soils', *Journal of Hydrology (N.Z.)*, 13, 121–34.

O'Loughlin, C. and Pearce, A. J. (1976). 'Influence of Cenozoic geology on mass movement and sediment yield response to forest removal, north Westland, New Zealand', *Bulletin of the International Association of Engineering Geology*, 14, 41–6.

Orr, C. M. (1974). 'The geological description of *in situ* rock masses as input data for engineering design', *C.S.I.R. South Africa Report*, series MEG/344, No. ME1274.

Osipov, V. I. (1975). 'Structural bonds and the properties of clays', *Bulletin of the International Association of Engineering Geology*, 12, 13–20.

Owens, L. B. (1976). 'Rates of weathering and soil formation on granite in Rhodesia', *Dissertation Abstracts International*, B, 77-1756.

Owens, L. B. and Watson, J. P. (1979). 'Landscape reduction by weathering in small Rhodesian watersheds', *Geology*, 7, 281–4.

Packard, F. A. (1974). 'The hydraulic geometry of a discontinuous ephemeral stream on a bajada near Tucson, Arizona', Ph.D. dissertation University of Arizona (quoted by Bull, 1977).

Pacher, F. (1958). 'Kennziffern des Flächengefüges', *Geologie und Bauwesen*, 24, 223–7.

Pain, C. F. (1968). 'Geomorphic effects of floods in the Orere River catchment, eastern Hunua Ranges', *Journal of Hydrology (N.Z.)*, 7, 62–74.

Pain, C. F. (1969). 'The effect of some environmental factors on rapid mass movement in the Hunua Ranges, New Zealand', *Earth Science Journal*, 3, 101–7.

Pain, C. F. (1971). 'Rapid mass movement under forest and grass in the Hunua Ranges, New Zealand', *Australian Geographical Studies*, 9, 77–84.

Pain, C. F. (1972). 'Characteristics and geomorphic effects of earthquake-initiated landslides in the Adelbert Range, Papua-New Guinea', *Engineering Geology*, 6, 261–74.

Pain, C. F. (1975). 'The Kaugel diamicton – a late Quaternary mudflow deposit in the Kaugel Valley, Papua-New Guinea', *Zeitschrift für Geomorphologie*, 19, 430–42.

Pain, C. F. and Bowler, J. M. (1973). 'Denudation following the November 1970 earthquake at Madang, Papua-New Guinea', *Zeitschrift für Geomorphologie, Supplementband*, 18, 92–104.

Palmer, R. S. (1965). 'Waterdrop impact forces', *American Society of Agricultural Engineers*, 8, 70–2.

Parker, G. G. and Jenne, E. A. (1967). 'Structural failure of western highways caused by piping', *Highway Research Record*, 203, 57–76.

Parsons, A. (1976). 'An example of the application of deductive models to field measurement of hillslope form', *Zeitschrift für Geomorphologie, Supplementband*, 25, 145–53.

Patterson, S. H. (1971). 'Investigations of ferruginous bauxite and other mineral resources on Kanai and a reconnaissance of ferruginous bauxite deposits on Maui, Hawaii', *U.S. Geological Survey Professional Paper*, 656, 1–74.

Patton, F. D. (1966). 'Multiple modes of shear failure in rock and related materials', Ph.D. thesis, University of Illinois (University Microfilms No. 667786).

Patton, F. D. and Deere, D. U. (1971). 'Geologic factors controlling slope stability in open pit mines', pp. 23–48 in C. O. Brawner and V. Milligan (eds.), *Stability in Open Pit Mining* (New York: Society of Mining Engineers, American Institute of Mining, Metallurgical and Petroleum Engineers).

Pearce, A. J. (1976). 'Geomorphic and hydrologic consequences of vegetation destruction, Sudbury, Ontario', *Canadian Journal of Earth Sciences*, 13, 1358–73.

Pearce, A. J. and Elson, J. A. (1973). 'Postglacial rates of denudation by soil movement, free face retreat, and fluvial erosion, Mont St. Hilaire, Quebec', *Canadian Journal of Earth Sciences*, 10, 91–101.

Peck, R. B. (1969). 'Stability of natural slope', pp. 437–51 in *Stability and Performance of Slopes and Embankments* (Soil Mechanics and Foundation Division American Society of Civil Engineers).

Penck, W. (1924). *Die Morphologische Analyse: Ein Kapitel der Physikalischen Geologie*. Geographische Abhandlungen 2, Reihe, Heft, 2, Stuttgart.

Pereira, H. C. (1973). *Land Use And Water Resources* (London: Cambridge University Press).

Perla, R. I. and Martinelli, M. (1976). *Avalanche Handbook*. U.S. Department of Agriculture Forest Service, Agriculture Handbook, 489.

Persons, B. S. (1970). *Laterite Genesis, Location, Use* (New York: Plenum Press).

Piteau, D. R. (1971). 'Geological factors significant to the stability of slopes cut in rock', *Proceedings of the Open Pit Mining Symposium, South African Institute of Mining and Metallurgy*, 33–53.

Piteau, D. R. (1973). 'Characterizing and extrapolating rock joint properties in engineering practice', *Rock Mechanics, Supplement*, 2, 5–31.

Pitty, A. F. (1971). *Introduction to Geomorphology* (London: Methuen).

Plafker, G. and Ericksen, G. E. (1978). 'Nevados Huascarán avalanches, Peru' pp. 277–314 in B. Voight (ed.), *Rockslides and Avalanches* (Amsterdam: Elsevier).

Polynov, B. B. (1937). *Cycle of Weathering*, translated by A. Muir (London: Murby).

Potts, A. S. (1970). 'Frost action in rocks: some experimental data', *Transactions of the Institute of British Geographers*, 49, 109–24.

Prior, D. B. (1977). 'Coastal mudslide morphology and processes on Eocene clays in Denmark', *Geografisk Tidsskrift*, 77, 14–33.

Prior, D. B. and Suhayda, J. N. (1979). 'Application of infinite slope analysis to subaqueous sediment instability, Mississippi Delta', *Engineering Geology*, 14, 1–10.

Protodyakonov, M. M. (1960). 'New methods of determining mechanical properties of rock', *Proceedings International Conference on Strata Control, Paris*, Paper C2, 187–95.

Protodyakonov, M. M. (1969). *Mechanical Properties of Rocks* (in Russian), translated by Israel Program for Scientific Translations, Jerusalem.

Pugh, J. C. (1966). 'The landforms of low latitudes', pp. 121–38 in G. H. Dury (ed.), *Essays in Geomorphology* (London: Heinemann).

Radbruch-Hall, D. H. (1978). 'Gravitational creep of rock masses on slopes', pp. 607–57 in B. Voight (ed.), *Rockslides and Avalanches* (Amsterdam: Elsevier).

Ragan, R. M. (1968). 'An experimental investigation of partial area contributions', *International Association for Scientific Hydrology Publication*, 76, 241–9.

Rapp, A. (1960a). 'Recent development of mountain slopes in Kärkevagge and surroundings, northern Scandinavia', *Geografiska Annaler*, 42, 73–200.

Rapp, A. (1960b). 'Talus slopes and mountain walls at Tempelfjorden, Spitsbergen', *Norsk Polarinstitutt Skrifter*, 119, 1–96.

Rapp, A. (1975). 'Studies of mass wasting in the Arctic and in the tropics', pp. 79–103 in *Mass Wasting* (Norwich: Geo Abstracts).

Rapp, A., Berry, L., and Temple, P. (eds.) (1973). 'Studies of soil erosion and sedimentation in Tanzania', *Research Monograph*, No. 1 (Bureau of Resource Assessment and Land Use Planning, University of Dar es Salaam).

Reeves, C. C. (1976). *Caliche: Origin, Classification, Morphology and Uses* (Lubbock, Texas: Estacado Books).

Reiche, P. (1950). 'A survey of weathering processes and products', *New Mexico University Publications in Geology*, 3.

Rice, A. (1976). 'Insolation warmed over', *Geology*, 4, 61–2.

Rice, R. M. and Foggin, G. T. (1971). 'Effect of high intensity storms on soil slippage on mountainous watersheds in southern California', *Water Resources Research*, 7, 1485–96.

Roberts, A. (1977). *Geotechnology* (Oxford: Pergamon).

Robertson, A. MacG. (1971). 'The interpretation of geological factors for use in slope theory', *Proceedings of the Open Pit Mining Symposium, South African Institute of Mining and Metallurgy*, 55–71.

Robinson, R. (1979). 'Variation of energy release, rate of occurrence and b-value of earthquakes in the main seismic region, New Zealand', *Physics of the Earth and Planetary Interiors*, 18, 209-20.

Rogers, N. W. and Selby, M. J. (1980). 'Mechanisms of shallow translational landsliding during summer rainstorms: North Island, New Zealand', *Geografiska Annaler*, 62A, 11-21.

Roš, M. and Eichinger, A. (1928). 'Versuche zur Klärung der Frage der Bruchgefar. II. Nicht-metallische Stoffe Eidgenössische Materialprufungsanstalt der Eidgen', Zürich.

Rouse, W. C. (1975). 'Engineering properties and slope form in granular soils', *Engineering Geology*, 9, 221-36.

Rouse, W. W. and Farhan, Y. I. (1976). 'Threshold slopes in South Wales', *Quarterly Journal of Engineering Geology*, 9, 327-38.

Ruxton, B. P. and Berry, L. (1957). 'The weathering of granite and associated erosional features in Hong Kong', *Bulletin of the Geological Society of America*, 68, 1263-92.

Ruxton, B. P. and McDougall, I. (1967). 'Denudation rates in northeast Papua from potassium-argon dating of lavas', *American Journal of Science*, 265, 545-61.

Saito, M. and Uezawa, H. (1961). 'Failure of soil due to creep', *Proceedings, 5th International Conference on Soil Mechanics and Foundation Engineering*, 1, 315-18.

Samuels, S. G. (1950). 'The effect of base exchange on the engineering properties of soil', *Building Research Station of Great Britain*, Note C176.

Savigear, R. A. G. (1952). 'Some observations on slope development in South Wales', *Transactions of the Institute of British Geographers*, 18, 31-51.

Schattner, I. (1961). 'Weathering phenomena in the crystalline of the Sinai in the light of current notions', *Bulletin of the Research Council of Israel*, 10G, 247-266.

Scheiddeger, A. E. (1970). *Theoretical Geomorphology*, 2nd ed. (London: Allen & Unwin).

Scholz, C. H. and Engelder, J. T. (1976). 'The role of asperity indentation and ploughing in rock friction', *International Journal of Rock Mechanics, and Mining Science, and Geomechanics Abstracts*, 13, 149-63.

Schumm, S. A. (1962). 'Erosion on miniature pediments in Badlands National Monument, South Dakota', *Geological Society of American Bulletin*, 73, 719-24.

Schumm, S. A. and Chorley, R. J. (1964). 'The fall of threatening rock', *American Journal of Science*, 262, 1041-54.

Schuster, R. L. and Krizek, J. (eds.) (1978). *Landslides: Analysis and Control* (Washington: Transportation Research Board).

Scott, K. M. and Williams, R. P. (1978). 'Erosion and sediment yields in the Transverse Ranges, Southern California', *U.S. Geological Survey Professional Paper*, 1030, 1-38.

Seed, H. B., Mori, K., and Chan, C. K. (1977). 'Influence of seismic history on liquefaction of sands', *Journal of the Geotechnical Engineering Division, American Society of Civil Engineers*, 103 (GT4), Proceedings Paper, 12841, 257-70.

Seed, H. B. and Wilson, S. D. (1969). 'The Turnagain Heights landslide, Anchorage, Alaska', pp. 357-85 in *Stability and Performance of Slopes and Embankments* (Soil Mechanics and Foundations Division, American Society of Civil Engineers).

Seed, B., Woodward, R., and Lundgren, R. (1964). 'Clay mineralogical aspects of the Atterberg limits', *Journal of the Soil Mechanics and Foundations Division, Proceedings of the American Society of Civil Engineers*, SM4, 107-31.

Selby, M. J. (1967a). 'Aspects of the geomorphology of the greywacke ranges bordering the lower and middle Waikato Basins', *Earth Science Journal*, 1, 37-58.

Selby, M. J. (1967b). 'Erosion by high intensity rainstorms in the lower Waikato Basin', *Earth Science Journal*, 1, 153-6.

Selby, M. J. (1970). 'Design of a hand-portable rainfall-simulating infiltrometer, with trial results from the Otutira catchment', *Journal of Hydrology (N.Z.)*, 9, 117-32.

Selby, M. J. (1971a). 'Salt-weathering of landforms, and an Antarctic example', *Proceedings 6th Geography Conference, New Zealand*, 30-5.

Selby, M. J. (1971b). 'Slopes and their development in an ice-free arid area of Antarctica', *Geografiska Annaler*, 53A, 235-45.

Selby, M. J. (1972). 'Antarctic tors', *Zeitschrift für Geomorphologie, Supplementband*, 13, 73-86.

Selby, M. J. (1973). 'An investigation into the causes of runoff from a catchment of pumice lithology in New Zealand', *Hydrological Sciences Bulletin*, 18, 255-80.

Selby, M. J. (1974a). 'Slope evolution in an Antarctic oasis', *New Zealand Geographer*, 30, 18-34.

Selby, M. J. (1974b). 'Dominant geomorphic events in landform evolution', *Bulletin of the International Association of Engineering Geology*, 9, 85-9.

Selby, M. J. (1976). 'Slope erosion due to extreme rainfall: a case study from New Zealand', *Geografiska Annaler*, 58A, 131-8.

Selby, M. J. (1977a). 'Bornhardts of the Namib Desert', *Zeitschrift für Geomorphologie*, 21, 1-13.

Selby, M. J. (1977b). 'On the origin of sheeting and laminae in granitic rocks: evidence from Antarctica, the Namib Desert and the Central Sahara',

Madoqua, 171–9.

Selby, M. J. (1979). 'Slope stability studies in New Zealand', pp. 120–34 in D. L. Murray and P. Ackroyd (eds.), *Physical Hydrology: New Zealand Experience* (Wellington: N.Z. Hydrological Society).

Selby, M. J. (1980). 'A rock mass strength classification for geomorphic purposes: with tests from Antarctica and New Zealand', *Zeitschrift für Geomorphologie*, 24, 31–51.

Selby, M. J. and Hosking, P. J. (1971). 'Causes of infiltration into yellow-brown pumice soils', *Journal of Hydrology (N.Z.)*, 19, 113–19.

Serafim, J. L. (1964). 'Rock mechanics considerations in the design of concrete dams', pp. 611–45 in *Proceedings of the International Conference on the State of Stress in the Earth's Crust* (New York: Elsevier).

Serafim, J. L. (1968). 'Influence of interstitial water on the behaviour of rock masses', pp. 55–97 in O. C. Zienkiewicz and D. Stagg (eds.), *Rock Mechanics in Engineering Practice* (New York: Wiley).

Seyček, J. (1978). 'Residual shear strength of soils', *Bulletin of the International Association of Engineering Geology*, 17, 73–5.

Sharpe, C. F. S. (1938). *Landslides and Related Phenomena* (New Jersey: Pageant).

Shreve, R. L. (1968). 'Leakage and fluidisation in air-layer lubricated landslides', *Geological Society of America Bulletin*, 79, 653–8.

Siever, R. (1959). 'Petrology and geochemistry of silica cementation in some Pennsylvanian sandstones', pp. 56–76 in H. A. Ireland (ed.), *Silica in sediments: a symposium* (Society of Economic Paleontologists and Mineralogists, Special Publication 7).

Silvestri, T. (1961). 'Determinazione sperimentale de resistenza meccania del materiale constituente il corpo di una diga del tipo "Rockfill"', *Geotechnica*, 8, 186–91.

Simonett, D. S. (1967). 'Landslide distribution and earthquakes in the Bewani and Torricelli Mountains, New Guinea' pp. 64–84 in J. N. Jennings and J. A. Mabbutt (eds.), *Landform Studies from Australia and New Guinea* (Canberra: A.N.U. Press).

Simonett, D. S. (1970). 'The role of landslides in slope development in the high rainfall tropics', *Final Report, Office of Naval Research, Geographical Branch*, Contract Number, 583(11), Task No., 389-133, 1–23.

Skempton, A. W. (1948). 'The rate of softening of stiff, fissured clays', *Proceedings of the 2nd International Conference on Soil Mechanics and Foundation Engineering, Rotterdam*, 2, 50–3.

Skempton, A. W. (1953a). 'Soil mechanics in relation to geology', *Proceedings Yorkshire Geological Society*, 29, 33–62.

Skempton, A. W. (1953b). 'The colloidal activity of clays', *Proceedings 3rd International Conference on Soil Mechanics and Foundation Engineering, Switzerland*, 1, 57.

Skempton, A. W. (1964). 'Long-term stability of clay slopes', *Géotechnique*, 14, 77–101.

Skempton, A. W. (1970). 'First-time slides in over-consolidated clays', *Géotechnique*, 20, 320–4.

Skempton, A. W. (1977). 'Slope stability of cuttings in brown London clay', *Proceedings of the 9th International Conference on Soil Mechanics and Foundation Engineering, Tokyo*, 3, 261–70.

Skempton, A. W. and Bishop, A. W. (1950). 'The measurement of the shear strength of soils', *Géotechnique*, 2, 90–108.

Skempton, A. W. and De Lory, F. A. (1957), 'Stability of natural slopes in London Clay', *Proceedings 4th International Conference on Soil Mechanics and Foundation Engineering, London*, 2, 378–81.

Skempton, A. W. and Hutchinson, J. N. (1969). 'Stability of natural slopes and embankment foundations', *7th International Conference on Soil Mechanics and Foundation Engineering, Mexico City*, 291–340.

Skempton, A. W. and Weeks, A. G. (1976). 'The Quaternary history of the Lower Greensand escarpment and Weald Clay vale near Seven Oaks, Kent', *Philosophical Transactions of the Royal Society, London*, A283, 493–526.

Slaymaker, O. (1977). 'An overview of geomorphic processes in the Canadian Cordillera', *Zeitschrift für Geomorphologie*, 21, 169–86.

Smalley, I. (1976). 'Factors relating to the landslide process in Canadian quickclays', *Earth Surface Processes*, 1, 163–72.

Smalley, I. and Taylor, R. (1972). 'The quickclay enigma', *Scientific Era*, December.

Smith, B. J. (1978). 'The origin and geomorphic implications of cliff foot recesses and tafoni on limestone hamadas in the northwest Sahara', *Zeitschrift für Geomorphologie*, 22, 21–43.

Smith, G. N. (1978). *Elements of Soil Mechanics for Civil and Mining Engineers* (London: Crosby Lockwood Staples).

Smith, W. D. (1978a). 'Earthquake risk in New Zealand: statistical estimates', *New Zealand Journal of Geology and Geophysics*, 21, 313–27.

Smith, W. D. (1978b). 'Spatial distribution of felt intensities for New Zealand earthquakes', *New Zealand Journal of Geology and Geophysics*, 21, 293–311.

Sopper, W. E. and Lull, H. W. (eds.) (1967). *International Symposium on Forest Hydrology* (Oxford: Pergamon).

Sopper W. E. and Lynch, J. A. (1970). 'Changes in water yield following partial forest cover removal on an experimental watershed', *International Association for Scientific Hydrology*, Publication 96, 369-89.

Spangler, M. G. (1960). *Soil Engineering* (Scranton: International Textbook Company).

Spencer, E. (1967). 'A method of analysis of the stability of embankments assuming parallel inter-slice forces', *Géotechnique*, 17, 11-26.

Sridharan, A. and Venkatappa Rao, G. (1979). 'Shear strength behaviour of saturated clays and the role of the effective stress concept', *Géotechnique*, 29, 177-93.

Starkel, L. (1966). 'Post-glacial climate and the moulding of European relief', pp. 15-32 in *Proceedings of the International Symposium on World Climate 8000 to 0 B.C.* (London: Royal Meteorological Society).

Starkel, L. (1972a). 'The modelling of monsoon area of India as related to catastrophic rainfall', *Geographica Polonica*, 23, 153-73.

Starkel, L. (1972b). 'The role of catastrophic rainfall in the shaping of the relief of the Lower Himalaya (Darjeeling Hills)', *Geographica Polonica*, 21, 103-47.

Starkel, L. (1976). 'The role of extreme (catastrophic) meteorological events in contemporary evolution of slopes', pp. 203-46 in E. Derbyshire (ed.), *Geomorphology and Climate* (London: Wiley).

Starkel, L. (1979). 'The role of extreme meteorological events in the shaping of mountain relief', *Geographica Polonica*, 41, 13-20.

Statham, I. (1973). 'Scree development under conditions of surface particle movement', *Institute of British Geographers, Transactions*, 59, 41-54.

Steinbrenner, E. C. (1955). 'The effect of repeated tractor trips on the physical properties of forest soils', *Northwest Science*, 29, 155-9.

Stephens, C. G. (1971). 'Laterite and silcrete in Australia: a study of the genetic relationships of laterite and silcrete and their companion materials, and their collective significance in the formation of the weathered mantle, soils, relief and drainage of the Australian continent', *Geoderma*, 5, 5-52.

Stevenson, C. M. (1968). 'An analysis of the chemical composition of rainwater and air over the British Isles and Eire for the years 1959-1964', *Quarterly Journal of the Royal Meteorological Society*, 94, 56-70.

Stimpson, B. (1979). 'Simple equations for determining the factor of safety of a planar wedge under various groundwater conditions', *Quarterly Journal of Engineering Geology*, 12, 3-7.

Stipp, J. J. and McDougall, I. (1968). 'Geochronology of the Banks Peninsula volcanoes, New Zealand', *New Zealand Journal of Geology and Geophysics*, 11, 1239-60.

Stoddart, D. R. (1969). 'World erosion and sedimentation', pp. 43-64 in R. J. Chorley (ed.), *Water, Earth and Man* (London: Methuen).

Strakhov, N. M. (1967). *Principles of Lithogenesis*, Vol. 1 (translated from Russian edition of 1964) (Oliver and Boyd: Edinburgh).

Sturgul, J. R. and Scheidegger, A. E. (1967). 'Tectonic stresses in the vicinity of a wall', *Rock Mechanics and Engineering Geology*, 5, 137-49.

Sugden, M. B., Van Wieringen, M., and Knight, K. (1977). 'Slip failures in bedded sediments', *Proceedings 9th International Conference on Soil Mechanics and Foundation Engineering*, 2, 155-60.

Swanson, F. J. and Swanston, D. N. (1977). 'Complex mass-movement terrains in the western Cascade Range, Oregon', *Geological Society of America, Reviews in Engineering Geology*, 3, 113-24.

Swanston, D. N. (1970). 'Mechanics of debris avalanching in shallow till soils of southeast Alaska', *USDA Forest Service Research Paper*, PNW-103, 1-17.

Taylor, D. W. (1937). 'Stability of earth slopes', *Journal of the Boston Society of Civil Engineers*, 24, 337-86.

Taylor, D. W. (1948). *Fundamentals of Soil Mechanics* (New York: Wiley).

Taylor, D. K., Hawley, J. E., and Riddolls, B. W. (1977). 'Slope stability in urban development', *Department of Scientific and Industrial Research (New Zealand), Information Series*, 122, 1-71.

Taylor, R. K. and Spears, D. A. (1970). 'The breakdown of British Coal Measure rocks', *International Journal of Rock Mechanics and Mining Science*, 7, 481-501.

Tennessee Valley Authority (1964). 'Bradshaw Creek-Elk River, a pilot study in area stream factor correlation', *Research Paper No. 4, T.V.A.*

Ter-Stepanian, G. I. (1974). 'Depth creep of slopes', *Bulletin of the International Association of Engineering Geology*, 9, 97-102.

Ter-Stepanian, G. I. (1977). 'Deep-reaching gravitational deformation of mountain slopes', *Bulletin of the International Association of Engineering Geology*, 16, 87-94.

Terzaghi, K. (1936). 'Stability of slopes in natural clay', *Proceedings 1st International Conference on Soil Mechanics and Foundation Engineering*, 1, 161-5.

Terzaghi, K. (1950). 'Mechanism of Landslides', *Geological Society of America, Engineering Geology (Berkey) Volume*, 83-123.

Terzaghi, K. (1953). 'Some miscellaneous notes on creep', *Proceedings 3rd International Conference on Soil Mechanics and Foundation Engineering*, 3, 205–6.

Terzaghi, K. (1962). 'Stability of steep slopes in hard unweathered rock', *Géotechnique*, 12, 251–70.

Terzaghi, K. and Peck, R. B. (1948). *Soil Mechanics in Engineering Practice* (New York: Wiley).

Thomas, M. F. (1965). 'Some aspects of the geomorphology of domes and tors in Nigeria', *Zeitschrift für Geomorphologie*, 9, 63–81.

Thomas, M. F. (1966). 'Some geomorphological implications of deep weathering patterns in crystalline rocks in Nigeria', *Institute of British Geographers, Transactions*, 40, 173–93.

Thomas, M. F. (1974). 'Granite landforms: a review of some recurrent problems of interpretation', *Institute of British Geographers, Special Publication*, 7, 13–37.

Thomas, M. F. (1978). 'The study of inselbergs', *Zeitschrift für Geomorphologie, Supplementband*, 31, 1–41.

Thornes, J. B. (1971). 'State, environment and attribute in scree-slope studies', *Institute of British Geographers, Special Publication*, 4, 49–63.

Torrance, J. K. (1975). 'Leaching, weathering and origin of Leda Clay in the Ottawa area', pp. 105–16 in *Mass Wasting* (Norwich: Geo Abstracts).

Trendall, A. F. (1962). 'The formation of "apparent peneplains" by a process of combined laterisation and surface wash', *Zeitschrift für Geomorphologie*, 6, 183–97.

Tricart, J. (1956). 'Etude experimentale du probleme de la gelivation', *Biuletyn Peryglacjalny*, 4, 285–318.

Tricart, J. (1962). 'Mécanismes normaux et phénomènes catastrophiques dans l'évolution des versants du bassin du Guil (Hautes-Alpes, France)', *Zeitschrift für Geomorphologie*, 5, 277–301.

Trimble, S. W. (1977). 'The fallacy of stream equilibrium in contemporary denudation studies', *American Journal of Science*, 277, 876–87.

Turner, F. J. (1952). '"Gefügerelief" illustrated by schist tor topography in Central Otago, New Zealand', *American Journal of Science*, 250, 802–7.

Twidale, C. R. (1964). 'Domed Inselbergs', *Institute of British Geographers, Transactions*, 34, 91–113.

Twidale, C. R. (1971). *Structural Landforms* (Canberra: Australian National University Press).

Twidale, C. R. (1973). 'On the origin of sheet jointing', *Rock Mechanics*, 5, 163–87.

Twidale, C. R. (1976a). *Analysis of Landforms* (Sydney: Wiley).

Twidale, C. R. (1976b). 'On the survival of paleoforms', *American Journal of Science*, 276, 77–95.

Twidale, C. R. and Bourne, J. A. (1975a). 'Episodic exposure of inselbergs', *Geological Society of America Bulletin*, 86, 1473–81.

Twidale, C. R. and Bourne, J. A. (1975b). 'The subsurface initiation of some minor granite landforms', *Journal of the Geological Society of Australia*, 22, 477–84.

Ursic, S. J. and Dendy, F. E. (1965). 'Sediment yields from small watersheds under various land uses and forest covers', *Proceedings of the Federal Inter-Agency Sedimentation Conference 1963. U.S. Department of Agriculture Miscellaneous Publications*, 970, 47–52.

Van Burkalow, A. (1945). 'Angle of repose and angle of sliding friction: an experimental study', *Geological Society of America Bulletin*, 56, 669–708.

Van de Graaff, W. J. E., Crowe, R. W. A., Bunting, J. A., and Jackson, M. J. (1977). 'Relict early Cainozoic drainages in arid Western Australia', *Zeitschrift für Geomorphologie*, 21, 379–400.

Varnes, D. J. (1958). 'Landslide types and processes', *Highway Research Board Special Report* (Washington, D.C.), 29, 20–47.

Varnes, D. J. (1975). 'Slope movements in the Western United States', pp. 1–17 in *Mass Wasting* (Norwich: Geo Abstracts).

Vickers, B. (1978). *Laboratory Work in Civil Engineering: Soil Mechanics* (London: Crosby Lockwood Staples).

Vincent, P. J. and Clarke, J. V. (1976). 'The terracette enigma – a review', *Biuletyn Peryglacjalny*, 25, 65–77.

Voight, B. (1973). 'Correlation between Atterberg plasticity limits and residual shear strength of natural soils', *Géotechnique*, 23, 265–7.

Voight, B. and Pariseau, W. G. (1978). 'Rockslides and avalanches: an introduction', pp. 1–67 in B. Voight (ed.), *Rockslides and Avalanches* (Amsterdam: Elsevier).

Vutukuri, V. S., Lama, R. D., and Saluja, S. S. (1974). *Handbook on Mechanical Properties of Rocks*, 4 vols. (Clausthal: Trans Tech Publications).

Walcott, R. I. (1972). 'Late Quaternary vertical movements in eastern North America: quantitative evidence of glacio-isostatic rebound', *Revue of Geophysics and Space Physics*, 10, 849–84.

Waldron, L. J. (1977). 'The shear resistance of root-permeated homogeneous and stratified soil', *Soil Science Society of America Journal*, 41, 843–9.

Walker, P. H. (1963). 'Soil history and debris avalanche deposits along the Illawarra Scarpland', *Australian Journal of Soil Research*, 1, 223–30.

Walker, P. H., Hutka, J., Moss, A. J., and Kinnell, P. I. A. (1977). 'Use of a versatile experimental system for soil erosion studies', *Soil Science Society of America Journal*, 41, 610–12.

Walling, D. E. (1978). 'Reliability considerations in the evaluation and analysis of river loads', *Zeitschrift für Geomorphologie, Supplement-band*, 29, 29–42.

Wasson, R. J. (1978). 'A debris flow at Reshun, Pakistan Hindu Kush', *Geografiska Annaler*, 60A, 151–9.

Watkins, J. R. (1967). 'The relationship between climate and the development of landforms in Cainozoic rocks of Queensland', *Journal of the Geological Society of Australia*, 14, 153–68.

Waylen, M. J. (1979). 'Chemical weathering in a drainage basin underlain by Old Red Sandstone', *Earth Surface Processes*, 4, 167–78.

Weaver, C. E. (1978). 'Mn-Fe coatings on saprolite fracture surfaces', *Journal of Sedimentary Petrology*, 48, 595–610.

Wellman, H. W. (1967). 'Reports on studies related to Quaternary diastrophism in New Zealand', *Quaternary Research (Japan)*, 6, 34–6.

Wentworth, C. K. (1943). 'Soil avalanches on Oahu, Hawaii', *Geological Society of American Bulletin*, 54, 53–64.

Wesley, L. D. (1977). 'Shear strength properties of halloysite and allophane clays in Java, Indonesia', *Géotechnique*, 27, 125–36.

Whalley, W. B. (1974). 'The mechanics of high magnitude, low-frequency rock failure', *Geographical Papers* (University of Reading), 27.

Whipkey, R. Z. (1965). 'Subsurface storm flow from forested slopes', *International Association for Scientific Hydrology, Bulletin*, 10, 74–85.

White, S. E. (1976). 'Is frost action really only hydration shattering? A review', *Arctic and Alpine Research*, 8, 1–6.

Whitman, R. V. and Bailey, W. A. (1967). 'Use of computers for slope stability analysis', *Proceedings American Society of Civil Engineers*, 93 (SM4), 475–98.

Wickham, G. E., Tiedemann, H. R., and Skinner, E. H. (1972). 'Support determinations based on geologic predictions', *Proceedings of the First North American Rapid Excavation and Tunnelling Conference, New York*, 1, 43–64.

Wilhelmy, H. (1964). 'Cavernous rock surfaces (tafoni) in semi-arid and arid climates', *Pakistan Geographical Review (Lahore)*, 19, 9–13.

Wilson, L. (1973). 'Variations in mean annual sediment yields as a function of mean annual precipitation', *American Journal of Science*, 273, 335–49.

Wilson, L. (1977). 'Sediment yield as a function of climate in United States rivers', *International Association of Hydrological Sciences*, Publication 122, 82–92.

Wiltshire, G. R. (1960). 'Rainfall intensities in New South Wales', *Journal of the Soil Conservation Service of New South Wales*, 16, 54–69.

Winkler, E. M. (1966). 'Important agents of weathering for building and monumental stone', *Engineering Geology*, 381–400.

Winkler, E. M. (1970). 'The importance of air pollution in the corrosion of stone and metals', *Engineering Geology*, 4, 327–34.

Winkler, E. M. and Wilhelm, E. J. (1970). 'Salt burst by hydration pressures in architectural stone in urban atmosphere', *Geological Society of America Bulletin*, 81, 567–72.

Wischmeier, W. H. (1959). 'A rainfall erosion index for a universal soil-loss equation', *Proceedings of the Soil Science Society of America*, 25, 246–9.

Wischmeier, W. H. (1976). 'Use and misuse of the universal soil-loss equation', *Journal of Soil and Water Conservation*, 31, 5–9.

Wischmeier, W. H. (1977). 'Soil erodibility by rainfall and runoff', pp. 45–56 in T. J. Toy (ed.), *Erosion, Research Techniques, Erodibility and Sediment Delivery* (Norwich: Geo Books).

Wischmeier, W. H., Johnson, C. B., and Cross, B. V. (1971). 'A soil erodibility nomograph for farmland and construction sites', *Journal of Soil and Water Conservation*, 26, 189–93.

Wischmeier, W. H. and Mannering, J. V. (1969). 'Relation of soil properties to its erodibility', *Proceedings of the Soil Science Society of America*, 33, 131–7.

Wischmeier, W. H. and Smith, D. D. (1965). 'Predicting rainfall-erosion losses from cropland east of the Rocky Mountains', *U.S. Department of Agriculture Handbook*, 282, 1–47.

Wischmeier, W. H. and Smith, D. D. (1978). 'Predicting rainfall erosion losses – a guide to conservation planning', *U.S. Department of Agriculture Handbook*, 537, 1–58.

Wischmeier, W. H., Smith, D. D., and Uhland, R. E. (1958). 'Evaluation of factors in the soil-loss equation', *Agricultural Engineering*, 39, 458.

Wolfe, J. A. (1977). 'Large Holocene low-angle landslide, Somar Island, Philippines', pp. 149–53 in D. R. Coates (ed.), *Landslides, Reviews in Engineering Geology*, vol. 3 (Boulder, Col. Geological Society of America).

Wolman, M. G. (1967). 'A cycle of sedimentation and erosion in urban river channels', *Geografiska Annaler*, 49A, 385–95.

Wolman, M. G. and Gerson, R. (1978). 'Relative scales of time and effectiveness of climate in watershed geomorphology', *Earth Surface Processes*, 3, 189–208.

Wolman, M. G. and Miller, J. P. (1960). 'Magnitude and frequency of forces in geomorphic processes', *Journal of Geology*, 68, 54–74.

Wood, B. L. (1969). 'Periglacial tor topography in southern New Zealand', *New Zealand Journal of Geology and Geophysics*, 12, 361–75.

Wopfner, H. and Twidale, C. R. (1967). 'Geomorphological history of the Lake Eyre Basin' pp. 118–42 in J. N. Jennings and J. A. Mabbutt (eds.), *Landform Studies from Australia and New Guinea* (Canberra: Australia National University Press).

Wu, T. H. (1966). *Soil Mechanics* (Boston: Allyn and Bacon).

Yee, C. S. and Harr, R. D. (1977). 'Influence of soil aggregation on slope stability in the Oregon Coast Ranges', *Environmental Geology*, 1, 367–77.

Yong, R. N. and Warkentin, B. P. (1966). *Introduction to Soil Behaviour* (New York: Macmillan).

Yoshikawa, T. (1974). 'Denudation and tectonic movement in contemporary Japan', *Bulletin of the Department of Geography, University of Tokyo*, 6, 1–14.

Young, A. (1963a). 'Deductive models of slope evolution', *Nachrichten der Akademie der Wissenschaften in Göttingen, II. Mathematisch-Physikalische Klasse*, 5, 45–66.

Young, A. (1963b). 'Soil movement on slopes', *Nature*, 200, 120–1.

Young, A. (1972). *Slopes* (Edinburgh: Oliver and Boyd).

Young, A. (1974). 'The rate of slope retreat', *Institute of British Geographers, Special Publication*, 7, 65–78.

Young, A. (1978). 'A twelve-year record of soil movement on a slope', *Zeitschrift für Geomorphologie, Supplementband*, 29, 104–10.

Zalesskii, B. V. (ed.) (1964). *Physical and Mechanical Properties of Rocks*. (Academy of Sciences of the U.S.S.R., translated 1967, Israel Program for Scientific Translations, Jerusalem).

Zaruba, Q. and Mencl, V. (1969). *Landslides and their Control* (Amsterdam: Elsevier).

Zaruba, Q. and Mencl, V. (1976). *Engineering Geology* (Amsterdam: Elsevier).

Zingg, A. W. (1940). 'Degree and length of land slope as it affects soil loss in runoff', *Agricultural Engineering*, 21, 59–64.

Index

LANCASTER UNIVERSITY LIBRARY
POPULAR LOAN

Issued for SEVEN DAYS during Term or for a Vacation

Due for return by end of service on date below